Microbial Pathogenomics

Genome Dynamics

Vol. 6

Microbial Pathogenomics

Volume Editors

Hilde de Reuse Paris
Stefan Bereswill Berlin

39 figures, 30 in color, and 12 tables, 2009

KARGER

Basel · Freiburg · Paris · London · New York · Bangalore · Bangkok · Shanghai · Singapore · Tokyo · Sydney

Dr. Hilde de Reuse
Institut Pasteur
Helicobacter Pathogenesis Group
Microbiology Department
28 rue du Docteur Roux
75724 Paris (France)

Prof. Dr. Stefan Bereswill
Charité-Universitätsmedizin Berlin
Institut für Mikrobiologie und Hygiene
Robert-Koch-Forum, Campus Charité Mitte (CCM)
Dorotheenstrasse 96
10117 Berlin (Germany)

Library of Congress Cataloging-in-Publication Data

Microbial pathogenomics / volume editors, Hilde de Reuse, Stefan Bereswill.
p. ; cm. -- (Genome dynamics, ISSN 1660-9263 ; vol. 6)
Includes bibliographical references and indexes.
ISBN 978-3-8055-9192-8 (hard cover : alk. paper)
1. Bacterial genomes. 2. Pathogenic bacteria. I. Reuse, Hilde de. II. Bereswill, Stefan. III. Series: Genome dynamics, v. 6. 1660-9263 ;
[DNLM: 1. Bacteria--genetics. 2. Bacteria--pathogenicity. 3. Genome, Bacterial. W1 GE336DK v.6 2009 / QW 51 M62687 2009]
QH434.M53 2009
616.9'201--dc22

2009027454

Bibliographic Indices. This publication is listed in bibliographic services, including Current Contents®

www.karger.com
Printed in Switzerland on acid-free and non-aging paper (ISO 9706) by Reinhardt Druck, Basel
ISSN 1660–9263
ISBN 978–3–8055–9192–8
e-ISBN 978–3–8055–9193–5

Contents

Editorial

The book series 'Genome Dynamics' aims to provide readers with an up-to-date overview on genome structure and diversity. Such knowledge is of particular interest for human health, as already demonstrated in the first volume of the series entitled 'Genome and Disease'. In this volume, we discussed the different mechanisms of genetic instability affecting our genes and leading to human disease. Importantly, genome analysis can also tell us how human pathogens impair health, how we interact with them and fight against their harmful effects. More than a decade after the publication of the genome sequence of *Haemophilus influenzae* and just before entering into a new era of genome analysis opened by the 'next generation' sequencing technologies, it is time to review our current knowledge of pathogen genomics and its contribution to the understanding and treatment of infectious diseases. Therefore, we have invited two reputed microbiologists, Hilde de Reuse (Institut Pasteur, Paris) and Stefan Bereswill (Charité University Medicine Berlin), to provide us with their view on the current status, medical impact and future developments of 'Microbial Pathogenomics'. As you will see, the result is very impressive. Many thanks to both guest editors for this very informative volume on key aspects and novel trends in this major field of research.

Jean-Nicolas Volff
Lyon, February 2009

Preface

The rapid and ongoing process of functional and comparative genome analysis has revealed novel aspects of microbial biology and evolution, as well as of pathogenicity. In this book on 'Pathogenomics', we focus on the genomics aspects of pathogenic bacteria because of their importance and their unique host-adaptation strategies.

Genomes from each important human bacterial pathogen have now been sequenced. For many of them multiple sequences of different strains and of closely related species (non-pathogenic or animals pathogens) are available. Population genomics of pathogenic bacteria have metamorphosed epidemiology and provided astonishing information on the mechanisms related to bacterial persistence or host adaptation. In addition, 'Pathogenomics' has also shed new light on the forces that shape the evolutionary history of bacterial pathogenesis and virulence acquisition in some cases through co-evolution with the host. Even more spectacular, bacterial genome information was used successfully to retrace the ancient human population migrations, as is illustrated in this book by the gastric pathogen *Helicobacter pylori.*

More generally, multiple genomic sequences provide insights into the evolutionary processes that have shaped bacterial genomes and generated their diversity. Analysis of genome plasticity and the bacterial gene pools have led to new concepts such as the core genome (genes in common to all sequenced strains) and the pan-genome (the sum of the core and of dispensable genomes shared by all sequenced strains). The overwhelming quantity of information couldn't have resulted in answers to biologically relevant questions without a concomitant revolution in the development of bioinformatics approaches and high throughput experimental technologies (functional genomics).

This book intends to summarize these different aspects and novel trends in bacterial pathogenomics by presenting a unique collection of reviews written by leading researchers in the field. The contributions were peer-reviewed by a panel of international experts.

The current technologies including computational tools and functional approaches for genome analysis are presented in illustrated chapters. This includes visualization tools for genome comparison, databases, *in silico* metabolic reconstructions and function prediction, as well as interactomics for the study of protein-protein interactions. Contributions dealing with pan-genomics and reverse vaccinology introduce the reader to the actual strategies used by genomics researchers to face the problems generated by bacterial diversity in the prevention and treatment of infectious diseases.

Taking individual bacterial pathogens as examples, the authors discuss the evolutionary forces that accompany human–pathogen interactions in the light of bacterial ecology. Most important frameworks of host-adaptation are illustrated by *Helicobacter pylori* and *Mycobacterium tuberculosis* that are human-specific and highly persistent.

Other chapters outline how bacterial pathogens have evolved through several mechanisms with one major role for horizontal gene transfer. Bacteria with different pathogenic strategies have been shaped. Some, like *Escherichia coli* have acquired the capacity to rapidly adapt to changing environments in order to enhance the spectrum of sites within the host that can be infected. For *Pseudomonas aeruginosa*, the strategies allow versatility for the occupation of a wide range of different environmental niches in addition to the human host. Others, like *Legionella* manipulate and subvert host mechanisms by synthesizing eukaryotic-like proteins that mimic specific cellular functions. Most fascinating are the signatures or possibility to deduce the life style of a bacterium as illustrated by a host-restricted organism such as *Bartonella* or by the versatile *Pseudomonas*. In the case of other pathogens such as *Helicobacter pylori* or *Campylobacter*, genome evolution through loss, gain and mutation of genes is also discussed.

In conclusion, the unique combination of topics dealing with technology, pathogenesis and evolution provides the reader with a global view of current and future trends in bacterial genomics. Teachers and lecturers will make use of the illustrative presentation to optimize knowledge transfer and learning strategies.

Hilde de Reuse, Paris
Stefan Bereswill, Berlin
February 2009

de Reuse H, Bereswill S (eds): Microbial Pathogenomics.
Genome Dyn. Basel, Karger, 2009, vol 6, pp 1–20

Genome Comparison of Bacterial Pathogens

T.M. Wassenaar[a,b] · J. Bohlin[c] · T.T. Binnewies[a,d] · D.W. Ussery[a]

[a]Center for Biological Sequence Analysis, Department of Systems Biology, Technical University of Denmark, Lyngby, Denmark; [b]Molecular Microbiology and Genomics Consultants, Zotzenheim, Germany; [c]Norwegian School of Veterinary Science, Epi-Center, Department of Food Safety and Infection Biology, and National Veterinary Institute, Section of Epidemiology, Oslo, Norway; [d]Roche Diagnostics Ltd., Rotkreuz, Switzerland

Abstract

Bacterial pathogens are being sequenced at an increasing rate. To many microbiologists, it appears that there simply is not enough time to digest all the information suddenly available. In this chapter we present several tools for comparison of sequenced pathogenic genomes, and discuss differences between pathogens and non-pathogens. The presented tools allow comparison of large numbers of genomes in a hypothesis-driven manner. Visualization of the results is very important for clear presentation of the results and various ways of graphical representation are introduced.

The first complete sequence of a bacterial genome was published in 1995 [1]. Since then, more than 800 bacterial and archaeal genomes have been fully sequenced and published, and in addition for more than a thousand genomes a near-to complete sequence has become publicly available. The rate at which completed bacterial genome sequences are added to the public domain is increasing with time (fig. 1, left panel). These statistics were obtained from the NCBI Genome Project web pages [2]. Pathogens comprise a large fraction of the sequenced bacterial genomes and since many of these belong to the Proteobacteria, this and a few other bacterial phyla are highly overrepresented in the available genome sequences (fig. 1, right panel). This should be borne in mind when interpreting BLAST E-values, as that program assumes an equal chance for any homology to be found by chance, whereas that chance greatly increases when searching with genes from, e.g., Proteobacteria or Firmicutes.

In this chapter we compare the sequenced genomes of pathogenic bacteria amongst each other and with non-pathogenic bacteria, using some common and relatively simple methods of comparison. Instead of zooming in on a single given genome sequence, we use tools to compare genomes within a well-defined group of related organisms, such as bacteria sharing a particular life style, or belonging to a particular species,

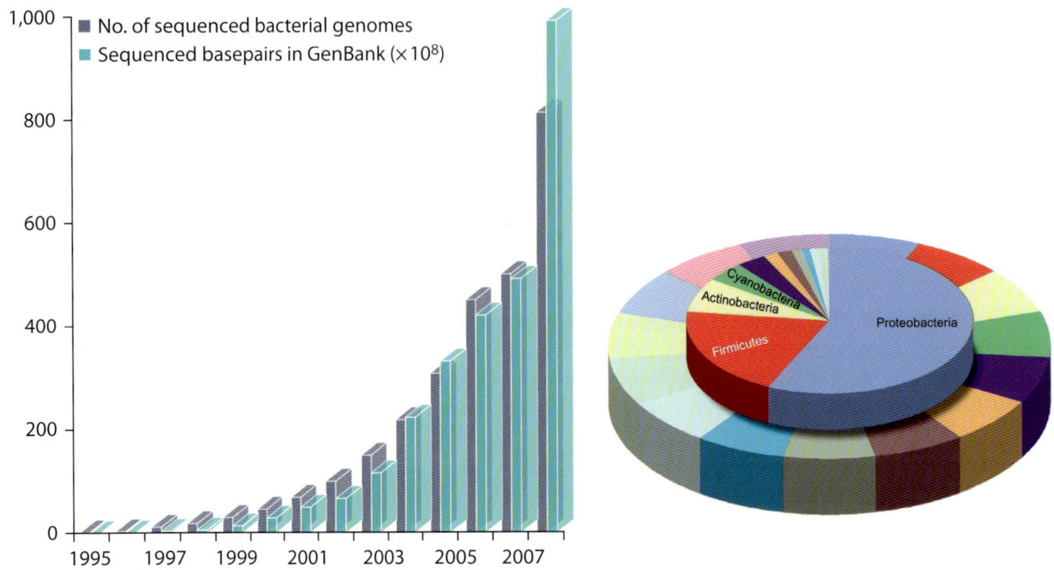

Fig. 1. To the left the increase in number of sequenced bacterial genomes (including archaeal genomes) and stored nucleotide sequences in GenBank are represented. To the right two pie charts represent a hypothetical equal proportion of 15 bacterial phyla (bottom chart) and the observed proportion of sequenced bacterial phyla, with Proteobacteria and Firmicutes being highly overrepresented (top chart).

genus or even phylum. Such comparisons are possible and doable despite the vast amount of data that is comprised in each individual genome sequence. Comparisons of many sequenced (pathogenic) bacterial genomes envisage the true genomic diversity of the Kingdom of bacteria. When performing phylogeny with high numbers of complete genome sequences computational time becomes an issue. Capturing the results in a meaningful (graphical) representation, and making sense of the observations are other challenges. Here we provide some simple examples of graphics that illustrate results based on complex data.

There are many methods to compare bacterial genomes [3] and it is not our intention to extensively cover all. The interested reader is directed towards a textbook produced by our group [4]. Instead, we will use tools to test some clearly defined hypotheses that deal with general features of bacterial pathogens, to illustrate this kind of hypothesis-driven bioinformatic analysis.

For the analyses presented here we have grouped all bacteria for which a genome sequence is listed at NCBI [2] according to their typical lifestyle, creating four groups: pathogenic, commensal/symbiotic, intracellular and free-living bacteria. Bacteria that are pathogenic to plants or cold-blooded animals were grouped together with pathogens causing disease in humans or other warm-blooded animals. All obligate intracellular bacteria were grouped as such, irrespective of their pathogenic potential. In this respect our grouping did not always follow the organism annotation given in the genome projects, provided by the authors who submitted the sequences. The reason why we preferred to keep all intracellular bacteria together is that such bacteria have genomes that are different in a number of ways from other bacteria, and we aimed to specifically analyze this group. Also note that some bacteria may be adapted to a free-living state but can also cause human (opportunistic) infections, in which case they were listed as 'pathogens'. As a consequence, the grouping is biased towards (human) pathogens (unless such organisms very rarely cause infections, in which case we grouped them as free-living).

Using these criteria 37%, or 253 out of the 675 genomes we used in our reference set were from pathogens, of which 31 were plant pathogens, 8 were insect pathogens and 5 were pathogens of cold-blooded animals including fish. 76 genomes (11%) were from benign organisms living with a host, including 20 plant symbionts, and 80 genomes (12%) were from intracellular organisms. 256 (38%) of the genomes were from organisms inhabiting either terrestrial or marine environments and for 10 bacteria insufficient information was available, so these genomes were removed. The resulting dataset of 665 bacterial genomes was used to address a number of questions as presented below.

Do Pathogens More Frequently Have Multiple DNA Replicons than Non-Pathogens?

The hypothesis tested here is based on the notion that free-living bacteria possibly need to have a more extensive adaptation potential, reflected by a larger genome, as they may encounter more variable situations during their life compared to pathogens. Multiple DNA replicons can exist in bacteria. By definition, a genome includes all chromosomes and, when applicable, plasmids that constitute an organism's total DNA. Chromosomes are independently replicating DNA molecules that are essential and present in single copy in the cell, and should carry at least one ribosomal RNA unit. Although this requirement is part of the definition of a chromosome, ribosomal RNA genes are not always annotated on chromosomes and sometimes seem to be absent despite the fact that the DNA molecule is classified as a 'chromosome'. Out of the 665 bacterial genomes analyzed, 10 genomes have three chromosomes, whilst another 45 genomes have two chromosomes, resulting in about 8% of the genomes having more than one chromosome. Some species or isolates contain plasmids that can be essential or non-essential, and can be present in single or multiple copies. Plasmids are frequently strain-specific and are more variable in size, gene content and copy number than chromosomes. The word 'genome' is only synonymous to 'chromosome' for organisms that contain one single chromosome without plasmids, which is only 401 genomes, or 60% of the total. Many bacterial pathogens carry plasmids that

Table 1. Number of chromosomes in bacteria with various lifestyles

Bacterial lifestyle	No. of genomes analyzed	1 Chromosome	2 Chromosomes	3 Chromosomes
Pathogenic	149	131 (88%)	12 (8%)	6 (4%)
Commensals/symbionts	66	62 (94%)	4 (6%)	0
Intracellular	63	60 (95%)	3 (5%)	0
Free-living	222	208 (93.6%)	13 (5.9%)	1 (0.5%)
All bacteria[a]	500	461 (92.2%)	32 (6.4%)	7 (1.4%)

[a]Redundancy was removed, in that genome sequences of the same species with the same number of chromosomes and plasmids were included only once.

can partly, or even completely, be responsible for virulent potential. The hypothesis tested here is whether pathogenic bacteria carry more, or more frequently, plasmids or multiple chromosomes than bacteria with a different lifestyle.

The number of chromosomes and eventually plasmids for each sequenced genome was extracted from the NCBI website. Of the 253 pathogens, 222 had a single chromosome, 22 had two chromosomes and 9 had three. The latter were all members of the genus *Burkholderia*, several of which were of the same species (all *Burkholderia* sequenced so far have three chromosomes with the exception of *B. mallei* and *B. pseudomallei* which have two). Thus, the set of genomes we use is partially redundant, as some species are represented more than once. Removal of such redundancy is problematic in those cases where plasmid content varies between isolates, as with *E. coli* (the number of chromosomes is usually constant within a species, with one exception: *Rhodobacter sphaeroides*, a photosynthetic organism, can have either 1 or 2 chromosomes). We therefore removed duplicated species, ignoring subspecies, only when plasmid content, lifestyle and host type was constant. This shortened the list to 500 genomes of which 149 were pathogens (table 1). Of these, 131 had a single chromosome, 12 had two chromosomes (*Brucella*, *Leptospira* and *Vibrio* species amongst others) and 6 *Burkholderia* genomes had three chromosomes. A comparison to bacteria with different lifestyles (all corrected for redundancy) is given in table 1. Intracellular pathogens have significantly more often a single chromosome ($p < 0.001$).

We next analyzed plasmid content, irrespective of chromosome counts. Although it is not guaranteed that plasmids are always sequenced along with the chromosome of an organism, the presence of a plasmid is generally well checked for pathogens, so that if anything, we could expect an under-reporting of plasmid content for bacteria with alternative lifestyles. Of the 149 non-redundant genomes from pathogenic bacteria, 78 (52%) did not have any plasmids reported. 35 had one plasmid, 20 had two plasmids and 16 had three or more, with the record holder *Borrelia burgdorferi* (strain

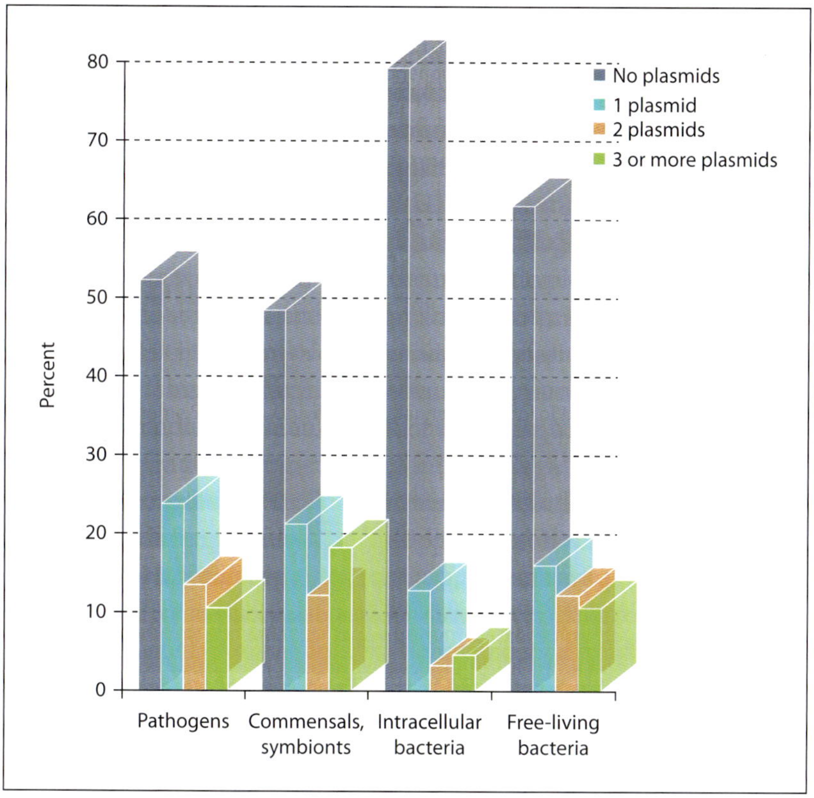

Fig. 2. Frequency of plasmids in bacteria with various lifestyles, corrected for redundancy.

B31), which has 21 plasmids. The results for the other bacteria are summarized in figure 2. The number of plasmids did not significantly correlate with lifestyle.

Do Pathogens Have a Genome Size or AT Content Different from Non-Pathogens?

A simple method to compare multiple genomes is to use a property that can be captured in a single numerical value. This can be its base composition (for example, the GC content as %GC), genome size, the number of ribosomal RNA units, protein-coding genes, repeat sequences, or any other property that can be expressed as a numerical value. Once that value is extracted, comparing the data for multiple genomes is relatively straightforward. We will illustrate such an analysis by comparing genome size of bacteria, to test if pathogens have a more strictly defined genome size than non-pathogens, notably free-living bacteria. The hypothesis is based on the notion that free-living bacteria possibly need to have a more extensive adaptation potential, reflected by a larger genome, as they may encounter more variable situations during their life than pathogens do.

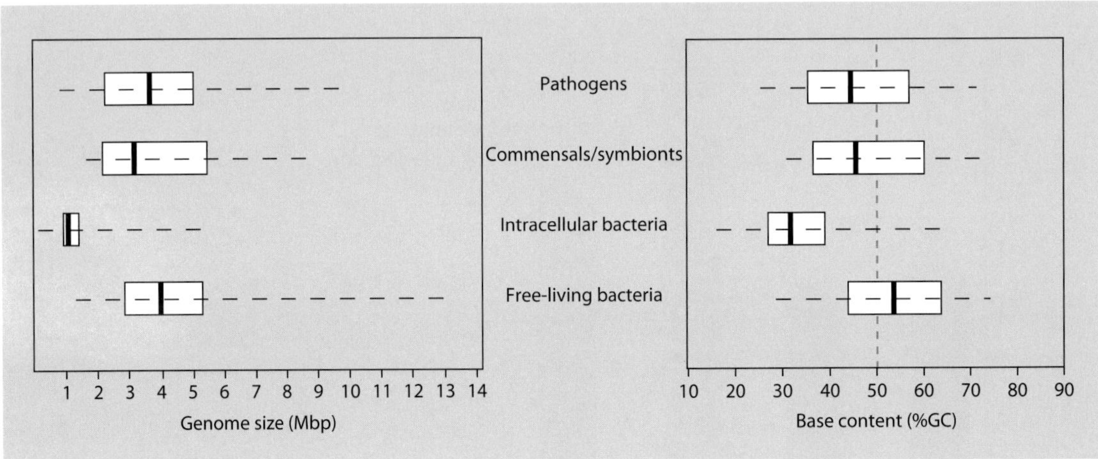

Fig. 3. To the left, the genome size distribution for 675 bacterial chromosomes is shown in a box and whiskers plot, grouped by life style of the organism. To the right, the base content is given as %GC for the same groups of organisms. The total spread of each data is given by a dotted line, the box represents the 25–75% distribution and the bar within the box gives the median. When the data distribution is skewed towards one end, the median will not be in the middle of the box, as can be seen for the commensals/symbionts.

At the time of writing (though this is a moving target), the largest complete bacterial genome sequenced was that of *Sorangium cellulosum* (strain 'So ce 56'), a myxobacterium belonging to the δ-Proteobacteria. It consists of a single chromosome of 13 Mb (13×10^6 bp). The biggest pathogenic bacterial genome sequenced to date is that of *Burkholderia xenovorans* LB400 (this member of the β-Proteobacteria is an opportunistic pathogen for cystic fibrosis patients) whose three chromosomes amount to 9.77 Mb. The smallest bacterial genome so far sequenced is that of *Carsonella ruddii* (PV), a γ-Proteobacteria that is an obligate endosymbiont of *Pachypsylla venusta* (a plant sap-feeding insect), having a mere 159,662 bp, or 0.159 Mb. The genome is believed to have undergone massive genome erosion [5]. The smallest genome of a pathogen known to date belongs to the obligate parasitic *Mycoplasma genitalium* G37, with 0.58 Mb, which happened to be the second bacterial genome to have been fully sequenced. Since this is an intracellular organism, it is not represented in the pathogenic group in our analysis.

As the mentioned record holders illustrate for Proteobacteria, genome size is not necessarily conserved within a bacterial phylum. The Actinobacteria are also vastly spread out between approximately 0.9 and 9.6 Mb. In contrast, 11 sequenced Chlamydiae genomes all fall within 1 and 1.2 Mb.

To visualize the variation in genome size for the groups of bacteria with a different lifestyle, a box and whiskers plot was constructed (fig. 3, left). Such a plot is suitable to compare and visualize a single numeric variable in large numbers of genomes, as it captures the commonality and spread of the findings. Figure 3 shows that indeed the largest genomes are observed for free-living bacteria, and the smallest genomes are reserved for intracellular bacteria. However, overlapping genome sizes are observed for the majority of pathogens, commensals/symbionts and the free-living bacteria. The most striking group to differ is that of the intracellular organisms, for which half of the genomes are around 1 Mb. An association between a small genome size and an intracellular lifestyle was found as statistically significant ($p < 0.001$). The originally proposed hypothesis that free-living bacteria would have a larger genome was slightly less significant ($p < 0.01$). These analyses were done by regression analysis using a multinomial model.

Another example of a box and whiskers plot is given in figure 3, right panel, where the base content (expressed as %GC) is plotted. Again, the most striking group is that of intracellular bacteria, which generally have a low GC content. The correlation between low GC content and intracellular lifestyle is again highly significant ($p < 0.001$) whereas free-living bacteria more frequently contain genomes with a higher GC content ($p < 0.001$).

Next, we searched for statistically significant correlations between GC content and either genome size, lifestyle or host type. We found that genome size is significantly associated with GC content, in that a higher GC content is more often observed for larger genomes and a low GC content is frequent in small genomes ($p < 0.001$). A weaker association ($p < 0.01$) was found for bacteria living in association with plants, which tend to have genomes with a lower GC content; this was the only host type that significantly correlated with any of the other investigated parameters. A correlation between pathogenic bacteria and either GC content or genome size could not be identified. A highly significant association was found, however, between genome size and plasmid content: larger plasmid counts were found for larger genomes. Although this finding does not seem surprising (as genome size includes plasmids plus chromosomes), plasmids usually contribute only marginally to the complete size of the genome. In fact, it seems that some bacteria need more DNA than others, and if that is the case, this DNA is more often distributed on multiple plasmids.

From this analysis we conclude that pathogenic bacteria do not generally have a shorter genome or a different overall base composition than other bacteria with the exception of the obligate intracellular bacteria, many of which happen to be pathogenic to their host.

Can Local Variation in Base Content Identify DNA that is Horizontally Acquired?

The next hypothesis we tested is not explicit for pathogens, as bacteria in all environments can partake in horizontal DNA uptake. Nevertheless, for pathogens it is known that virulence genes and antibiotic resistance genes spread by way of DNA uptake, with or without the action of mobile elements such as plasmids, transposons,

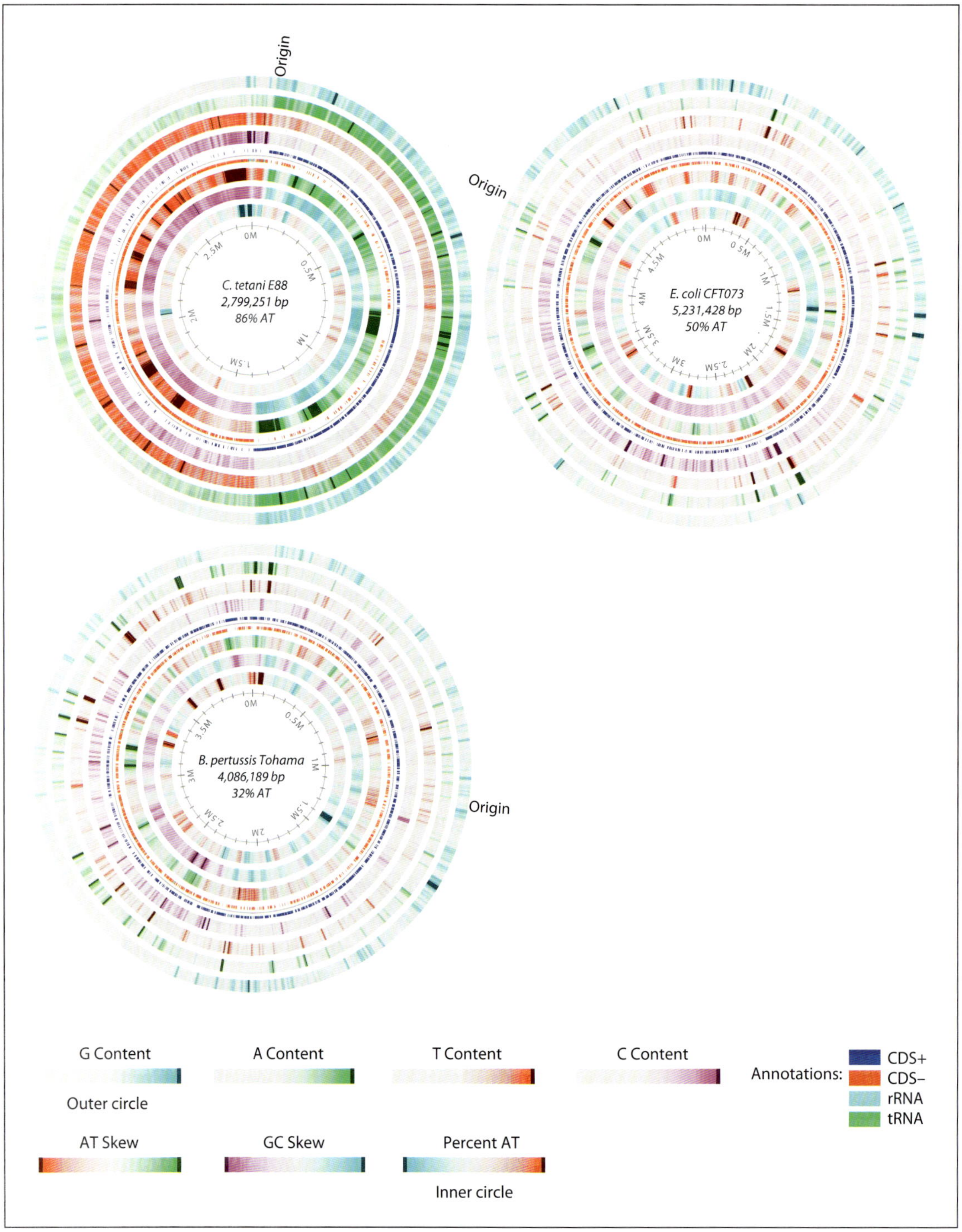

Fig. 4. Base Atlases for the chromosome of three pathogens whose genomes differ in AT content. The origins of replication are indicated. The color scales have been adjusted for each genome for maximum visualization. All color scales represent fixed averages with the exception of the %AT (innermost circle) which is depicted as deviation from the mean. Further explanation of Base Atlases is provided in [4] and [8].

integrons or gene cassettes. We do not aim to prove or disprove that DNA acquisition exists, but here we question the frequently expressed view that (recently) acquired DNA presumably has a different base composition and can be recognized by this property. (The theory predicts that differences in base composition will eventually be ameliorated by mutations [6]). We will only consider non-replicating DNA, which has to be incorporated into the chromosome. For such DNA to be recognizable by AT content, two things have to apply: (i) the AT content of the acceptor DNA has to be more or less constant for all its endogenous DNA that is not horizontally acquired; and (ii) the donor and acceptor DNA have to differ in AT content.

Let us first consider the first requirement. When examining the variation of AT content within a given genome, a general trend can be observed in that a large region containing the origin of DNA replication tends to be more GC-rich (i.e. less AT-rich), and the region around the replication terminus is more AT-rich (described in [7], and further explored in [4]). AT-rich sequences melt more easily than GC-rich sequences, due in part to the extra hydrogen bond present in a GC base pair. As a consequence it seems that, contra-intuitively, the origin of replication is the least likely to start replication. However, the 'large region' around the replication origin is approximately 5% of the total length of the chromosome, flanking either side of the origin, up to hundreds of kb. Within this region there is indeed a short stretch of a few bp, right around where the replication origin bubble opens up, that is significantly more AT-rich and will melt easily. Nevertheless, the average, or global AT content is not necessarily that which is observed locally along a chromosome, depending on the position.

How can one make such observations? In order to calculate relative or local %AT, a window is defined (say, investigating 100 bp) for which the %AT is calculated. This window is then moved step-by-step all along the genome, and for each step (of a single nucleotide shift) the obtained local %AT is written down. These scores can then be graphically represented as a graph of an artificially opened chromosome, or on a circular map, which we call an atlas (fig. 4) in which %AT (and the relative abundance of individual bases) can be visualized by color codes.

A web-based tool for Base Atlases is available at the Genome Atlas Website [9] which plots a variety of data by color intensity in two ways: either absolute values are represented (which, in case of %AT would mean the more AT-rich, the darker red the lane would appear at that location), or relative values as degree of standard deviation. A Base Atlas is a specific type of atlas that is designed to show variation in base composition (see [4] for further explanations). In the case of AT content, we would color a genome that would have the global average AT content all over its genome as grey. As already discussed, a genome contains regions that have more, or less AT compared to its global average, and these are colored as red (for more AT) or blue (for less AT) relative to the global average. That way a genome of a highly AT-rich organism can still have blue patches (as a GC-rich organism can have red regions) as can be seen in the inner circle of the left and right-hand atlases in figure 4.

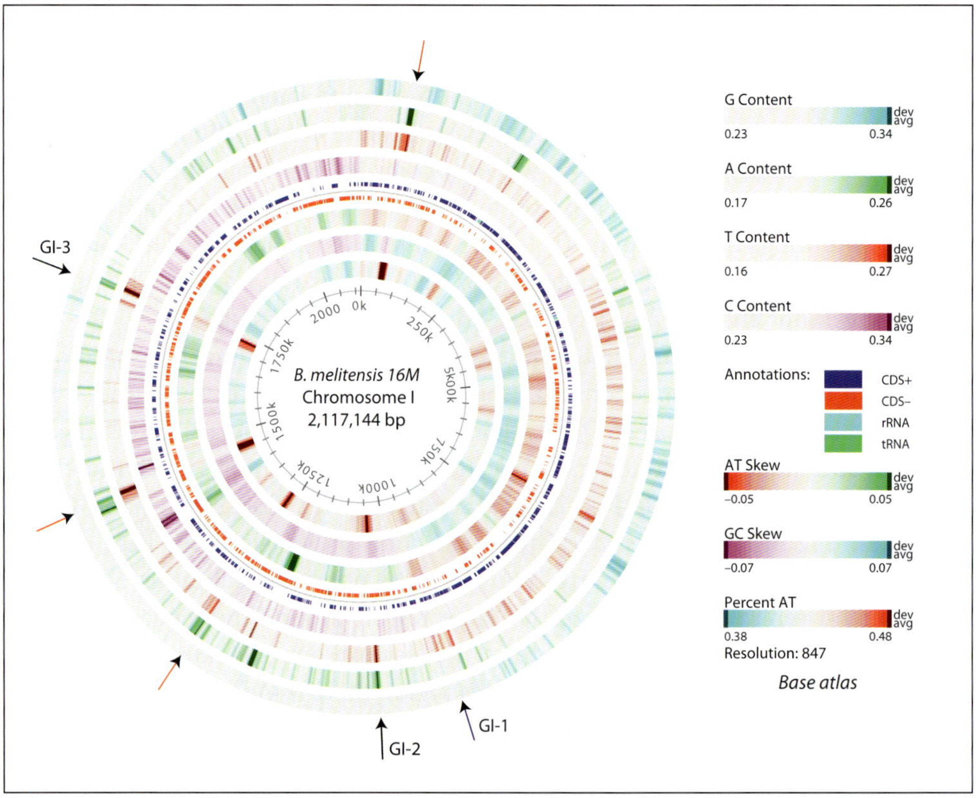

Fig. 5. Base Atlas of *Brucella melitensis* strain 16M, Chromosome 1. The genomic islands (GI) 1, 2 and 3 as identified in [10] are indicated by black arrows. Red arrows indicate other regions with striking AT content, whereas the blue arrow of GI-1 indicates that the AT content of this GI is not strikingly different from the rest of the genome.

Another example of a Base Atlas is given in figure 5, for *Brucella melitensis*, causing brucellosis (only chromosome 1 is shown). In this atlas some regions stick out as much richer in AT than the rest of the DNA. Two of these regions have been proven to be genomic islands (GIs) [10], however the regions around 50, 1250 and 1450 kb were not identified as such. Conversely GI-1 does not show up for having exceptional base content. Thus, AT content is not a reliable predictor to identify GIs.

When others compared base composition of many bacterial species (not only pathogens), it was observed that global AT content more or less associated with the ecological niche it occupies [11, 12]. Based on a genome's bias in codon usage, it is possible to predict with reasonable accuracy its likely environmental niche [13]. This would imply that 'neighboring' bacteria are likely to have similar base composition. Thus, those organisms that are most likely to exchange DNA (as they occupy the same ecological niche) also are more likely to have similar base compositions. It could be speculated that DNA exchange is one drive behind this diversification. The consequence would be that exchanged DNA might not at all be so different in base composition, weakening the second requirement.

There is an explanation why a stretch of endogenous DNA, not horizontally acquired, has a base composition different from the local AT content. Since AT content is related to codon usage (see below) and thus gene expression (or vice versa, as the cause and effect cannot be stated), genes that are expressed at extremely high or low levels will frequently differ in AT content from other, more moderately expressed genes. In addition, particular mutational events can drive a gene towards being more AT-rich, and not all genes of a genome undergo the same selection pressures to fixate such mutations in the population. In all, it cannot be taken for granted that an aberrant AT content of a gene or a gene locus means that this DNA was (recently) horizontally acquired. Additional evidence is needed in order to make such a statement, such as inverted repeats flanking the identified gene or locus, or (remnants of) genes that are involved in DNA mobilization located in direct vicinity. The presence and position of repeats can also be visualized in an atlas. In addition, particular physical properties of the DNA that depend on base composition can be visualized on a Genome Atlas, and these are independent indicators of mobile DNA. Figure 6 shows the Genome Atlas of *B. melitensis*; a Genome Atlas combines lanes from a Structure Atlas, a Base Atlas and a Repeat Atlas and from years of experience in comparative genomics we can say that this combination gives a good overview of the main features of a given chromosome. The Genome Atlas of *B. melitensis* clearly shows the presence of repeat sequences, structural features and aberrant base composition for GI-3 whereas repeats are absent for GI-2 and base composition of GI-1 is relatively normal. All mentioned atlas types are online available from our website [9].

In conclusion, atypical base composition can be an indication of horizontally acquired DNA, but additional evidence is needed to support such a prediction, as not all genes with a 'strange' base composition signature are actual strangers to the genome. As an extreme example rRNA genes can have a highly aberrant GC content and seem striking by many other parameters as visualized on a Genome Atlas, but they hardly ever undergo horizontal transfer (if at all). Conversely not all horizontally acquired DNA will have a DNA composition that can be recognized as 'different' to the recipient genome, and it can be quite difficult to identify horizontally acquired DNA as such. There are instances where the amino acid sequence of the proteins indicate horizontal transfer, whilst the DNA sequence appears 'normal' compared to the chromosomal background [14].

How Can DNA Base Composition Vary?

Since most of the DNA in a bacterial genome codes for genes, the coding region has the most effect on global base composition of a genome. Nearly all bacteria use the same genetic code, and redundancy in this code means that various codons (from 1 to 6) code for a single amino acid. By preferential use of particular codons, the total base

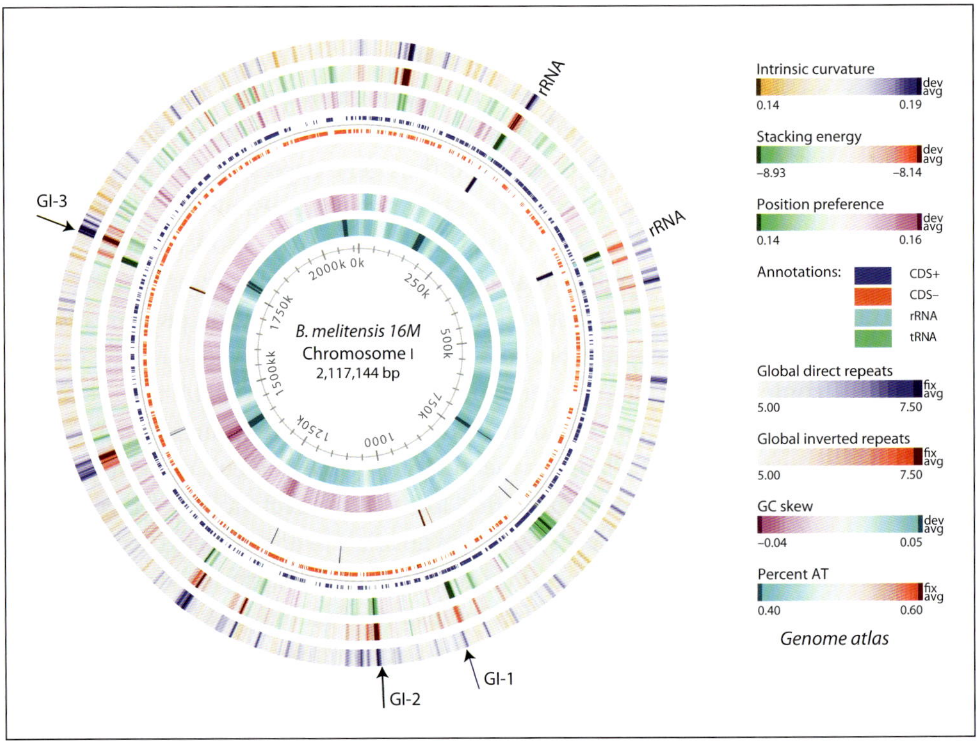

Fig. 6. Genome Atlas of chromosome 1 of *Brucella melitensis*. The outer three lanes represent physical properties of the DNA (intrinsic curvature, stacking energy and position preference). Following the two lanes with annotated genes for the positive and negative strand, two lanes show the presence of repeats, and the last two lanes are taken from a Base Atlas. For further explanation, see [4] and [8].

composition of a genome can be influenced. Since most of the variation in codons coding for a single amino acid is in the third base, here is most of the signal that ultimately defines the global AT content.

We aimed to compare two pathogens resulting in similar clinical outcome that differed significantly in AT content; however, when browsing the list at NCBI, an interesting observation was made in that most gastroenteric infections are caused by medium to AT-rich organisms, whereas pneumonic infections are rarely caused by AT-rich pathogens (except for the intracellular mycoplasmas) and far more frequently by GC-rich organisms. We selected the less extreme examples of *Francisella philomiragia* (32.6% GC), which can cause pneumonia in near-drowning victims, and *Burkholderia mallei* (68.5% GC, chromosome 1 is shown only), which causes glanders and rapid-onset pneumonia.

The preferential codon use of these bacteria with contrasting base composition is illustrated in figure 7. In the figure, the codon usage is arranged around a wheel plot with the third position base grouped together. From these wheel plots it is apparent

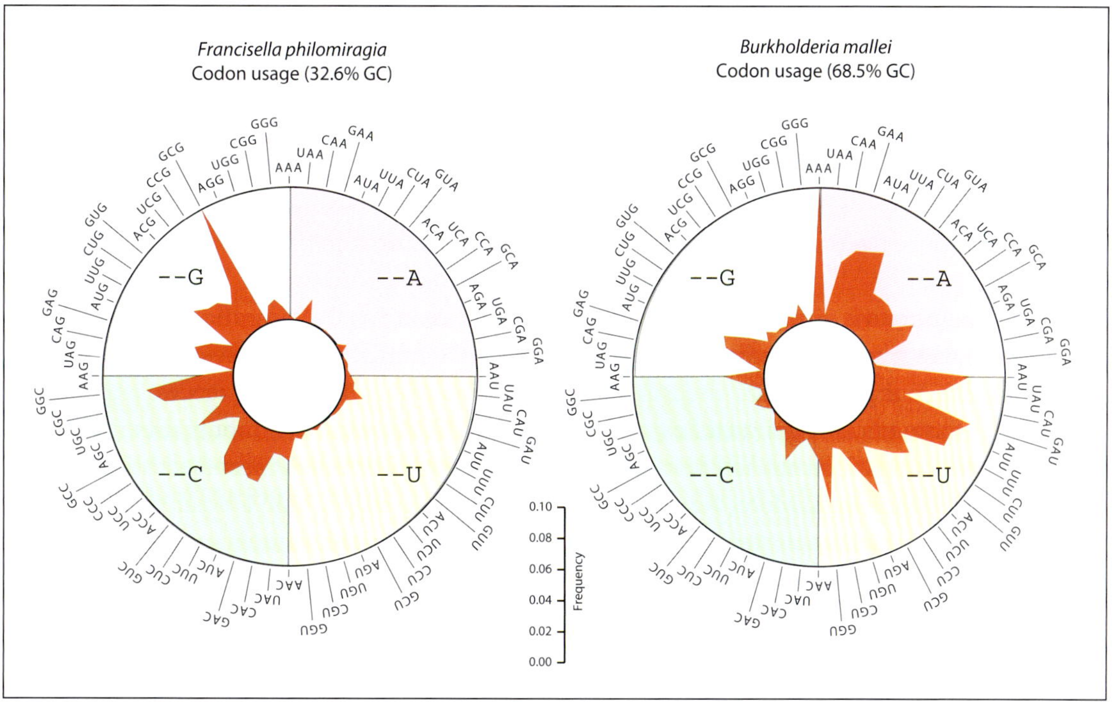

Fig. 7. Codon usage wheel plot of *Francisella philomiragia* and of *Burkholderia mallei*. The red spikes represent the relative frequencies of the codons, using the scale indicated in the middle. Their use of codons is clearly different, largely due to the last nucleotide of the triplets. The analysis is modified from [15].

that the preferred third base differs extensively between the two organisms. This drives (or is driven by) base composition, which extensively differs between these pathogens. Nevertheless, it appears that pathogens living in the same environment will have similar %GC composition, and hence also similar codon usage. At this point, although it has not been proven, it looks as though environmental limiting conditions affect the relative ease with which certain nucleotides can be made, and this in turn is what drives the base composition and codon usage.

Codon usage and the availability of tRNAs can affect the efficiency of translation, notably for those amino acids that depend on more than one tRNA (variation in the third base is usually overcome by the third base wobble). Thus, highly expressed genes would more frequently use codons for which high numbers of tRNAs are available, and conversely production of a protein that uses codons for which tRNAs are in limited supply will be slowed down during translation. For this reason, expression of 'foreign' DNA from a different environment, such as cloned DNA, can be problematic when the codon usage does not match the host strain, and naturally acquired foreign DNA is no exception. The implication is that DNA with a very different base composition is less likely to be efficiently expressed. Additional structural constraints likely decrease the probability that foreign DNA is efficiently incorporated in a genome of largely different base composition. Indeed, similarity in base composition is one of the strongest predictors of successful gene transfer [16].

How to Recognize DNA Insertions if Not by Base Composition?

Alternative methods have been developed to identify DNA insertions resulting from DNA transfer that are more sophisticated than just looking at base composition [17]. DNA alignments are used to investigate similarity between sequences, and BLAST (Basic Local Alignment Search Tool) [18, 19], is the most commonly used alignment tool. BLAST is not automatically suitable for large DNA input segments such as complete genomes. Moreover, the standard representation of BLAST results as text alignments is impractical when using complete genomes. Specific tools have been designed to align and visualize genome sequences of which the Artemis Comparison Tool (ACT) is worth mentioning. ACT comes in two versions. The program can be downloaded and used on a local computer [20] or remotely used as a web-based version of ACT with pre-computed comparisons between several hundred bacterial genomes [21].

Sequence alignments are frequently leading to statements such as: 'gene *x* in organism XX probably originated from organism YY by horizontal gene transfer'. The reasoning being that gene *x* has most similarity to gene *y* of organism YY, which happened to be present in the GenBank database. A word of caution is needed before one would accept such a statement. First of all, similarity of two genes is no evidence of direct genetic lineage. In the stated example, gene *y* could have been derived from organism XX (so gene *y* went from XX to YY instead of the other way round). Without additional evidence, the direction of gene flow cannot be stated. Another possibility is that both genes *x* and *y* come from an ancestral gene which has not been sequenced yet.

What additional evidence would be needed to confidentially state that indeed our gene of interest was inserted into a genome? How can we be certain a gene is inserted in one genome, and not deleted instead in the other genome? When this question is not relevant, such an event is neutrally called an indel (for INsertion/DELetion), which leaves both options open. Only when more genomes are available for comparison, one can begin to envisage the insertion, deletion and recombination events that shape a genome. After all, a genome sequence is a snapshot in evolutionary time and genomes are not static. The best way forward is to compare the region where our gene of interest is found between multiple members of the species or genus. If most related genomes are lacking the gene and only a few contain it, it becomes more likely that the gene was an insertion. Obviously, sampling bias can heavily influence the results of such comparisons.

The view in older textbooks of biological diversity and evolution often envisions clonal bacteria, which slowly evolve through the gradual accumulation of single-nucleotide changes. Occasionally a gene might be duplicated or a novel gene added by DNA transfer, but in general it has been commonly perceived that if one were to sequence two different strains of a species, the sequences would for the most part be similar and the two strains would share most of their genes. The currently available genome sequence data tell a different story. At the time of writing there were 32 *E. coli/Shigella* genomes sequenced with a coverage of at least 99%. One of the surprising observations is the diversity between these genomes. The size of the chromosome ranges from just over 3 to 5.6 Mb – that is, more than a million bp is present in some *E. coli* strains and missing in others. This very large variation represents mainly coding sequences, and the consequence of this diversity within a species is considerable.

One aspect we have ignored is the difference in selection pressures that genes in a genome may undergo. Selection can be positive, negative or neutral, but due to space limitations their consequences are not discussed here. The reader is referred to key publications on this subject provided for *Streptococcus* [22], *E. coli* [23, 24] and from a general perspective [25, 26].

Once a genome is sequenced and its genes are identified and annotated, one can BLAST each individual gene of that genome against a set of genomes derived from related organisms. This produces an enormous amount of information, even for just comparing two genomes against each other. For comparison of many genomes, the results can be summarized in a BLAST matrix [27]. Such a matrix reports the numbers of significant BLAST hits found for all individual genes in each genome, when compared to the next genome, and presents a wealth of information in a single table.

It would be even more informative if one could see which genes were actually found present or absent in each genome. The problem is that genes are not static, so that a particular gene may be present at a 9 o'clock position in one genome, only to be found at a 5 o'clock position in the next (the convention is to put the origin of replication, which every chromosome has, at 12 o'clock but this rule is not always obeyed). Thus, visualization becomes problematic if we want to maintain the information on gene location for each genome.

As a compromise, we have developed the BLAST Atlas. This is a graphical representation of genome-wise BLAST comparisons whereby all BLAST hits are plotted with reference to gene location of one reference genome [28]. A zoomable version of this tool is now available online [29]. An example of a BLAST Atlas is given in figure 8, using the *E. coli* isolate 53638 (believed to be intermediate between *E. coli* and *Shigella*) as the reference genome compared to 20 other *E. coli/Shigella* predicted proteomes (as we are only assessing protein-coding amino acid sequences here, their genomes are no longer completely represented). For each gene present in 53638, its presence in the other genomes is indicated by color. This produces a gap if the gene is absent in another genome, and as can be seen many gaps are shared by a number of strains. The genomes are sorted around the reference genome by their pathogenic potential, and colored accordingly. Naturally, the plot would look different with another genome selected as a reference, and it is generally better to assess at least

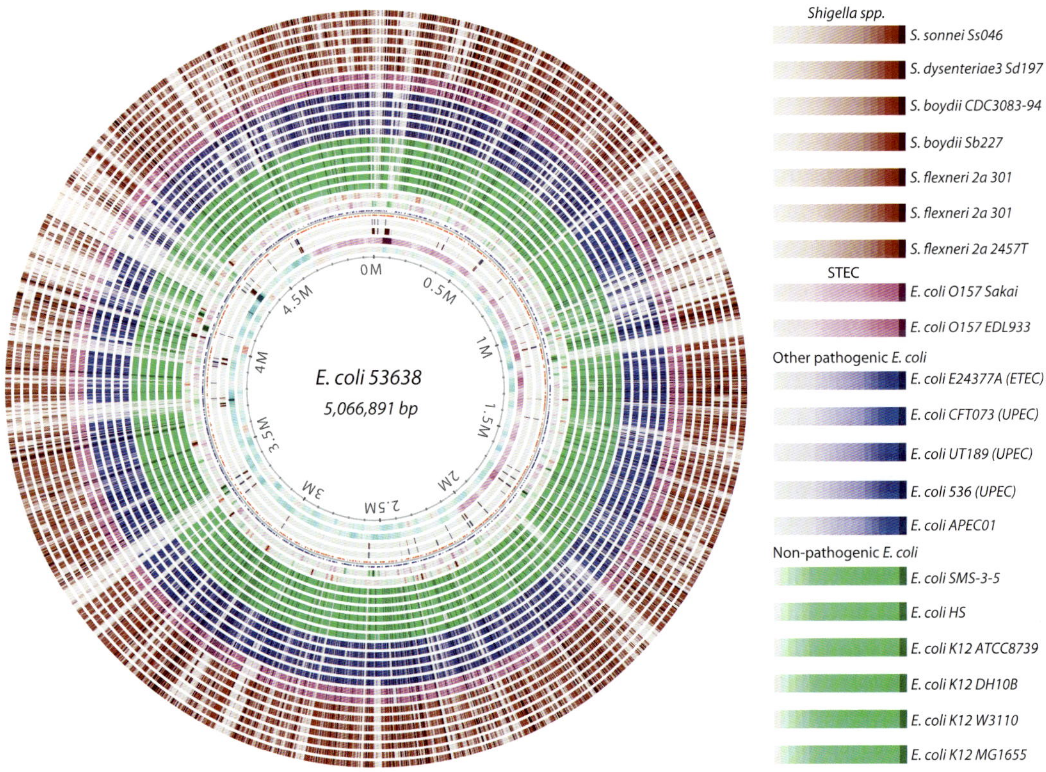

Fig. 8. Genome Blast Atlas of enteroinvasive *E. coli* (isolate 53638) as the reference strain, compared to a set of 13 sequenced *E. coli* and 7 *Shigella* genomes. The legend indicates which genome is represented in the lanes. The lanes inside the green BLAST lanes represent the Genome Atlas of *E. coli* 53638. Blast Atlases are described in [28].

two BLAST Atlases, with two reference genomes that are as different to each other as possible. It should once more be stressed that the location is plotted with reference to the genome in the middle, so a BLAST Atlas tells you whether a gene is present in a genome, but not where that gene is.

Phylogeny of Bacterial Genomes

The value of complete bacterial genome sequences is no longer doubted, and can address questions that would otherwise remain unanswered. The 'anthrax case' in the USA, where letters were posted that had been deliberately contaminated with *Bacillus anthracis*, would not have been solved if fractions of genome sequences from various

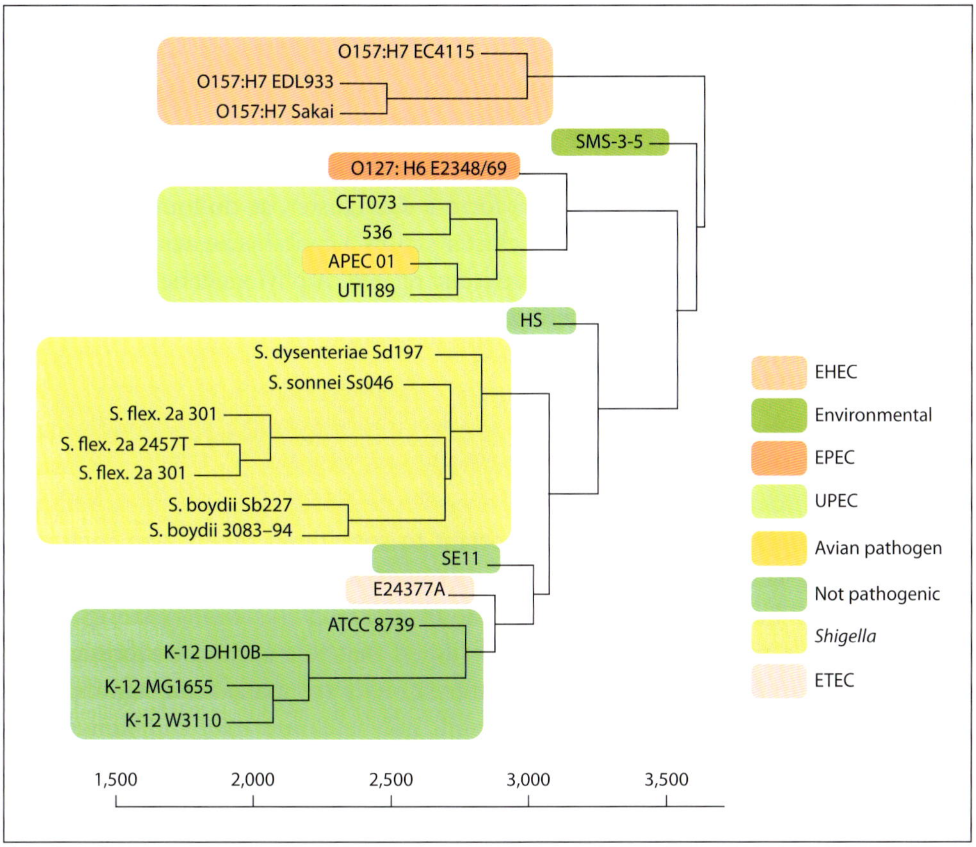

Fig. 9. Dendrogram based on complete genome sequences of 16 *E. coli* isolates and seven *Shigella* species. The color codes identify source or pathogenic properties of the isolates. EHEC = Enterohemolytic *E. coli*; EPEC = enteropathogenic *E. coli*; UPEC = uropathogenic *E. coli*; ETEC = enterotoxic *E. coli*. S. flex = *Shigella flexneri*.

isolates had not been generated [30]. For this organism, multilocus sequence typing (MLST), a frequently used typing method based on partial sequences of a few household genes, would have been useless as the investigated isolates were too similar. At the other end of the spectrum, the diversity within the species can be so large that MLST would provide an incorrect impression of similarity, or, when horizontal gene transfer is frequent, phylogenetic signal is lost in the investigated MLST genes. Only complete genome sequences can reveal the true variation in such cases. A phylogenetic tree based on complete genome sequences compares all those genes that are shared by two or more of the investigated isolates [31]. Figure 9 provides an example of such a tree, based on shared gene families within the genomes. The 'Manhattan distance' can be interpreted as a measure of the distance between two genomes – in this context it is the number of gene families where the two genomes differ, e.g. the number of gene families present in one but not the other genome. Thus, for example, the three *E. coli* K-12 genomes should have very small distances, as they do in figure 9. Since the total number of gene families varies from population to population, this can be corrected for by dividing all distances with the size of the sample pan-genome.

Notice in figure 9 that all Shigella genomes cluster within *E. coli* [32]. The three enterohemolytic *E. coli* isolates (EHEC) form a sub-cluster, as do four of five non-pathogenic isolates. The uropathogenic cluster (UPEC) contains an avian pathogenic strain, which reveals that the two are genetically related. A phylogenetic tree based on single genes or a combination of a few genes would be different, and less robust than this whole-genome tree.

Know Your Sequenced Pathogen

In order to compare genomes, it is important to sometimes take a step back, and make sure that we really know what it is that we are comparing. For example, the first sequenced bacterial genome was that of *Haemophilus influenza* [1]. Since *H. influenza* is a pathogen, most people assumed that this sequence represented a pathogenic strain, and many sequence comparisons were made (and many papers published) using this as a 'pathogenic' genome, maybe contrasting it to 'non-pathogenic' genomes. However, the *H. influenza* Rd genome sequenced was from a rough strain (KW20) of serotype d, and is non-pathogenic. About 10 years later, another *H. influenza* genome sequence (strain 86–028NP) was published, this time from a non-typeable pathogenic isolate [33].

In a similar manner, the first sequenced *Campylobacter jejuni* isolate (a common causative of enteritis) is described as a human clinical isolate, but its history of storage and multiple passage has resulted in some atypical phenotypes such as a poor motility that was not described in the genome publication [34, 35]. For *C. jejuni* subsp. *doylei* strain 269.97 it is stated that this organism causes bacteremia. True, this strain was isolated from a bacteremic patient, but *C. jejuni doylei* most frequently causes enteric infections (like *C. jejuni* subsp. *jejuni*) and it is not known if the sequenced strain has any property that makes it more prone to cause bacteremia than other *C. jejuni* strains. Factors independent of the bacteria, such as the immune status of the host, the infection dose, its residual microflora etc. all play a role in the outcome of disease. The pathogenic nature of a bacterium is dictated by its genome but also by its gene expression, protein modification, secretion efficiency and other factors that cannot be easily predicted from genome sequences.

Did you know that *Clostridium botulinum* does not cause disease in humans? At least, such is stated for strain Eklund 17B for which this information is of course correct, but it only applies to that strain. Of the 14 listed strains of *Staphylococcus aureus* subsp. *aureus* for which a genome sequence is available, 9 are listed to cause toxic shock syndrome and staphylococcal scarlet syndrome, whereas one strain causes mastitis, one causes a variety of infections and two strains cause septicemia and pneumonia. Clearly, this reflects either the origin of the isolate, or the interest of the researcher that filed the sequence, but it is highly questionable that these clinical outcomes of infection are reflected by the individual genomes listed here. A healthy dose of common sense (and relevant microbiological knowledge) is needed to interpret the filed meta-data of sequenced genomes.

For *Helicobacter pylori* (a human pathogen living in the stomach) it has been suggested that multiple laboratory passage (as the first sequenced strain 26695 had undergone) may have induced multiplication of repeat sequences, compared to a fresh clinical isolate J99 subsequently sequenced [36]. For some organisms it is known that their genome can change depending on growth conditions, as was shown for the *Bacillus cereus* complex [37]. In such a case, knowledge of the growth conditions for the cells from which the sequenced DNA was derived is essential to interpret the observed variation. As stated above, genome sequence is like a snapshot in evolutionary history, and one must be cautious about making conclusions of an organism's life from only a single snapshot.

Concluding Remarks

With hundreds of genomes available for analysis, there's a real need for tools to quickly and efficiently compare, visualize and analyze many genomes. It is likely that in the near future it will become commonplace to compare thousands of genomes, especially in the light of newer and faster sequencing technologies, which are currently under development. Statistical methods of calculation and visualization, such as box and whiskers plots will be necessary, as well as the development of new tools to be able to handle the huge amount of sequence information.

References

1 Fleischmann RD, Adams MD, White O, Clayton RA, Kirkness EF, et al: Whole-genome random sequencing and assembly of *Haemophilus influenzae* Rd. Science 1995;269:496–512.

2 http://www.ncbi.nlm.nih.gov/genomes/lproks.cgi

3 Binnewies TT, Motro Y, Hallin PF, Lund O, Dunn D, et al: Ten years of bacterial genome sequencing: comparative-genomics-based discoveries. Funct Integr Genomics 2006;6:165–185.

4 Ussery DW, Borini S, Wassenaar TM: Computing for Comparative Microbial Genomics: Bioinformatics for Microbiologists (Computational series). Springer Verlag London, 2009.

5 Toh H, Weiss BL, Perkin SA, Yamashita A, Oshima K, et al: Massive genome erosion and functional adaptations provide insights into the symbiotic lifestyle of *Sodalis glossinidius* in the tsetse host. Genome Res 2006;16:149–156.

6 Baran RH, Ko H: Detecting horizontally transferred and essential genes based on dinucleotide relative abundance. DNA Res 2008;15:267–276.

7 Ussery DW, Hallin PF: AT content in sequenced prokaryotic genomes. Microbiol 2004;150:749–752.

8 Jensen LJ, Friis C, Ussery DW: Three views of microbial genomes. Res Microbiol 1999;150:773–777.

9 http://www.cbs.dtu.dk/services/GenomeAtlas

10 Rajashekara G, Glasner JD, Glover DA, Splitter GA: Comparative whole-genome hybridization reveals genomic islands in *Brucella* species. J Bacteriol 2004; 186:5040–5051.
11 Musto H, Naya H, Zavala A, Romero H, Alvarez-Valin F, Bernardi G: Genomic GC: level, optimal growth temperature, and genome size in prokaryotes. Biochem Biophys Res Commun 2006;347:1–3.
12 Foerstner KU, von Mering C, Hooper SD, Bork P: Environments shape the nucleotide composition of genomes. EMBO Rep 2005;6:1208–1213.
13 Willenbrock H, Friis C, Friis AS, Ussery DW: An environmental signature for 323 microbial genomes based on codon adaptation indices. Genome Biol 2006;7:R114.
14 Podell S, Gaasterland T, Allen EE: A database of phylogenetically atypical genes in archaeal and bacterial genomes, identified using the DarkHorse algorithm. BMC Bioinformatics 2008;9:419.
15 Ussery DW, Hallin PF, Lagesen K, Wassenaar TM: Genome update: tRNAs in sequenced microbial genomes. Microbiol 2004;150:1603–1606.
16 Medrano-Soto A, Moreno-Hagelsieb G, Vinuesa P, Christen JA, Collado-Vides J: Successful lateral transfer requires codon usage compatibility between foreign genes and recipient genomes. Mol Biol Evol 2004;21:1884–1894.
17 Bohlin J, Skjerve E, Ussery DW: Investigations of oligonucleotide usage variance within and between prokaryotes. PLoS Comput Biol 2008;4:e1000057.
18 Altschul SF, Gish W, Miller W, Myers EW, Lipman DJ: Basic local alignment search tool. J Mol Biol 1990;215:403–410.
19 http://blast.ncbi.nlm.nih.gov/Blast.cgi
20 Carver TJ, Rutherford KM, Berriman M, Rajandream MA, Barrell BG, Parkhill J: ACT: the Artemis Comparison Tool. Bioinformatics 2005;21: 3422–3423.
21 http://www.webact.org/WebACT/home
22 Anisimova M, Bielawski J, Dunn K, Yang Z: Phylogenomic analysis of natural selection pressure in *Streptococcus* genomes. BMC Evol Biol 2007;7: 154.
23 Chen SL, Hung CS, Xu J, Reigstad CS, Magrini V, et al: Identification of genes subject to positive selection in uropathogenic strains of *Escherichia coli*: a comparative genomics approach. Proc Natl Acad Sci USA 2006;103:5977–5982.
24 Petersen L, Bollback JP, Dimmic M, Hubisz M, Nielsen R: Genes under positive selection in *Escherichia coli*. Genome Res 2007;17:1336–1343.
25 Lynch M, Conery JS: The origins of genome complexity. Science 2003;302:1401–1404.
26 Ochman H, Davalos LM: The nature and dynamics of bacterial genomes. Science 2006;311:1730–1733.
27 Binnewies TT, Hallin PF, Staerfeldt HH, Ussery DW: Genome Update: proteome comparisons. Microbiology 2005;151:1–4.
28 Hallin PF, Ussery DW: CBS Genome Atlas Database: a dynamic storage for bioinformatic results and sequence data. Bioinformatics 2004;20:3682–3686.
29 http://www.cbs.dtu.dk/services/gwBrowser
30 Keim P, Pearson T, Okinaka R: Microbial forensics: DNA fingerprinting of *Bacillus anthracis* (anthrax). Anal Chem 2008;80:4791–4799.
31 Henz SR, Huson DH, Auch AF, Nieselt-Struwe K, Schuster SC: Whole-genome prokaryotic phylogeny. Bioinformatics 2005;21:2329–2335.
32 Snippen LG, Kiil K, Almøy T, Ussery D: Manuscript in preparation.
33 Harrison A, Dyer DW, Gillaspy A, Ray WC, Mungur R, et al: Genomic sequence of an otitis media isolate of nontypeable *Haemophilus influenzae*: comparative study with *H. influenzae* serotype d, strain KW20. J Bacteriol 2005;187:4627–4636.
34 Parkhill J, Wren BW, Mungall K, Ketley JM, Churcher C, et al: The genome sequence of the food-borne pathogen *Campylobacter jejuni* reveals hypervariable sequences. Nature 2000;403:665–668.
35 Gaynor EC, Cawthraw S, Manning G, MacKichan JK, Falkow S, Newell DG: The genome-sequenced variant of *Campylobacter jejuni* NCTC 11168 and the original clonal clinical isolate differ markedly in colonization, gene expression, and virulence-associated phenotypes. J Bacteriol 2004;186:503–517.
36 Alm RA, Ling LS, Moir DT, King BL, Brown ED, et al: Genomic-sequence comparison of two unrelated isolates of the human gastric pathogen *Helicobacter pylori*. Nature 1999;397:176–180.
37 Carlson CR, Kolstø AB: A small (2.4 Mb) *Bacillus cereus* chromosome corresponds to a conserved region of a longer (5.3 Mb) *Bacillus cereus* chromosome. Mol Microbiol 1994;13:161–169.

Trudy M. Wassenaar
Molecular Microbiology and Genomics Consultants
Tannenstrasse 7
DE–55576 Zotzenheim (Germany)
Tel. +49 6701 8531, Fax +49 6701 901803, E-Mail trudy@mmgc.eu

de Reuse H, Bereswill S (eds): Microbial Pathogenomics.
Genome Dyn. Basel, Karger, 2009, vol 6, pp 21–34

In silico Reconstruction of the Metabolic and Pathogenic Potential of Bacterial Genomes Using Subsystems

L.K. McNeil[a] · R.K. Aziz[b]

[a]National Center for Supercomputing Applications, University of Illinois, Urbana, Ill., USA; [b]Department of Microbiology and Immunology, Faculty of Pharmacy, Cairo University, Cairo, Egypt

Abstract

Whole genome sequencing has revolutionized biological sciences, and is leading to a paradigm shift in microbiology. As more microbial genomes are sequenced, and more bioinformatics tools are developed, it has become possible to predict the metabolism of an organism from genomic data. In contrast, predicting the pathogenic potential of parasitic microbes and their interactions with their hosts is still a challenge, especially as the definition of pathogenesis itself is still evolving. In this review, we introduce the subsystem-based technology for genome annotation and analysis, and we discuss some subsystem-based tools available in the National Microbial Pathogen Data Resource (NMPDR, http://www.nmpdr.org) and their potential application in comparative genomics and pathogenomics.

Two centuries ago, the origin of infectious diseases was still obscure, and infection was more of a mythological than a scientific issue. Even though Anthony van Leeuwenhoek (1632–1722) observed the first microbes, which he called 'living animalcules', under his prototypic microscope, it was not until the work of Louis Pasteur (1822–1895) and Robert Koch (1843–1910) that a paradigm shift was realized in understanding the etiology of infectious diseases [1]. This paradigm shift was mainly driven by new technology that allowed humans to see microbes under the microscope, to culture them, and to detect their reactions biochemically. It marked the advent of the novel science, microbiology, and the start of the germ theory of disease causation.

It has been suggested that another paradigm shift is in the making as we enter the post-genomic era, in which we can detect living forms without the need for microscopy, culture, or classical biochemistry [2]. There were signs of a radical change coming as microbiologists started accepting a unique DNA sequence as proof of the presence of a microorganism [1]. Technology has moved quickly from whole-genome sequencing of cultured bacteria [3, 4], to sequencing metagenomes without culture or even DNA cloning steps [5, 6]. It has become possible to sequence and assemble the complete genome of a microbe, partially reconstruct its metabolic networks, and predict – solely from sequence data – how the microbe would obtain food and energy, without the need to see or grow that microbe. It has become possible to sequence the metagenome of a particular ecosystem and collect – again exclusively from sequence data – a large amount of information about that ecosystem and the relative contribution of different organisms in it, again without the need to grow, isolate, or even identify any of the living forms in that ecosystem [7]. Like the nineteenth century's first microbiological revolution, today's revolution is driven by novel technologies that have totally changed the way microbiology is practiced. We have moved from the study of single genes and single phenotypes to the study of genomes, transcriptomes, proteomes, and metabolomes. Focus has shifted from culture-based and biochemical methods for bacterial isolation and detection, to sequence-based methods for decoding the information that genomes carry to better understand microbial life [2].

When it comes to decoding the sequence information within a microbial genome, all genes are not equally decipherable. When the first bacterial genome was annotated in 1995, nothing could be said about the functions of 42% of its genes because they had no match in the database, or they matched an entry labeled 'hypothetical' [3]. Growth of the databases and ongoing curation of the sequence of *Haemophilus influenzae* Rd KW20 increased the proportion of functionally categorized genes from 58% to 62% by May 2008 [8] (fig. 1). Genes that encode information-transfer and metabolic reactions and pathways are well conserved among different living forms and are well defined. The great advances in biochemistry and molecular biology in the past century have resulted in very accurate maps of these central metabolic pathways. Consequently, it is now possible to predict the primary metabolic patterns of a newly sequenced organism. In May 2008, for example, the automated annotation of the genome of the large (4.7 Mb) gamma-proteobacterium *Yersinia pseudotuberculosis* YPIII by the RAST server [9] resulted in only 22% of genes having no assigned function. The automated metabolic reconstruction found that 44% of genes played a role in complete subsystems, distributed among 20 broad categories of biological processes. In the virulence category, 12 complete subsystems were automatically identified. Subsystems are groups of proteins with related functions, such as pathways of metabolism, complex structures, or phenotypes.

The successful, automated annotation of a known pathogen is made possible by comparative analysis with a database of subsystems and functional annotations curated by human experts. But how close is this to a complete picture? Have all the genes that play a role in the pathogenic potential of that organism been identified? Is it possible to sequence an entirely new organism and predict whether it will be pathogenic or not? And if an organism has a pathogenic potential, is it possible to predict to which host it is specific, or whether it has the potential to switch or broaden its host specificity? And suppose a microbe is capable of causing a specific disease

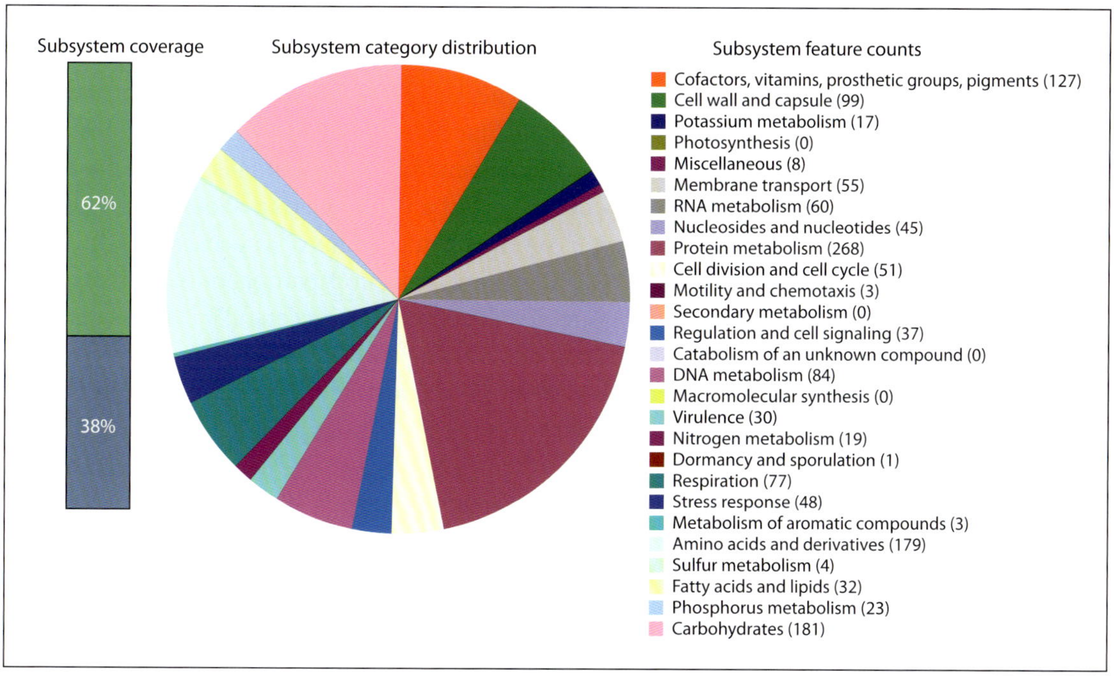

Fig. 1. Metabolic reconstruction, or subsystems summary, of the genome of *Haemophilus influenzae* Rd strain KW20 from the National Microbial Pathogen Data Resource. Subsystems comprise genes grouped together to describe an active biological process, such as a metabolic pathway, complex, or phenotype.

in a particular host; is it possible to predict whether it will ever get in contact with that host? A clear understanding and comprehensive definition of pathogenicity are required to answer these questions.

What is a Pathogen and what is a Virulence Factor?

The first definition of a pathogen was developed in the 1880s by Robert Koch, who set criteria for establishing the causality of infectious diseases (reviewed in [1, 10]). A pathogen, according to Koch's postulates, is a microbe isolated in pure culture from every individual suffering from the disease, but not from healthy counterparts. Subsequent inoculation of a healthy individual with the isolated organism should then cause the same disease. Koch himself realized the limitations of these postulates, but they provided a rigorous framework for experimental microbiology, which advanced our understanding of diseases such as anthrax, cholera, and tuberculosis. Koch's postulates have been revised extensively, and several other postulates or guidelines have been developed to establish disease causality and set boundaries between what is a pathogenic organism and what is not [1]. A century later, in the era of molecular microbiology, experimental focus switched from entire organisms to individual genes. To define a virulence gene, i.e., a gene whose product contributes to the pathogenic potential of an organism, Stanley Falkow paralleled Koch's postulates with his molecular postulates for virulence gene identification [11]. Again the first of these postulates set an exclusive condition that a virulence trait should be 'associated with pathogenic members of a genus or pathogenic strains of a species' [11], implying a clear-cut demarcation between a pathogenic and a non-pathogenic organism.

Like Koch's postulates, Falkow's molecular postulates successfully lead the quest for virulence gene discovery, which has been a rising theme in literature in the past two decades. However, it has become evident in the post-genomic era – even to Falkow himself [12] – that these postulates have several limitations as well. For example, many actual virulence genes/proteins are present in both pathogenic and non-pathogenic bacteria, but still play a role in causing human diseases [13]. Additionally, a number of proteins are bifunctional, having one biochemical role conserved among a large number of taxa and a second, host-specific role with virulence potential, e.g., streptococcal GAPDH is also a plasmin(ogen)-binding protein [14, 15]. Another factor that hinders virulence gene discovery by genetic methods is that phenotypes often result from the expression of multiple genes; thus, knocking out one or two bacterial genes might not result in a mutant totally unable to survive within the host environment. Further confusing the issue is the fact that horizontal gene transfer often leads to multiple paralogs in the same genomes. Although these paralogs might not be functionally redundant, it is very likely that they could complement each other when one is deleted. The concept of pathogenesis becomes even more complicated as we take the host into consideration.

Five years ago, the American Academy of Microbiology (AAM) convened a colloquium to discuss the application of genomics to the development of a comprehensive understanding of pathogenesis [16]. The panel defined pathogenesis in terms of the survival and evolution of disease-causing organisms, labeling pathogens as obligate, opportunistic, or accidental. Obligate pathogens evolve strictly according to their ability to cause disease. Opportunistic pathogens do not rely on the disease state to survive, but are subject to the evolutionary pressure of their pathology. Accidental pathogens may cause disease but are not spread by means of the disease, thus disconnecting evolution from pathogenicity. Pathogenicity may drive the evolution of an organism, co-evolve with an organism, or arise independently from the evolution of an organism. Along with this three-part definition of pathogenicity, the same committee recognized two virulence strategies: attacking the host with toxins, or subverting host factors to cause disease [16]. It is not always obvious how to neatly apply these definitions to a given disease-causing species.

Take, for example, Group A *Streptococcus* (GAS). GAS is an obligate human pathogen that can be carried harmlessly by a human host, can cause mild pharyngitis, necrotizing fasciitis, or even fatal bacteremia. GAS secretes toxins and subverts the host immune response both in the primary infection and in causing the post-infection sequelae rheumatic fever and acute glomerulonephritis. Recent experiments designed to define a pathogenic profile, or set of genes associated with disease and predictive of invasiveness, have failed. Only one of 266 virulence factors was found to be reliably associated with invasive GAS infection rather than mild pharyngitis [17]. All isolates tested in that study caused either mild or severe disease, so perhaps it is not surprising that most virulence factors tested were found in most isolates. When a similar study of a limited set of virulence factors was performed [18], this time with carriage isolates as controls, no clear association was found between emm-type or superantigen and disease. In fact, strains of serotype M12 were significantly associated with invasive disease and, at the same time, were predictive of carriage [18]. This result is congruent with clinical observations [19] and with the report that different strains of inbred mice respond very differently to GAS challenge, while different individuals of the same strain respond similarly [20]. The resolution of the interaction between host factors and bacterial virulence factors will require large-scale studies and correspondingly large data sets.

Systematic explorations of the relationship between host genetics and severity of disease have recently been made possible by the availability of a panel of advanced recombinant inbred (ARI) mice with defined genetic variation [21]. To identify the important differences in host response to GAS, mice from 33 isogenic ARI strains were challenged with identical inocula. While all mice developed bacteremia, differences in disease severity, bacterial dissemination and mortality rates were significantly correlated with strain when age was held constant [22]. An analysis of disease phenotypes in the context of mouse genotypes identified a quantitative trait locus (QTL) on chromosome 2 that strongly predicted disease severity. This QTL harbors genes encoding synthesis pathways for interleukin 1-alpha and prostaglandin E, which are known to play a role in the regulation of host immune responses to bacterial infections [23]. Results of such large-scale investigations will be crucial for unraveling host-pathogen interactions. Genome-wide studies of virulence factors are needed, and results must be integrated into genomic databases so that they may be easily analyzed in an intuitive way by experimental, not only computational, biologists.

The First 1,000 Genomes

In November 2003, the AAM colloquium on genomics and pathogenesis made the following recommendations for advancing the field of pathogenomics: 'The sequences of many hosts, pathogens, their nonpathogenic relatives, commensals, as well as a diverse array of microorganisms, are all needed to complete the picture of pathogenesis and provide a phylogenetic framework for understanding the phenomenon. Moreover, improvements are needed in the two most important tools of genomics: annotation methodologies and sequence databases' [16]. The panel recognized that annotation was the bottle neck of genomics and that new tools should be both high-throughput and user-friendly. At the time, 125 bacterial genomes were complete and published. Of these, 84 were classified as pathogenic, and 65 were known to cause disease in humans. Thirteen other genomes represented commensal or symbiotic bacteria, with the remaining 27 classified as environmental [16].

Almost simultaneously, in December 2003, the Fellowship for Interpretation of Genomes (FIG) launched the Project to Annotate 1000 Genomes to develop the strategy and tools for accurate, high-throughput annotation in preparation for an expected onslaught of sequence data [24]. FIG developed the SEED annotation environment to support the vertical annotation of genes in a comparative context across multiple genomes, using subsystems to provide a multidimensional framework for capturing the knowledge of subject experts. An expert defines a subsystem as a set of functional roles that act together in a biological pathway, process, or structure, which is supported in one or a few genomes by experimental evidence. Based on experimental evidence and first-hand knowledge, the expert manually annotates at least one gene in an exemplar genome for each functional role in the subsystem. These known genes are then analyzed in the SEED environment, which provides one-click tools to compare chromosomal regions surrounding a focus gene, to align and build a phylogenetic tree of selected orthologs, and to locate chromosomal clusters containing the focus gene in other genomes. The expert curator assigns the functional annotation to genes in other genomes based on an integration of evidence including sequence similarity, functional clustering, phylogenetic profiling, and metabolic context. The subsystem is displayed as a spreadsheet with functions in columns and genomes in rows. Cells of the spreadsheet are populated by the gene or genes that encode each function in each organism. All genes in one column play the same functional role and are assigned a consistent, meaningful annotation. Each column in the spreadsheet also represents a protein family, called a FIGfam, and the collection of columns in a subsystem spreadsheet represents a set of functionally related protein families. Subsystems annotation provides both a means to improve consistency and accuracy of annotations, as well as a framework for characterizing functional variants of biological systems, such as alternative metabolic pathways.

Shortly after the development of SEED began, the National Institute of Allergy and Infectious Diseases (NIAID) announced a new bioinformatic venture to integrate genomic and other biological data for biodefense research. In cooperation with investigators at the University of Chicago, Argonne National Laboratory, and the University of Illinois, FIG responded with a proposal to build the National Microbial Pathogen Data Resource (NMPDR) based on the new SEED environment. In July 2004, NMPDR became one of eight Bioinformatics Resource Centers for Biodefense and Emerging/Re-Emerging Infectious Disease [25]. NMPDR was originally focused on the food- and water-borne, Category C pathogens *Campylobacter, Listeria, Staphylococcus, Streptococcus*, and *Vibrio* [26]. Recently, the sexually transmitted pathogens *Chlamydia, Haemophilus, Mycoplasma, Neisseria, Treponema*, and

Ureaplasma were added to our mandate. Because NMPDR is based on the comparative analysis tools in SEED, all essentially complete, public genomes are available for analysis in NMPDR.

As anticipated by the AAM colloquium and the Project to Annotate 1000 Genomes (P1K), a need quickly arose for an accurate, automated, user-friendly annotation service to process new genomes prior to including them in the SEED for manual extension of subsystems, and subsequently, into NMPDR. According to data listed in the Genomes Online Database (GOLD [27]) in May 2008, 2,040 bacterial genomes were either completed or in the process of being sequenced. Of these, 1,004 are pathogenic, with 875 reported to cause disease in humans. Commensal and symbiotic bacteria number 289, with the remaining 874 classified as environmental. The efforts of the International Human Microbiome Consortium will continue to increase the number of human commensal and pathogenic bacterial genomes needing annotation and analysis. Likewise, the number of environmental genomes will soon be increased by the Genomic Encyclopedia of Bacteria and Archaea project, a large-scale collaboration between the DOE Joint Genome Institute (JGI) and the Deutsche Sammlung von Mikroorganismen und Zellkulturen (DSMZ) to sequence genomes systematically selected from the tree of life.

That sequencing has outpaced annotation is evident by the small proportion of genomes, 19%, associated with a reference to the scientific literature. Another 36% have been made available in public databases either as finished (closed) or draft assemblies without a published analysis. P1K has recently culminated with the release of the Rapid Annotation server based on Subsystem Technology, or RAST [9]. RAST identifies protein-encoding, rRNA and tRNA genes, uses FIGfams to assign functions to the genes, predicts which subsystems are completely populated by the genome, and provides a partial metabolic reconstruction based on complete, functional subsystems. The result is easily downloaded in several formats. The user may also view the result in the context of the genomes available in SEED while maintaining the privacy of the new sequence. The SEED-viewer environment developed for RAST became the template for a new, menu-driven, intuitive user-interface for NMPDR.

Value Added by Curated Subsystems

FIGfams

P1K resulted in a growing collection of more than 500 functional subsystems from which FIGfams are computed. There are two types of FIGfams. The original concept of a FIGfam is a protein family extracted from a column of a populated subsystem, which represents an expert assertion of function. An extension of that concept resulted in a set of FIGfams that are computed from the combination of shared sequence homology and genomic context. These automated FIGfams lack an expert assertion, but they provide a pre-computed starting point for experts to explore further, using bioinformatic or experimental techniques. All FIGfams are available in NMPDR and SEED in an interactive environment that may be accessed from a protein annotation page, or may be searched with a keyword, identifier, or protein sequence. A FIGfam page presents the FIGfam id, a list of sequence ids of proteins that belong to the family, the subsystem(s) (if any) that the FIGfam was extracted from, the average sequence length of the member proteins, and an interactive graphic. The graphic depicts genomic regions centered on the focus FIGfam. Several genomes are depicted, each in a different row, and sets of proteins that share similar sequences in different genomes are labeled with the same number and color. This allows the visual comparison of genomic context of the focus FIGfam. The identities of individual proteins and genomes are displayed in pop-up boxes when pointed to, and clicking will open the annotation overview page for that protein. The genomes shown in the display may be selected by the user from an ordered taxonomy of available organisms, and the size of the region shown may be reset by the user. The selected sequences are downloadable in FASTA format.

Metabolic Reconstructions for in silico Systems Biology

The collection of functional subsystems curated by subject experts provides a partial metabolic reconstruction of any individual genome as a step toward creating a reaction network describing the metabolic capabilities encoded in a genome [28]. Subsystems that represent a metabolic process are presented with links to information about the enzyme-catalyzed reactions associated with the functional roles in the subsystem. Links to defined reactions in KEGG (http://www.genome.ad.jp/kegg/) and to the Gene Ontology database, AmiGO (http://www.geneontology.org/), which has downstream links to a variety of other pathway databases, are added to the table of functional roles by the subsystem curator. For example, the Glycolysis and Gluconeogenesis subsystem contains functional roles for glucokinase, phosphofructokinase, etc. These functional roles are associated with reactions representing the breakdown of glucose into pyruvate (and the reverse process). Another set of curated links to KEGG reactions is provided by a team of collaborators at Hope College. These 'Hope Reactions' are used to define metabolic scenarios, which are coherent subnetworks of reactions that specify input and output metabolites (e.g., glucose and pyruvate in the case of glycolysis) as well as the stoichiometry of the metabolic process represented [29]. Reactions are curated for 145 subsystems that cover most of central and intermediate metabolism. The set of curated reactions present in each genome is automatically identified, then a path-finding algorithm determines whether this set of reactions is capable of transforming the input metabolites into the output metabolites for each scenario in these subsystems. The scenarios can be linked together across subsystems by matching output metabolites from one scenario to input metabolites from another scenario, to get a bigger picture of the metabolic capabilities of the organism. The complete set of scenarios for each genome will soon be available for download from NMPDR. The ultimate goal is to automatically generate substantially complete, genome-scale metabolic networks for all genomes in NMPDR and to provide the set of scenarios for each organism packaged as a network or stoichiometric matrix for metabolic flux analysis with a tool such as FluxAnalyzer [30].

Subsystems Generate Testable Hypotheses

Subsystems may be used as a matrix for the generation of testable hypotheses because they point out gaps in our knowledge, even of well studied systems. Folate synthesis and salvage, for example, are pathways that have been studied for decades in model organisms from all domains of life. The first functional role of the de novo tetrahydrofolate biosynthetic pathway in bacteria, fungi, and plants has long been known to be played by GTP cyclohydrolase I (GCYH-I; EC 3.5.4.16), encoded in *Escherichia coli* by the *folE* gene. That gene and subsequent functional roles played by the *folB-KPCA* genes were used to define a subsystem. When sequence similarity was used as the basis for extending the subsystem from *E. coli* to other bacterial genomes, orthologs of *folE* could not be identified in about 30 bacterial species that did contain orthologs of all the other folate biosynthesis genes. This suggested that an alternate, unrecognized protein performs the function in those genomes. Evidence other than sequence similarity was considered in an effort to identify candidate genes, such as phylogenetic profiling and clustering. The Signature Genes tool at NMPDR was used to find the set of genes present in the diverse organisms that perform de novo folate biosynthesis without a recognized FolE homolog, but absent from *E. coli* K12. Among these genes, a candidate of unknown function was located in the context of other genes in the pathway, for example, in the immediate vicinity of *folK* and/or *folP* in *Thermotoga, Xanthomonas*, and *Methylococcus*, and near *folM* in *Nitrosomonas*. In the *Neisseria*, the candidate is adjacent to dihydrobiopterin reductase. The GCYH-I activity of the candidates from *Thermotoga maritima, Bacillus subtilis, Acinetobacter baylyi*, and *Neisseria gonorrhoeae* was experimentally verified. This new GCYH-I, annotated as type 2, is found in about 20% of sequenced bacteria, including the pathogenic *Staphylococci* and *Neisseria* [31].

Continuing exploration of the Folate Biosynthesis subsystem (fig. 2) with as many as 400 genomes across all domains revealed more new discoveries [32]. The populated subsystem had empty cells in most of the rows, or genomes, for *folQ*, which encodes dihydroneopterin triphosphate (DHNTP) pyrophosphatase activity in *E. coli* [33, 34]. By integrating evidence of gene similarity, clustering, fusion, and phylogenetic distribution, candidate genes were predicted to fill the role of *folQ* in some bacteria and plants, but the identity of the protein that plays the role is still an open question in most bacteria. While *folQ* represents a globally missing gene, other empty cells in the subsystem spreadsheet indicated locally missing genes for almost every step of the synthesis pathway. Candidates for such missing genes in bacteria and plants were then predicted using comparative genomic context, and representative candidates were experimentally confirmed.

Organism haemophilus	Domain Bac	Variant [?] = all	*folE	*folQ	*DHNA	FolB2	FolX	FolK	*FolP	*DHFS	*AUX	*DHFR	*fpgs
Haemophilus ducreyi 35000HP	Bacteria	111	945_1	503_41	835_5	264		511	545_9	1244_10_16	644_29	958_12	1244_10_16
Haemophilus influenzae 86-028NP	Bacteria	111	364_1	135_41	1681_5	1455		596	277_9	1514_10_16	1691_29	767_12	1514_10_16
Haemophilus influenzae R2846	Bacteria	111	774_1	494									1072_10_16
Haemophilus influenzae R2866	Bacteria	111	896_1	1513_41	1839_5	567		1186	913_9	988_10_16	89_29	359_12	988_10_16
Haemophilus influenzae Rd KW20	Bacteria	2.101	1381_1	297_41	250_5	1081		63	1272_9, 1396_9	1206_10_16	671_29	866_12	1206_10_16
Haemophilus somnus 129PT	Bacteria	111	1059_1	1580_41	868_5	1303		623	389_9	1029_10_16	697_29	1291_12	1029_10_16
Haemophilus somnus 2336	Bacteria	111	302_1	688_41	930_5	1470		1245	2105_9	331_10_16	181_29	1613_12	331_10_16

*folQ
3: folQ Dihydroneopterin triphosphate pyrophosphohydolase
4: folQ3 Putative DHNTP pyrophosphatase
41: folQ2 Dihydroneopterin triphosphate pyrophosphohydolase type 2

Fig. 2. The Folate Biosynthesis subsystem spreadsheet, focused on *Haemophilus*. Columns represent functional roles, which may be played by different proteins in different organisms, as is the case for *folQ*. Rows represent different genomes, and the cells of the spreadsheet are populated by genes responsible for the function. Within rows, background colors represent genes that are clustered on the chromosome. The complete subsystem includes a separate table of functional roles with reactions, and a diagram.

The extrapolation of this strategy to pathogenic reconstruction awaits improvements in virulence subsystems. NMPDR curators are actively seeking collaborations with subject experts with the goal of building subsystems that define virulence pathways for different aspects of pathogenesis, e.g., evasion of host defenses, adhesion, toxigenesis, host-cell invasion, etc. [35]. From these subsystems, virulence protein families will be defined, virulence motifs will be determined, and it will be possible to predict candidate pathogenesis genes in newly sequenced genomes. This will not spare the need to verify these functions experimentally, just as predicting roles in metabolic pathways does not spare the need to experimentally confirm the activity. What this will do is to accelerate medical microbiology research in emerging or re-emerging pathogens (e.g., *Legionella pneumophila* and *Streptococcus pyogenes*), biothreats (e.g., *Bacillus anthracis* and *Francisella tularensis*), unculturable or slowly growing organisms (e.g., *Mycobacterium tuberculosis, M. leprae*, and *Treponema pallidum*), and pathogens for which no genetic manipulation system has been developed (e.g., Chlamydiae).

Comparative Pathogenomics Tools in NMPDR

Increasingly sophisticated analyses of the whole genomes, core genomes, pangenomes, dispensable genomes, and pathogenomes of various groups of pathogens have been published as the number of available genomes has expanded (reviewed in [2]). When few fully sequenced genomes of the same species were available, biologists used experimental rather than computational techniques to estimate the relatedness of many strains of a given serotype or phenotype, for example, comparative genomic hybridization on whole genome microarrays [36] and PCR screening for the presence of prophages [37] or other regions of diversity [38]. These studies provided estimates of the complement of genes shared by all members of a given species, the core genome or chromosomal backbone. These studies also provided an estimate of dispensable genomes or pathogenomes corresponding to a species or to a defined serotype or disease phenotype. The practical utility of these results is limited, however, by the availability of clinical strains or computational tools used to generate the data sets, as well as by the format of the data sets, which are frequently provided as supplemental tables of gene id numbers in PDF format on the web sites of journal publishers. While it is certainly possible to use these id numbers to retrieve the nucleotide or amino acid sequence from the corresponding database, it is a tedious task for most wet-bench, experimental biologists. In response, NMPDR has developed user-friendly tools to empower biologists to make use of genomic data that is regularly updated.

NMPDR provides several tools for whole genome comparison on the basis of sequence similarity or functional annotation. One is the Signature Genes tool, which may be used to compute a core genome or to define a signature associated with a limited group of genomes that display an interesting phenotype. This tool uses precomputed BLASTP results to compare the sequences of all proteins in a selected reference genome to all those in a set of genomes selected in the comparison, or inclusion, set. The user may set the stringency and the scope of the comparison. Stringency is determined by the E-value of the BLASTP similarity, which is set to 1e-10 by default, and the scope is controlled by a commonality factor, which is set to 0.8 (80% of comparison genomes) by default. For example, with reference to the genome of *Streptococcus mutans* there are 850 proteins shared with an E-value of less than 1e-10 and commonality of 1.0 by all 24 finished (closed) streptococcal genomes in version 23 of NMPDR (3 *S. agalactiae*, 1 *S. equi*, 1 *S. mitis*, 1 *S. mutans*, 3 *S. pneumoniae*, 11 *S. pyogenes*, 1 *S. sanguinus*, 2 *S. thermophilus*, 1 *S. uberis*). Another use of the Signature Genes tool is to compare a reference genome with genomes selected in an inclusion set, and contrast these with genomes in an exclusion set. Users may find the answers to questions such as, which genes are found in two strains of GAS that are associated with rheumatic fever, but not in the other strains of GAS? This allows users to find the set of genes that represent the signature of a phenotype or serotype. The entire results table may be downloaded, and the protein or DNA sequences are also downloadable in FASTA format. For each protein found, the results table links to pages describing and providing evidence for the annotation, as well as to pages describing the subsystems for those proteins that are included in a subsystem. These links allow the user to immediately explore the physical and functional context of any protein that matches the search criteria. Comparative analysis of proteins in common to organisms with a shared phenotype but absent from other closely related organisms that lack the phenotype will inform experimental science and move the field of pathogenomics forward.

Conclusion

Pathogenomics arises at the intersection of genomics and microbial pathogenesis. This new field has been defined as the study of host and pathogen genomes [6, 13] and as the study of pathogenomes [39, 40], i.e., the large sections of genomes encoding virulence genes and driving intra-species diversification within microbial genomes. The tools for generating whole genome sequences and annotating them have improved dramatically since the genome sequence of the first bacterial pathogen was published. Tools for comparative analysis of whole genome sequences are becoming more powerful and easy to use. The future of pathogenomics research will be to explore newly sequenced genomes and, ideally, to predict the lifestyle of the organism and its potential interactions with other organisms in its habitat, notably eukaryotic hosts. Metabolic reconstruction from genomic data alone has become possible thanks to the achievements of biochemists, who cataloged pathways involved in the central machinery of life. Additional 'omic' data, some existing in the literature but not yet accessible in the sequence databases, and much data still to be collected, will be needed to catalog the disease-causing potential and virulence pathways of known pathogens. Pathogenic reconstruction is the challenge for microbiologists in the post-genomic era.

Acknowledgements

The authors thank Andrei Osterman and the editors for the opportunity to contribute to this volume. We also gratefully acknowledge the enormous effort of curators and developers at FIG, Argonne National Laboratory, University of Chicago, and University of Illinois. Special thanks to Matt De Jongh of Hope College for productive discussions about metabolic reaction networks. This work was supported with Federal funds from the National Institute of Allergy and Infectious Diseases, National Institutes of Health, Department of Health and Human Services, USA, under Contract HHSN266200400042C.

References

1 Fredericks DN, Relman DA: Sequence-based identification of microbial pathogens: a reconsideration of Koch's postulates. Clin Microbiol Rev 1996;9:18–33.

2 Medini D, Serruto D, Parkhill J, Relman DA, Donati C, et al: Microbiology in the post-genomic era. Nat Rev Microbiol 2008;6:419–430.

3 Fleischmann RD, Adams MD, White O, Clayton RA, Kirkness EF, et al: Whole-genome random sequencing and assembly of *Haemophilus influenzae* Rd. Science 1995;269:496–512.

4 Fraser CM, Gocayne JD, White O, Adams MD, Clayton RA, et al: The minimal gene complement of *Mycoplasma genitalium*. Science 1995;270:397–403.

5 Tyson GW, Chapman J, Hugenholtz P, Allen EE, Ram RJ, et al: Community structure and metabolism through reconstruction of microbial genomes from the environment. Nature 2004;428:37–43.

6 Crossman L, Cerdeno-Tarraga A, Bentley S, Parkhill J: Pathogenomics. Nat Rev Microbiol 2003;1:176–177.

7 Dinsdale EA, Edwards RA, Hall D, Angly F, Breitbart M, et al: Functional metagenomic profiling of nine biomes. Nature 2008;452:629–632.

8 National Microbial Pathogen Data Resource [database on the Internet]. Version of March 24, 2008. Chicago: Computation Institute, University of Chicago/Argonne National Laboratory/Fellowship for Interpretation of Genomes; 2004- [cited 2008 May 10]. Available from: http://www.nmpdr.org//FIG/seedviewer.cgi?pattern = 71421.1;page = SearchResult

9 Aziz RK, Bartels D, Best AA, DeJongh M, Disz T, et al: The RAST Server: rapid annotations using subsystems technology: BMC Genomics 2008;9:75–89.

10 Inglis TJ: Principia aetiologica: taking causality beyond Koch's postulates. J Med Microbiol 2007;56:1419–1422.

11 Falkow S: Molecular Koch's postulates applied to microbial pathogenicity. Rev Infect Dis 1988;10:S274–S276.

12 Falkow S: Molecular Koch's postulates applied to bacterial pathogenicity–a personal recollection 15 years later. Nat Rev Microbiol 2004;2:67–72.

13 Pallen MJ, Wren BW: Bacterial pathogenomics. Nature 2007;449:835–842.

14 Winram SB, Lottenberg R: The plasmin-binding protein Plr of group A streptococci is identified as glyceraldehyde-3-phosphate dehydrogenase. Microbiology 1996;142:2311–2320.

15 Gase K, Gase A, Schirmer H, Malke H: Cloning, sequencing and functional overexpression of the *Streptococcus equisimilis* H46A *gapC* gene encoding a glyceraldehyde-3-phosphate dehydrogenase that also functions as a plasmin(ogen)-binding protein. Purification and biochemical characterization of the protein. Eur J Biochem 1996;239:42–51.

16 Buckley M: The genomics of disease-causing organisms: mapping a strategy for discovery and defense. American Academy of Microbiology 2004 (http://academy.asm.org/index.php?option = com_content&task = blogcategory&id = 22&Itemid = 57).

17 McMillan DJ, Beiko RG, Geffers R, Buer J, Schouls LM, et al: Genes for the majority of group A streptococcal virulence factors and extracellular surface proteins do not confer an increased propensity to cause invasive disease. Clin Infect Dis 2006;43:884–891.

18 Rogers S, Commons R, Danchin MH, Selvaraj G, Kelpie L, et al: Strain prevalence, rather than innate virulence potential, is the major factor responsible for an increase in serious group A *Streptococcus* infections. J Infect Dis 2007;195:1625–1633.

19 Kotb M, Norrby-Teglund A, McGeer A, El-Sherbini H, Dorak MT, et al: An immunogenetic and molecular basis for differences in outcomes of invasive group A streptococcal infections. Nat Med 2002;8:1398–1404.

20 Medina E, Goldmann O, Rohde M, Lengeling A, Chhatwals GS: Genetic control of susceptibility to group A streptococcal infection in mice. J Infect Dis 2001;184:846–852.

21 Peirce JL, Lu L, Gu J, Silver LM, Williams RW: A new set of BXD recombinant inbred lines from advanced intercross populations in mice. BMC Genet 2004;5:7–23.

22 Aziz RK, Kansal R, Abdeltawab NF, Rowe SL, Su Y, et al: Susceptibility to severe streptococcal sepsis: use of a large set of isogenic mouse lines to study genetic and environmental factors. Genes Immun 2007;8:404–415.

23 Abdeltawab NF, Aziz RK, Kansall R, Rowe SL, Su Y, et al: An unbiased systems genetics approach to mapping genetic loci modulating susceptibility to severe streptococcal sepsis. PLoS Pathogens 2008;4:e1000042.

24 Overbeek R, Begley T, Butler RM, Choudhuri JV, Chuang HY, et al: The subsystems approach to genome annotation and its use in the project to annotate 1000 genomes. Nucleic Acids Res 2005;33:5691–5702.

25 Greene JM, Collins F, Lefkowitz EJ, Roos D, Scheuermann RH, et al: National Institute of Allergy and Infectious Diseases bioinformatics resource centers: new assets for pathogen informatics. Infect Immun 2007;75:3212–3219.

26 McNeil LK, Reich C, Aziz RK, Bartels D, Cohoon M, et al: The National Microbial Pathogen Database Resource (NMPDR): A genomics platform based on subsystem annotation. Nucleic Acids Res 2007;35: D347–D353.

27 Liolios K, Mavromatis K, Tavernarakis N, Kyrpides NC: The Genomes OnLine Database (GOLD) in 2007: status of genomic and metagenomic projects and their associated metadata. Nucleic Acids Res 2008;36:D475–D479.

28 Palsson B: Two-dimensional annotation of genomes. Nat Biotechnol 2004;22:1218–1219.

29 De Jongh M, Formsma K, Boillot P, Gould J, Rycenga M, Best A: Toward the automated generation of genome-scale metabolic networks in the SEED. BMC Bioinformatics 2007;8:139–155.

30 Klamt S, Stelling J, Ginkel M, Gilles ED: FluxAnalyzer: exploring structure, pathways, and flux distributions in metabolic networks on interactive flux maps. Bioinformatics 2003;19:261–269.

31 El Yacoubi B, Bonnett S, Anderson JN, Swairjo MA, Iwata-Reuyl D, de Crecy-Lagard V: Discovery of a new prokaryotic type I GTP cyclohydrolase family. J Biol Chem 2006;281:37586–37593.

32 de Crécy-Lagard V, El Yacoubi B, de la Garza RD, Noiriel A, Hanson AD: Comparative genomics of bacterial and plant folate synthesis and salvage: predictions and validations. BMC Genomics 2007;8: 245–249.

33 Klaus SM, Wegkamp A, Sybesma W, Hugenholtz J, Gregory JF 3rd, Hanson AD: A nudix enzyme removes pyrophosphate from dihydroneopterin triphosphate in the folate synthesis pathway of bacteria and plants. J Biol Chem 2005;280:5274–5280.

34 Gabelli SB, Bianchet MA, Xu W, Dunn CA, Niu ZD, et al: Structure and function of the *E. coli* dihydroneopterin triphosphate pyrophosphatase: a Nudix enzyme involved in folate biosynthesis. Structure 2007;15:1014–1022.

35 Curtis MA, Slaney JM, Aduse-Opoku J: Critical pathways in microbial virulence. J Clin Periodontol 2005;32:28–38.

36 Smoot JC, Barbian KD, Van Gompel JJ, Smoot LM, Chaussee MS, et al: Genome sequence and comparative microarray analysis of serotype M18 group A *Streptococcus* strains associated with acute rheumatic fever outbreaks. Proc Natl Acad Sci USA 2002;99:4668–4673.

37 Banks DJ, Porcella SF, Barbian KD, Beres SB, Philips LE, et al: Progress toward characterization of the group A *Streptococcus* metagenome: complete genome sequence of a macrolide-resistant serotype M6 strain. J Infect Dis 2004;190:727–738.

38 Green NM, Zhang S, Porcella SF, Nagiec MJ, Barbian KD, et al: Genome sequence of a serotype M28 strain of group A *Streptococcus*: potential new insights into puerperal sepsis and bacterial disease specificity. J Infect Dis 2005;192:760–770.

39 Collyn F, Guy L, Marceau M, Simonet M, Roten CA: Describing ancient horizontal gene transfers at the nucleotide and gene levels by comparative pathogenicity island genometrics. Bioinformatics 2006;22: 1072–1079.

40 Yoon SH, Park YK, Lee S, Choi D, Oh TK, et al: Towards pathogenomics: a web-based resource for pathogenicity islands. Nucleic Acids Res 2007;35: D395–D400.

Leslie K. McNeil
National Center for Supercomputing Applications
1205 W. Clark St.
Urbana, IL 61801 (USA)
Tel. +1 217 244 0597, Fax +1 217 244 2909, E-Mail lkmcneil@yahoo.com

de Reuse H, Bereswill S (eds): Microbial Pathogenomics.
Genome Dyn. Basel, Karger, 2009, vol 6, pp 35–47

The Bacterial Pan-Genome and Reverse Vaccinology

H. Tettelin

Institute for Genome Sciences, University of Maryland School of Medicine, Baltimore, Md., USA

Abstract

The whole genome sequence of most human bacterial pathogens is available and the advent of next-generation sequencing technologies will result in a large number of sequenced isolates per pathogenic species. The study of multiple genome sequences of a given bacterium provides insights into its evolution, pathogenic potential and diversity. The pathogen's pan-genome, defined as the sum of the core genome shared by all sequenced strains and the dispensable genome present only in a subset of the isolates, can be analyzed to assess the size and diversity of the gene repertoire that the species has access to. This information is then used to better inform the reverse vaccinology approach whereby vaccine candidates are identified and prioritized in silico based on genomic data. Bioinformatics integration of genome sequence data with functional genomics results and clinical meta-data is essential to maximize the use of this large amount of information to answer biologically relevant questions.

We have come a long way since the release of the first complete genome sequence of a bacterial pathogen, *Haemophilus influenzae* [1] thirteen years ago. The whole genome shotgun approach, then revolutionary, is now the standard for genome sequencing. Its application has led to the availability of one or more genome sequences for most of the major human pathogens, as well as other bacteria. As of January 2009, the Genomes Online Database (GOLD v2.0, http://www.genomesonline.org) lists 766 complete published bacterial genomes and another 2,262 ongoing ones. The advent of next generation sequencing technologies [2] will significantly increase these numbers to the point where genome sequence data for most known bacterial species will eventually become available.

This wealth of information provides a solid framework to interrogate intra-species bacterial diversity. This type of diversity can be mediated by spontaneous mutations, recombination, and/or lateral gene transfer. Among other outcomes, these mechanisms result in gene acquisition and loss, and therefore contribute to variation in gene content between isolates of a species. It has been shown, for instance, that strains of the O157 serotype of *Escherichia coli* share a 4.1-Mb genome backbone with the non-pathogenic laboratory strain K-12 but also harbor an additional 1.4 Mb of sequence encoding 1,387 genes, many of which are involved in virulence [3]. Further analysis of enterohemorrhagic (O157:H7) and uropathogenic strains of *E. coli* revealed extensive gene content variation mainly in the form of pathogenicity islands [4]. Similar studies in streptococci [5, 6], staphylococci [7], and other pathogens [8] revealed significant gene content variation across isolates and a significant fraction of this diversity was encoded by mobile genetic elements such as pathogenicity islands, bacteriophages or plasmids.

When searching for new candidates for the development of effective vaccines against pathogens, it is important to understand their gene content diversity. Indeed, designing vaccines against potent antigens such as CagA in *Helicobacter pylori* [9] or the newly discovered pilus in *Streptococcus pneumoniae* [10, 11] would not lead to broadly protective vaccines given the limited presence of these antigens among strains of the species. Knowledge of the genomic diversity of the species of interest better informs the identification and prioritization of vaccine candidates and often provides several candidates to consider and characterize simultaneously. The use of genome sequence information to identify vaccine candidates has been termed reverse vaccinology [12]. Over time, there has been an evolution of the approaches used in vaccine research. Stanley Plotkin recently described his view of the six revolutions in vaccinology [13]. The first five have already happened: attenuated organisms, inactivated organisms, cell culture and reassortment, genetic engineering, and induction of cellular immunity. In his view, there are many contenders for the sixth revolution including combination vaccines, new adjuvants, proteomics, vaccines against noninfectious diseases, and reverse vaccinology. Among these only proteomics and reverse vaccinology focus, at least in part, on the identification of new candidates; and these two techniques are complementary. For instance, proteomics can be used to confirm predictions on candidate proteins made during the genome-mining step of reverse vaccinology. Most importantly, both of these approaches will be affected by the variation of gene content occurring among isolates of the pathogen studied.

This chapter covers the impact that genomic diversity has on the prediction of vaccine candidates using reverse vaccinology.

Bacterial Diversity and the Pan-Genome Concept

The genome sequence of multiple strains of most of the major human pathogens is currently available where one or more genomes are complete and free of gaps while others are draft whole genome sequences with or without partial closure of gaps. The availability of the genome sequence of a single strain of a pathogenic species provides genetic information about its metabolic capabilities, lifestyle, pathogenic potential and genomic structure. In many instances, the entire gene repertoire deciphered from the first genome sequence of a pathogen has been represented on a DNA microarray that is then used to interrogate the species diversity by microarray-based comparative genomic hybridizations (mCGH) (examples of bacterial mCGH studies include [14–24]).

Very little was known about the genetic diversity of *Streptococcus agalactiae* (group B *Streptococcus,* GBS), a major pathogen and a leading cause of disease in newborn infants and the elderly [25–27], when its first genome sequence was published in 2002 [28]. A microarray was constructed based on this genome and 19 human isolates of GBS representing the major disease-causing serotypes were hybridized. This experiment revealed major islands of genomic diversity distributed across the reference genome [28]. In total, 18% of the genes from the reference strain, including important virulence determinants and surface proteins, were not detected in at least one of the 19 strains. Of these variable genes, 91% were clustered in 15 genomic regions of five or more contiguous genes, most of which displayed characteristics of potentially mobile or foreign DNA based on their nucleotide composition and flanking repeats. While the mCGH experiments revealed extensive diversity within GBS strains when compared to the reference genome, highlighting reference regions that are divergent elsewhere, they did not interrogate the genomic fragments of the 19 strains that were not shared with the reference. To overcome this limitation, the complete sequences of six additional GBS strains representing the major disease-causing serotypes were generated and added to the two GBS genomes that were publicly available at the time [28, 29]. Comparison of the eight GBS genomes confirmed the diversity identified by mCGH but also unraveled the sequence and coding potential of many genomic islands that were not shared with the reference genome [30]. Overall, the eight isolates shared a high degree of synteny interrupted by 69 interspersed genomic islands that were absent in one or more genomes.

The high degree of diversity exhibited by the GBS species leads to two important questions: how large is the gene repertoire accessible to this species and how many genomes should be sequenced to characterize this repertoire? In order to address these questions, the GBS core genome was defined as the ~1,800 genes shared by all of the eight strains and the rest of the genes were scrutinized. By analyzing all permutations of adding a new genome to N genomes already considered, where N ranges from 1 to 7, it was determined that each GBS sequence contributes an average of 33 new genes that had not been identified in previous genomes [30]. Mathematical extrapolation of the average number of new genes provided by each genome in all permutations revealed a curve that does not cross the X axis, suggesting that a large number of genomes would have to be sequenced in order to characterize the GBS pan-genome (fig. 1). The pan-genome is defined as the sum of the core genome shared by all isolates and the dispensable genome that is composed of genes shared by only a subset of the strains together with genes specific to individual strains [31]. In general, the core genome encodes functions related to the basic biology and phenotypes of the species

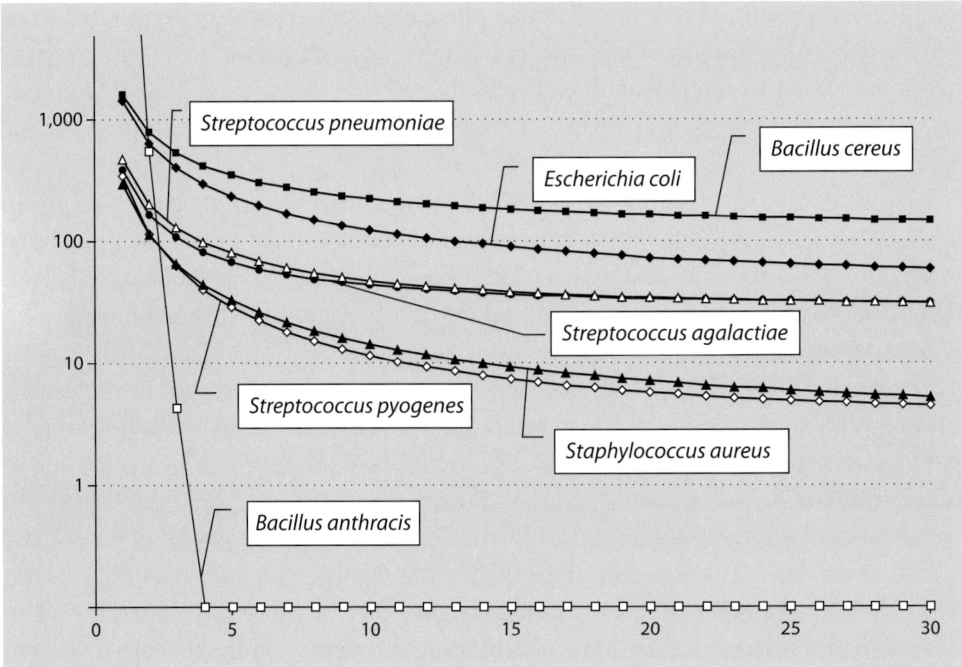

Fig. 1. Pan-genome analysis of seven bacterial species. The average number of new genes (y-axis) discovered with the availability of an additional whole genome sequence is represented in logarithmic scale as a function of the number of genomes already analyzed (x-axis). The curves are power-law regressions calculated based on all permutations of adding a new genome sequence to N genomes (for details on the pan-genome analysis and regression see [30, 31]). The species depicted were chosen because they are important pathogens and there exist at least seven whole genome sequences publicly available for each of them. This number of genomes provides sufficient statistical power for the regressions. Unfortunately, only five genome sequences are publicly available for *Neisseria meningitidis* so this species could not be analyzed. The total number of genome sequences used for each species was as follows: *Staphylococcus aureus* 7, *Streptococcus pyogenes* 7, *Streptococcus agalactiae* 8, *Streptococcus pneumoniae* 10, *Bacillus anthracis* 10, *Escherichia coli* 13, and *Bacillus cereus* 14. However, for simplicity of the display only the theoretical mathematical extrapolation to 30 genomes is depicted. All but one of the species exhibit a curve that does not reach zero, indicating that their pan-genome is open, while that of *Bacillus anthracis* is closed.

while the dispensable genome contributes to the diversity and likely provides functions that are not essential for the basic life cycle but some of which confer selective advantages including niche adaptation, antibiotic resistance and the ability to colonize new hosts [32].

The pan-genome analysis was conducted on the genomes from other species, including *S. pneumoniae, Streptococcus pyogenes, Staphylococcus aureus, E. coli, Bacillus cereus* and *Bacillus anthracis.* As indicated in figure 1, the pan-genome from all of these species except *B. anthracis* appears to be very large as well, leading to the concept of an open pan-genome species where the entire gene repertoire has yet to be defined and the pan-genome is much larger than the genome of any individual strain. In contrast, an average of four *B. anthracis* genomes is sufficient to characterize its pan-genome, likely reflecting the higher clonality of this organism. Indeed, *B. anthracis* is considered to be a clone of *B. cereus* and while the *B. anthracis* pan-genome is closed, the *B. cereus* pan-genome is open. Recently, the *Haemophilus influenzae* and *S. pneumoniae* supragenomes, another denomination for a concept similar to the pan-genome, have been studied and confirmed to be much larger than any individual genome [33–35]. The pan-genome and supragenome concepts constitute an attempt at estimating the size of the species gene repertoire and its content. They build upon the concept of species genome put forward by Lan and Reeves [8] where the species' coding potential was partitioned in core and auxiliary genes. With the fairly limited number of genome sequences per species available to date, we are not yet in a position to predict the actual size of open pan-genomes, that is to 'close' them. It is clear that a large number of additional genome sequences for many species will have to be generated but the exact number is unknown.

The issue of sampling exists when studying pan-genomes since only the data from sequenced isolates can be used, and these isolates are never chosen randomly. The availability of next generation sequencing technologies [2] that provide higher throughput and decrease costs will enable the sequencing of many more genomes of all pathogenic species and provide a framework for a more representative sampling of isolates to study. The main technologies currently on the market include the Roche/454 Life Sciences pyrosequencing method (www.roche-applied-science.com), the Illumina/Solexa reversible terminator chemistry and clonal single molecule array approach (www.illumina.com), the ABI SOLiD sequencing by sequential ligation system (www.appliedbiosystems.com) and the Helicos Biosciences single molecule sequencing platform (www.helicosbio.com). Many more technologies are under development and hold promise to further increase throughput [36]. These next generation platforms come with drawbacks that in most cases consist of a somewhat lower accuracy than Sanger sequencing and shorter read lengths. The former is usually compensated by achieving higher sequence coverage than with the classical approach, while the latter remains an inherent problem, especially in the case of de novo sequencing of a genome for which no reference genome is available. Nevertheless, it is foreseeable that these technologies will lead to the availability of multiple genome sequences for most of the bacterial species known to date. As a consequence, mCGH, especially when used to assay bacterial diversity as described above, will progressively be replaced by whole genome sequencing that overcomes its limitations [36].

The conclusion of this section is that the genomic diversity of many bacterial species, including pathogens, can be quite extensive. Species with an open pan-genome exhibit remarkably high levels of diversity, have access to a large gene repertoire, and therefore harbor the potential of being extremely versatile and adaptable. Such abilities raise concerns for disease treatment given that these pathogens possess a more extensive tool set to evade immunity and vaccination, and develop multi-drug resistance. It is therefore important to consider the entire gene repertoire from the pan-genome when searching for new protein candidates for vaccine development. Knowledge of the distribution of specific proteins will help inform identification and prioritization: do they belong to the core, dispensable, or strain-specific subsets?; are they associated with invasive or carriage isolates?; are they over-represented in isolates from endemic geographical areas?, etc. In the next section reverse vaccinology is discussed, an example of the use of genome sequence information for the identification of vaccine candidates.

Reverse Vaccinology

Reverse vaccinology was pioneered on *Neisseria meningitidis* and proved successful with the use of the genome sequence from a single isolate [37]. As of January 2009, there are still only five complete genome sequences of *N. meningitidis* publicly available in GenBank. It is therefore not possible to perform meaningful regressions to determine whether the species' pan-genome is open or closed. It is known, however, that significant genomic differences exist between the sequenced strains, including large variations in gene content [38].

Reverse vaccinology inverts the steps of classical approaches to vaccine research that involve one of two methods: generation of live-attenuated strains by serial passages in vitro or isolation of protective antigens from the cultured organism by biochemical, serological or genetic techniques [39]. These methods only work for organisms that can be cultured, are time consuming, and only identify abundant antigens. In the case of serogroup B strains of *N. meningitidis,* 40 years of classical vaccine research led to a few antigens that were highly variable and only conferred protection against the strain they were isolated from. Generation of a successful vaccine was further stymied by the inability to use the serogroup B capsular polysaccharide as an antigen due to the fact that it is identical to a polysialic acid present in many of our tissues and therefore constitutes a risk of autoimmunity [12]. To circumvent these shortcomings, the whole genome sequence of a serogroup B strain of the meningococcus was generated and analyzed. All the proteins predicted to be encoded by the genome were submitted to an in silico pipeline geared at the identification of proteins likely to be exposed at the surface of the bacteria and therefore accessible to antibodies [40]. Criteria for selection included proteins known to carry out functions at the surface of the cell and proteins harboring amino acid motifs characteristic of: targeting to the membrane (signal peptides), anchoring in the lipid bilayer (lipoproteins), anchoring in the outer-membrane of Gram-negative bacteria or the cell wall of Gram-positive bacteria, and interaction with host proteins or structures (e.g. integrin binding domains) [41]. Proteins known to be cytoplasmic or likely to be embedded in the cell's membrane and inaccessible to antibodies were systematically excluded. This analysis identified 570 potential surface antigens within the genome of *N. meningitidis.* These candidate antigens were subjected to experimental characterization to assess their antigenicity, accessibility at the surface, and conservation across strains [42]. All candidate genes were cloned in *E. coli* expression vectors and 350 recombinant proteins were successfully purified in sufficient amounts for mouse immunizations. Sera recovered from these mice were then used for characterization of the candidates. Expression of the proteins by the meningococcus was assayed by western blot on both whole cell extracts and outer-membrane vesicles. Surface exposure and accessibility was tested by enzyme-linked immunosorbent assay (ELISA) and flow cytometry on whole cells. Finally, the probability that the antigens constitute viable vaccine candidates was evaluated based on the bactericidal assay where the complement-mediated bacterial killing activity of the antibodies is tested on whole cells. Of the 350 proteins available, 28 were positive in all of these experimental assays. Given the high degree of antigen variability in *N. meningitidis*, it was important to evaluate the level of conservation of these 28 candidates across a panel of diverse strains of *Neisseria*, including *N. meningitidis* strains of the five disease-causing serogroups (A, B, C, Y and W135) and other species of *Neisseria: N. cinerea, N. lactamica,* and *N. gonorrhoeae.* Amplification of the genes by PCR and sequencing revealed eight novel vaccine candidates that were highly conserved and therefore likely to confer broad protection when used for vaccine development. These antigens were tested individually and in combination for protection in the animal model as well as in human clinical trials [43]. A cocktail of five of the antigens (composed of a surface lipoprotein, a phospholipid-binding domain lipoprotein, a YceI family protein of unknown function, factor H-binding protein fHBP and the invasin NadA) has been successfully taken through phase I and II clinical trials in infants and has recently entered phase III trials [44]. This example underscores the power of reverse vaccinology in unraveling new protective antigens that could not be identified through four decades of classical vaccine research and in accelerating the delivery of new vaccines on the market.

Subsequently, the reverse vaccinology approach was applied to the GBS pan-genome [45]. Given the diversity encountered within this open pan-genome species and the failure to identify broadly protective individual antigens from the first genome sequence available (which by definition harbors all genes from the core genome), it was decided not to restrict the in silico predictions to proteins encoded by the core genome. Although core proteins are more likely to confer broad protection, our experience with a single GBS genome suggested that a combination of core and dispensable proteins would be necessary to achieve the desired levels of protection. The failure to use only core proteins may be due to the fact that only a fraction of the core surface proteins are expressed during infection, or the fact that expressed proteins are not accessible to antibodies, for example because they do not protrude far enough of the cell surface and are masked by capsular polysaccharides. A total of 589 proteins were predicted to be surface exposed from the pan-genome, 396 of which belonged to the core genome. Cloning and expression of the 589 candidates in *E. coli* resulted in 357 recombinant proteins that were successfully recovered in solution and used for mouse immunizations. Because one of the major problems with GBS is the infection of infants during delivery, the mouse model of disease consists of immunization of adult female mice followed by challenge of their pups with the pathogen within 48 h. Systematic screening of the purified candidates using this model revealed four antigens (a LysM domain protein involved in cell envelope functions and three cell-wall anchored proteins) capable of significantly protecting infant mice from challenge with a GBS strain known to carry the antigen [45]. Only one of these antigens, the Sip protein, was part of the core genome and yet it only provided partial protection. Sip was initially described as a universal vaccine candidate [46] yet its accessibility to antibodies was impaired by the presence of the polysaccharide capsule [45]. As expected, non-core antigens did not confer any protection against strains lacking the gene. In some instances, no or little protection was observed even when the challenge strain carried the gene, again suggesting an issue with antigen accessibility. Flow cytometry confirmed this hypothesis by demonstrating variable levels of antibody binding that correlated with animal protection results. A cocktail composed of the four antigens was used in the animal model and tested against a panel of diverse GBS challenge strains representing the major pathogenic serotypes. This resulted in high levels of protection ranging from 59 to 100%. This antigen combination also displayed a bactericidal effect, suggesting that it constituted a good candidate for vaccine development in humans.

The fact that the best cocktail of vaccine candidates contains only one protein from the core genome appears counterintuitive. Common sense would dictate that the best way to reach a broadly protective vaccine is to use antigens present on all strains. The GBS study demonstrated that some core antigens are not suitable for vaccine development. The problem of accessibility at the surface, for instance due to masking by a polysaccharide capsule as described for Sip [45] or the leucin-rich repeat GBS antigen Blr [47], needs to be considered and is not readily predictable in silico. The timing and level of expression of the antigens is also crucial and can be studied by transcriptomics and proteomics. The antigenicity of the candidates also varies and predictions based on epitope modeling or structural genomics can help prioritize antigens and guide vaccine development. Knowledge of the pan-genome enables classification of candidates into bins of various levels of conservation (core vs. dispensable) or impact (invasive vs. carriage) across isolates, and prioritization based on current vaccine needs. For instance, if a core antigen provides protection against 80% of isolates and the 20% not covered share dispensable genes, novel candidates should be searched in that subset of shared genes.

Data Integration

Integration of genome sequencing and functional genomics data is necessary for proper identification and prioritization of vaccine candidates. The development of bioinformatics tools to achieve this goal has become critical and several efforts are underway. The comparative genomics package Strepneumo (strepneumo-sybil.igs.

umaryland.edu) was recently released and as of January 2009 enables the detailed comparison of seventeen genomes of *Streptococcus pneumoniae*. The system is based on the public relational database schema GMOD (gmod.org) and the open source web-based genome comparison tool Sybil (sybil.sourceforge.net). Sybil allows users to search for genes or gene clusters of interest and visualize their genomic context. All of the views in Sybil are interactive and allow the user to browse the data seamlessly, for instance moving from a whole genome comparison to a local genome view to an individual gene report to the interrogation of that gene's cluster of orthologs. In the context of reverse vaccinology, Strepneumo in its present form enables detailed characterization of vaccine candidates in the context of multiple genomes (pan-genome) but does not provide a bonafide vaccine candidate prediction pipeline. Future enhancements of the system include the implementation of such a pipeline together with the incorporation of relevant publicly available data including microarray analyses (transcriptomics and mCGH), proteomics data and the new RNA-Seq approach for transcriptional profiling and RNA discovery [48]. The ultimate goal of the package is to answer high-level biological questions such as 'Display the list of all proteins that are shared by at least 70% of all sequenced strains, are located in genomic islands exhibiting an atypical nucleotide composition indicative of selective pressure or potential lateral transfer, are expressed upon adherence to epithelial cells and harbor structures predicted to be accessible epitopes.' We still have a long way to go before the system can handle such queries but they are feasible and the key is to integrate many data types in a single uniform database structure accompanied by powerful and user-friendly interfaces. The Strepneumo system will be updated with new genome data and functional genomics data as they become available over time. Similar systems will also be implemented for other species as the number of genome sequences per species continues to increase.

It is not possible to list all public tools available to perform biologically meaningful interrogations of genomics and functional genomics data. Some databases like the Comprehensive Microbial Resource (cmr.jcvi.org) aim at providing comparative power across a comprehensive list of completely sequenced species. Other databases target a subset of species like the Bioinformatics Resource Centers (www.brc-central.org) or MaGe (www.genoscope.cns.fr/agc/mage) [49]. The Bioinformatics Links Directory (bioinformatics.ca/links_directory) features a long list of links to molecular resources, tools and databases [50]. This directory provides an excellent starting point for users to get acquainted with the most useful and powerful publicly available tools for genomic data mining and analysis.

Conclusion and Perspectives

The reverse vaccinology approach has been applied to many bacterial species [e.g. 51–55]. With the availability of genomic data from most known human pathogens, it is almost inconceivable not to at least check antigens being considered for vaccine development against the DNA sequences to understand their distribution, diversity and characteristics. The rise of next-generation sequencing technologies will continue to flood databases with genome sequence information and will soon result in a fairly good representation of the pan-genome of virtually every pathogenic (and other) species known to date. The issue of strain selection for genome sequencing, which has been heavily biased towards a subset of invasive pathogenic isolates that most likely do not accurately represent the diversity of the species, will progressively be overcome owing to the ability to sequence hundreds of genomes cheaply and rapidly. Ideally, investigators will tackle all types of isolates including carriage strains, environmental relatives, fresh clinical isolates that have not been passaged in the laboratory, and multiple strains representing all the clades of the phylogeny of the species as it is currently known. This phylogeny might not be accurate but it will be refined as more genome sequences become available. In a perfect scenario, a large number of isolates should be selected randomly and sequenced but this might be limited by our ability to gain access to such random strains. An alternative is to conduct metagenomics studies where entire communities of pathogens are sequenced directly from their environment. This approach completely alleviates strain selection biases and tackles all species, including those that cannot be cultured in the laboratory. A large project currently underway aims at characterizing the human microbiome, the entire set of microbial species inhabiting our body [56] in order to understand the diversity of microbial communities in different cavities, how they vary in time within an individual, between individuals and how they affect our physiology as well as our predisposition to disease. The metagenomic approach will enhance our knowledge of the bacterial pan-genomes or pan-microbiomes if we operate at the community level. It is also possible to obtain the genome sequence of rare unculturable species thanks to the emerging field of single cell genomics [57]. Here individual cells of organisms of interest are isolated by dilution, separation or micro-manipulation techniques, and their genomic DNA is amplified by multiple displacement amplification [58] for further studies.

It is becoming increasingly important to integrate genome sequence data with functional genomics data, as well as clinical meta-data associated with the strains under study in order to maximize our ability to extract biologically relevant information from this flood of 'omics' information. The development of robust databases and powerful bioinformatics tools to interrogate them is a requisite and many projects are underway to achieve this goal. It is foreseeable that in silico analyses will provide more and more refined information, for instance on vaccine candidates by narrowing the number of proteins to study possibly by a log. But it is important to continue to use experimental validation of computer predictions and in turn to use these experimental results to refine computer prediction tools. The rapid advances in laboratory and bioinformatics technologies that we have observed recently paint a bright future for such feedback loop interactions...

Acknowledgements

I thank David Riley for help with pan-genome analyses and generation of figure 1.

References

1 Fleischmann RD, Adams MD, White O, Clayton RA, Kirkness EF, et al: Whole-genome random sequencing and assembly of *Haemophilus influenzae* Rd. Science 1995;269:496–512.

2 Shendure J, Ji H: Next-generation DNA sequencing. Nat Biotechnol 2008;26:1135–1145.

3 Perna NT, Plunkett G 3rd, Burland V, Mau B, Glasner JD, et al: Genome sequence of enterohaemorrhagic *Escherichia coli* O157:H7. Nature 2001;409: 529–533.

4 Welch RA, Burland V, Plunkett G 3rd, Redford P, Roesch P, et al: Extensive mosaic structure revealed by the complete genome sequence of uropathogenic *Escherichia coli*. Proc Natl Acad Sci USA 2002;99: 17020–17024.

5 Beres SB, Sylva GL, Sturdevant DE, Granville CN, Liu M, et al: Genome-wide molecular dissection of serotype M3 group A *Streptococcus* strains causing two epidemics of invasive infections. Proc Natl Acad Sci USA 2004;101:11833–11838.

6 Brochet M, Couve E, Glaser P, Guedon G, Payot S: Integrative conjugative elements and related elements are major contributors to the genome diversity of *Streptococcus agalactiae*. J Bacteriol 2008;190: 6913–6917.

7 Ben Zakour NL, Sturdevant DE, Even S, Guinane CM, Barbey C, et al: Genome-wide analysis of ruminant *Staphylococcus aureus* reveals diversification of the core genome. J Bacteriol 2008;190:6302–6317.

8 Lan R, Reeves PR: Intraspecies variation in bacterial genomes: the need for a species genome concept. Trends Microbiol 2000;8:396–401.

9 Torres J, Backert S: Pathogenesis of *Helicobacter pylori* infection. Helicobacter 2008;13(suppl 1):13–17.

10 Barocchi MA, Ries J, Zogaj X, Hemsley C, Albiger B, et al: A pneumococcal pilus influences virulence and host inflammatory responses. Proc Natl Acad Sci USA 2006;103:2857–2862.

11 Bagnoli F, Moschioni M, Donati C, Dimitrovska V, Ferlenghi I, et al: A second pilus type in *Streptococcus pneumoniae* is prevalent in emerging serotypes and mediates adhesion to host cells. J Bacteriol 2008; 190:5480–5492.

12 Rappuoli R: Reverse vaccinology. Curr Opin Microbiol 2000;3:445–450.

13 Plotkin SA: Six revolutions in vaccinology. Pediatr Infect Dis J 2005;24:1–9.

14 Aakra A, Nyquist OL, Snipen L, Reiersen TS, Nes IF: Survey of genomic diversity among *Enterococcus faecalis* strains by microarray-based comparative genomic hybridization. Appl Environ Microbiol 2007;73:2207–2217.

15 Hotopp JC, Grifantini R, Kumar N, Tzeng YL, Fouts D, et al: Comparative genomics of *Neisseria meningitidis*: core genome, islands of horizontal transfer and pathogen-specific genes. Microbiology 2006; 152:3733–3749.

16 Earl AM, Losick R, Kolter R: *Bacillus subtilis* genome diversity. J Bacteriol 2007;189:1163–1170.

17 Hu G, Liu I, Sham A, Stajich JE, Dietrich FS, Kronstad JW: Comparative hybridization reveals extensive genome variation in the AIDS-associated pathogen *Cryptococcus neoformans*. Genome Biol 2008;9:R41.

18 Lindroos HL, Mira A, Repsilber D, Vinnere O, Naslund K, et al: Characterization of the genome composition of *Bartonella koehlerae* by microarray comparative genomic hybridization profiling. J Bacteriol 2005;187:6155–6165.

19 Parker CT, Quinones B, Miller WG, Horn ST, Mandrell RE: Comparative genomic analysis of *Campylobacter jejuni* strains reveals diversity due to genomic elements similar to those present in *C. jejuni* strain RM1221. J Clin Microbiol 2006;44: 4125–4135.

20 Peng J, Zhang X, Yang J, Wang J, Yang E, et al: The use of comparative genomic hybridization to characterize genome dynamics and diversity among the serotypes of *Shigella*. BMC Genomics 2006;7:218.

21 Salama NR, Gonzalez-Valencia G, Deatherage B, Aviles-Jimenez F, Atherton JC, et al: Genetic analysis of *Helicobacter pylori* strain populations colonizing the stomach at different times post infection. J Bacteriol 2007;189:3834–3845.

22 Silva NA, McCluskey J, Jefferies JM, Hinds J, Smith A, et al: Genomic diversity between strains of the same serotype and multilocus sequence type among pneumococcal clinical isolates. Infect Immun 2006; 74:3513–3518.

23 Taboada EN, Acedillo RR, Carrillo CD, Findlay WA, Medeiros DT, et al: Large-scale comparative genomics meta-analysis of *Campylobacter jejuni* isolates reveals low level of genome plasticity. J Clin Microbiol 2004;42:4566–4576.

24 Zhang Y, Laing C, Steele M, Ziebell K, Johnson R, et al: Genome evolution in major *Escherichia coli* O157:H7 lineages. BMC Genomics 2007;8:121.
25 Farley MM, Harvey RC, Stull T, Smith JD, Schuchat A, et al: A population-based assessment of invasive disease due to group B *Streptococcus* in nonpregnant adults [see comments]. N Engl J Med 1993;328:1807–1811.
26 Doran KS, Nizet V: Molecular pathogenesis of neonatal group B streptococcal infection: no longer in its infancy. Mol Microbiol 2004;54:23–31.
27 Schuchat A, Wenger JD: Epidemiology of group B streptococcal disease. Risk factors, prevention strategies, and vaccine development. Epidemiol Rev 1994;16:374–402.
28 Tettelin H, Masignani V, Cieslewicz MJ, Eisen JA, Peterson S, et al: Complete genome sequence and comparative genomic analysis of an emerging human pathogen, serotype V *Streptococcus agalactiae*. Proc Natl Acad Sci USA 2002;99:12391–12396.
29 Glaser P, Rusniok C, Buchrieser C, Chevalier F, Frangeul L, et al: Genome sequence of *Streptococcus agalactiae*, a pathogen causing invasive neonatal disease. Mol Microbiol 2002;45:1499–1513.
30 Tettelin H, Masignani V, Cieslewicz MJ, Donati C, Medini D, et al: Genome analysis of multiple pathogenic isolates of *Streptococcus agalactiae*: Implications for the microbial 'pan-genome'. Proc Natl Acad Sci USA 2005;102:13950–13955.
31 Tettelin H, Riley D, Cattuto C, Medini D: Comparative genomics: the bacterial pan-genome. Curr Opin Microbiol 2008;11:472–477.
32 Medini D, Donati C, Tettelin H, Masignani V, Rappuoli R: The microbial pan-genome. Curr Opin Genet Dev 2005;15:589–594.
33 Hiller NL, Janto B, Hogg JS, Boissy R, Yu S, et al: Comparative genomic analyses of seventeen *Streptococcus pneumoniae* strains: insights into the pneumococcal supragenome. J Bacteriol 2007;189:8186–8195.
34 Hogg JS, Hu FZ, Janto B, Boissy R, Hayes J, et al: Characterization and modeling of the *Haemophilus influenzae* core and supragenomes based on the complete genomic sequences of Rd and 12 clinical nontypeable strains. Genome Biol 2007;8:R103.
35 Shen K, Antalis P, Gladitz J, Sayeed S, Ahmed A, et al: Identification, distribution, and expression of novel genes in 10 clinical isolates of nontypeable *Haemophilus influenzae*. Infect Immun 2005;73:3479–3491.
36 Coombs A: The sequencing shakeup. Nat Biotechnol 2008;26:1109–1112.
37 Tettelin H, Saunders NJ, Heidelberg J, Jeffries AC, Nelson KE, et al: Complete genome sequence of *Neisseria meningitidis* serogroup B strain MC58. Science 2000;287:1809–1815.
38 Bentley SD, Vernikos GS, Snyder LA, Churcher C, Arrowsmith C, et al: Meningococcal genetic variation mechanisms viewed through comparative analysis of serogroup C strain FAM18. PLoS Genet 2007;3:e23.
39 Rappuoli R, Del Giudice G: Identification of vaccine targets, in Paoletti LC, McInnes PM (eds): Vaccines: From Concept to Clinic. Boca Raton, CRC Press, 1999, pp 1–17.
40 Pizza M, Scarlato V, Masignani V, Giuliani MM, Arico B, et al: Identification of vaccine candidates against serogroup B meningococcus by whole-genome sequencing. Science 2000;287:1816–1820.
41 Tettelin H, Feldblyum TV: Genome sequencing and analysis; in Grandi G (ed): Genomics, Proteomics and Vaccines. London, John Wiley and Sons Ltd, 2004, pp 45–73.
42 Serruto D, Rappuoli R, Pizza M: Meningococcus B: from genome to vaccine; in Grandi G (ed): Genomics, Proteomics and Vaccines. London, John Wiley and Sons Ltd, 2004, pp 185–204.
43 Giuliani MM, Adu-Bobie J, Comanducci M, Arico B, Savino S, et al: A universal vaccine for serogroup B meningococcus. Proc Natl Acad Sci USA 2006;103:10834–10839.
44 Nicholls H: In silico vaccine. Nat Biotechnol 2008;26:597.
45 Maione D, Margarit I, Rinaudo CD, Masignani V, Mora M, et al: Identification of a universal group B streptococcus vaccine by multiple genome screen. Science 2005;309:148–150.
46 Brodeur BR, Boyer M, Charlebois I, Hamel J, Couture F, et al: Identification of group B streptococcal Sip protein, which elicits cross-protective immunity. Infect Immun 2000;68:5610–5618.
47 Waldemarsson J, Areschoug T, Lindahl G, Johnsson E: The streptococcal Blr and Slr proteins define a family of surface proteins with leucine-rich repeats: camouflaging by other surface structures. J Bacteriol 2006;188:378–388.
48 Graveley BR: Molecular biology: power sequencing. Nature 2008;453:1197–1198.
49 Vallenet D, Labarre L, Rouy Z, Barbe V, Bocs S, et al: MaGe: a microbial genome annotation system supported by synteny results. Nucleic Acids Res 2006;34:53–65.
50 Fox JA, McMillan S, Ouellette BF: Conducting research on the web: 2007 update for the bioinformatics links directory. Nucleic Acids Res 2007;35:3–5.
51 De Groot AS, Rappuoli R: Genome-derived vaccines. Expert Rev Vaccines 2004;3:59–76.
52 Serruto D, Rappuoli R: Post-genomic vaccine development. FEBS Lett 2006;580:2985–2992.

53 Yang HL, Zhu YZ, Qin JH, He P, Jiang XC, et al: In silico and microarray-based genomic approaches to identifying potential vaccine candidates against *Leptospira interrogans*. BMC Genomics 2006;7:293.

54 Graham SP, Honda Y, Pelle R, Mwangi DM, Glew EJ, et al: A novel strategy for the identification of antigens that are recognised by bovine MHC class I restricted cytotoxic T cells in a protozoan infection using reverse vaccinology. Immunome Res 2007;3:2.

55 Liu L, Cheng G, Wang C, Pan X, Cong Y, et al: Identification and experimental verification of protective antigens against *Streptococcus suis* serotype 2 based on genome sequence analysis. Curr Microbiol 2009;58:11–17.

56 Turnbaugh PJ, Ley RE, Hamady M, Fraser-Liggett CM, Knight R, Gordon JI: The human microbiome project. Nature 2007;449:804–810.

57 Walker A, Parkhill J: Single-cell genomics. Nat Rev Microbiol 2008;6:176–177.

58 Lasken RS: Single-cell genomic sequencing using Multiple Displacement Amplification. Curr Opin Microbiol 2007;10:510–516.

Hervé Tettelin, PhD, Associate Professor
Institute for Genome Sciences, Department of Microbiology and Immunology
University of Maryland School of Medicine BioPark II
Room 629, 801 West Baltimore Street
Baltimore, MD 21201 (USA)
Tel. +1 410 706 6764, Fax +1 410 706 1482, E-Mail Tettelin@som.umaryland.edu

de Reuse H, Bereswill S (eds): Microbial Pathogenomics.
Genome Dyn. Basel, Karger, 2009, vol 6, pp 48–61

'Guilty by Association' – Protein-Protein Interactions (PPIs) in Bacterial Pathogens

K. Schauer[a] · K. Stingl[b]

[a]Molecular Mechanisms of Intracellular Transport, UMR 144 CNRS, Institut Curie, Paris, France; [b]Institut für Allgemeine Zoologie und Genetik, Westfälische Wilhelms-Universität, Münster, Germany

Abstract

Protein-protein interaction (PPI) studies are frequently used as a starting point for the functional annotations of unknown proteins according to the principle of 'guilty by association'. Moreover, they deliver information for the understanding of specific virulence mechanisms. We provide an overview about the approaches used for the identification of PPIs in human bacterial pathogens, commenting on advantages and pitfalls of the methods. Furthermore, this review intends to show the impact of PPI studies on future research, taking *Helicobacter pylori*, one of the first sequenced human pathogens, as model organism.

Protein-Protein Interaction Networks Govern Biological Processes in Living Cells

Protein-protein interactions (PPIs) are operative at virtually any biological process. Research during the last decade revealed many multi-protein complexes and protein networks in prokaryotes as well as eukaryotes. In contrast to eukaryotes, which show a high compartmentalization of their cellular organization, bacteria are limited to 3–4 major compartments (cytoplasm, inner membrane and cell wall for Gram-positive bacteria and cytoplasm, inner membrane, outer membrane and periplasm for Gram-negative bacteria). Their cellular complexity is, in particular, provided by the interactions of macromolecules, among them PPIs, giving rise to numerous spatially and temporally defined sub-compartments. In these sub-compartments, PPIs can be stable, e.g. considering molecular machines like ribosomes, or transient when e.g. involved in signaling cascades. Therefore, PPIs can mediate the formation of a functional complex or they can be used to regulate a complex [1]. Since the spatio-temporal composition of protein complexes is decisive for protein function, PPIs can provide functional information far beyond sequence-based predictions. The availability of sequence data for numerous pathogenic bacteria together with the development

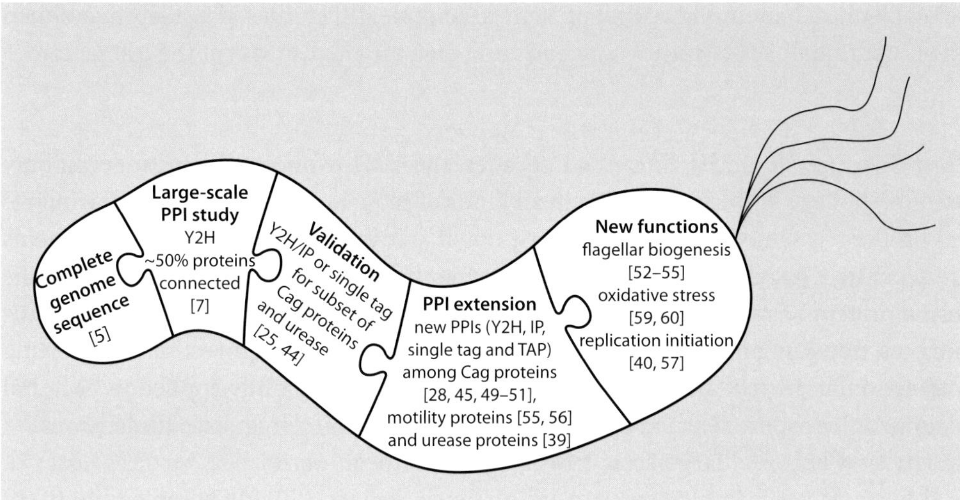

Fig. 1. 'Jigsaw pieces' leading to the understanding of new functions in *Helicobacter pylori*. PPI studies on *H. pylori* are illustrated, covering 10 years of research starting in 1997 when the complete genome sequence was published; corresponding references in brackets.

of powerful proteomic tools opened up new vistas for the exploration of the proteome-wide repertoire of PPIs, the interactome.

There are two principal goals to study PPIs in bacterial human pathogens. First, the understanding of PPIs aims at the discovery of protein function by the principle of 'guilty by association'. This means that the context of an unknown protein gives valuable information about its function. Second, PPIs can deliver information about molecular details of a complex and its regulation. In particular, virulence factors, essential for host colonization, are investigated. Both approaches often aim at the identification of new drug targets. Moreover, due to their usually relatively small genome sizes associated with host adaptation [2], bacterial pathogens present a manageable complexity to study protein networks that can help to reveal protein functions in more complex organisms. PPIs implicated in host-pathogen interactions benefit of increasing interest but are discussed elsewhere [3].

We will present an overview about the different techniques for the characterization of PPIs [4] that have successfully been applied to human bacterial pathogens. Furthermore, we will discuss the impact of PPI studies on future research, taking the gastric bacterium, *H. pylori*, one of the first human pathogens sequenced [5], as example (fig. 1).

PPI Assays Applied to Human Bacterial Pathogens

A variety of methods has been developed to study PPIs in bacterial pathogens. Commonly, they either detect binary interactions or multi-protein complexes (see below). For each method, we first present large-scale PPI studies, if available, and then go on with small-scale studies concentrating on a targeted subset of the interactome.

Binary PPIs

Yeast-Two Hybrid (Y2H). Since two decades, the Y2H is one of the most commonly used methods to study binary PPIs in all kinds of sequenced organisms. The principle lies on the reassembly of a split transcriptional activator in yeast [6], whose domains are separately fused to two proteins of interest. In case of physical interaction of the fusion proteins, a reporter gene is transcribed in the yeast nucleus. Hence, Y2H identifies both transient and stable interactions but only in the case of direct self-supporting interaction of the bait and the prey proteins. Y2H was frequently applied in bacterial pathogens (see some selected examples in table 1), even in large-scale dimensions.

The first bacterial large-scale PPI analysis has been performed for *H. pylori* [7]. In this study, 261 bait constructs were screened against a highly complex library of genome-encoded random polypeptides. Fifty *H. pylori* proteins with previously demonstrated PPIs were included for validation. This approach identified over 1,200 PPIs connecting nearly half of the *H. pylori* proteins. It permitted the assignment of unannotated proteins to biological pathways and the definition of interaction domains as putative drug targets (PIMrider = http://pim.hybrigenics.com). The first Y2H-based proteome-wide PPI map for pathogens was obtained for *Campylobacter jejuni* [8]. A pooled matrix approach was used in which over 89% of the predicted full length ORFs were chosen as bait and prey. Statistical methods were applied to generate confidence scores that identified 2,884 high confidence PPIs that covered 67% of the *C. jejuni* proteins. Surprisingly, comparison between *C. jejuni* and *H. pylori*, which are closely related ε-proteobacteria, did not show a significant overlap in conserved protein subnetworks. Recently, the first complete map for *Treponema pallidum* was published [9]. A subset of 991 high confidence PPIs linked 55% of the proteome. Annotations for at least 18 proteins have been improved and eight PPIs of a sub-network (DNA replication) have been confirmed by co-immunoprecipitation. When PPIs from this study were compared with the data from *C. jejuni*, *E. coli* and *H. pylori*, there was again only marginal overlap.

Low degree of overlap between Y2H studies in different organisms can principally stem from (i) artifacts produced by the analysis method (i.e. false-positives, false-negatives), (ii) 'sticky' or 'promiscuous' proteins, which bias the dataset, and whose biological impact has to be evaluated by the researcher and, (iii) biologically relevant species-specific PPIs. Usually, the error rate in Y2H large-scale datasets is estimated based on the data overlap with reliable small-scale studies. Likewise, it was estimated that e.g. 77% of the PPIs are missing in the large-scale study of *T. pallidum* (false-negatives). Similarly, all published large-scale PPI studies of *Saccharomyces cerevisiae* probably cover only 50% of the total interactome [10]. The false-positive rate for large-scale data, which is mainly caused by heterologous overexpression of the interacting proteins, was estimated to be 25–72% [10, 11]. Hence, due to the high false-positive rate of large-scale studies and the low PPI coverage, low overlap between different studies is inevitable and stresses the need for validation experiments.

Bacterial-Two Hybrid and Protein Fragment Complementation (PFC). Bacterial-two hybrid is based on a transcriptional activation that is similar to Y2H, but profits from a cytoplasmic localization of the PPIs [12, 13]. Protein fragment complementation (PFC) relies on the reconstitution of an essential activity of a bacterial cytoplasmic enzyme [14]. For both distinct methods, nuclear translocation of the interacting proteins is not required and, thus, membrane proteins can also be analyzed. Furthermore, the PPI study can be performed in the organism of interest or a close relative.

A standardized bacterial-two hybrid assay was performed for several selected ORF fragments of the type IV secretion system (T4SS) of *Rickettsia sibirica* using *E. coli* as a host [15]. Nearly half of the PPIs previously identified by Y2H of *Agrobacterium tumefaciens* T4SS were confirmed in this bacterial-two hybrid assay. However, nearly 50 PPI partners were identified on average for each T4SS subunit. This large network is supported by the fact that the majority of the positive preys was found to interact with more than one bait. Unfortunately, most of the interactions were only observed once. Validation studies are needed to highlight the physiologically relevant interactions.

PFC was used to study PPIs of genetically intractable mycobacteria, like *Mycobacterium tuberculosis* (M-PFC) [16]. The functional reconstitution of two murine dihydrofolate reductase (mDHFR) domains, conferring resistance to the antibiotic trimethoprim, was used as a reporter for PPIs. The M-PFC was successfully tested for *M. tuberculosis* membrane-spanning sensor histidine kinase DevS (Rv3132c) and its corresponding response regulator DevR (Rv3133c) [16]. After validation, the secreted antigen Cfp-10 was used as bait. Six proteins were identified as interactors, including Esat-6, a known partner of Cfp-10 [17]. Except for one PPI, all identified PPIs were validated by conventional Y2H and pull-down experiments. The applicability of M-PFC for high-throughput and the quantification of PPIs by growth analysis in the presence of trimethoprim indicates that M-PFC will be a powerful tool for future large-scale analyses. Recently, a third secreted virulence factor was found to interact with the Cfp-10/Esat-6 secretion system (ESX-1) in a bacterial-two hybrid study [18], suggesting that secretion of multiple substrates by ESX-1 contributes to virulence of *Mycobacterium.*

The split-Trp is another PFC assay that monitors the functional reconstitution of tryptophan biosynthesis in tryptophan auxotrophic organisms. Originally developed in *S. cerevisiae* [19], this method was introduced to prokaryotes [20]. Several well-characterized bacterial and eukaryotic interacting proteins were examined in tryptophan auxotrophic *E. coli* and *M. smegmatis* strains to demonstrate the feasibility of the approach. This method complements the M-PFC assay described above and awaits application for the identification of novel PPIs.

Far-Western Blotting. The far-Western (or gel overlay) analysis is based on the same principles as the classical Western blotting approach, thereby detecting stable binary PPIs. Instead of detecting a protein by the respective antibody, a labeled or antibody-

Table 1. Overview of selected protein-protein interaction studies performed in bacterial pathogens

Method	Pros	Contras	Organism	Reference
Yeast-two hybrid (Y2H)				
Large-scale datasets	feasible for every sequenced bacterium	heterologous overexpression (many false-positive PPI)	*Campylobacter jejuni* *Helicobacter pylori* *Treponema pallidum*	[8] [7] [9]
Small-scale datasets	sensitive for transient interactions detection of interacting domains	many false-negatives PPI occurs in nucleus of yeast cell, not suitable for membrane proteins detects only binary PPIs	*H. pylori* *Legionella pneumophila* *Mycobacterium tuberculosis* *Shigella flexneri* *Yersinia*	[25, 28, 40, 44, 51] [21, 66] [26, 29–32, 67–69] [27, 70, 71] [72–75]
Bacteria-two hybrid	sensitive for transient interactions detection of interacting domains	heterologous overexpression (many false-positive PPI)	*M. marinum/M. tuberculosis* *Rickettsia sibirica*	[18] [15]
M-PFC = mycobacterial protein fragment complementation	cytoplasmic environment of PPI PPI in original organism or close relative	detects only binary PPIs	*M. tuberculosis*	[16]
'Split-Trp'	also for membrane proteins		*E. coli/M. smegmatis* as host for PPIs	[20]
Far-Western blotting/ protein (print) overlay	easy handling	relies on specificity of antibody/purification grade of recombinant proteins detects only stable PPIs	*L. pneumophila* *M. tuberculosis*	[21] [22]
Surface plasmon resonance (SPR)	validation of PPIs and establishment of interaction kinetics (affinity, rates of association and dissociation)	in vitro interaction of purified (recombinant) proteins risk of protein inactivation by immobilization to the surface	*Y. pestis* *Pseudomonas aeruginosa*	[24] [23]
2D blue-native/SDS gel electrophoresis	no modification (tagging) of bait protein also applicable for membrane proteins multi-protein complexes	subjective identification of PPIs for complex protein samples co-migration of proteins not belonging to a complex detects only stable PPIs	*H. pylori*	[43]

Table 1. Continued

Method	Pros	Contras	Organism	Reference
Affinity purification (pull-down) Immunoprecipitation	mostly for validation of Y2H datasets genetic tools in the pathogen are dispensable no modification (tagging) of bait protein	strongly relies on specificity of antibody (usually high background of unspecific interactors) detects only stable PPIs	*H. pylori* *M. tuberculosis* *S. flexneri* *Y. pestis*	[25, 28] [29, 30] [27] [73]
Single tag	higher yield than for TAP	if homologue, pathogen has to be genetically manipulable tag might interfere with function and PPI mostly overexpressed detects only stable PPIs	*Brucella suis* *H. pylori* *L. pneumophila* *Mycobacterium* *S. flexneri* *Y. pestis*	[76, 77] [40, 44, 49–51, 55] [66] [17, 18, 26, 31, 32] [33, 71] [74]
Tandem-affinity purification (TAP)	high specificity physiological expression in original organism in combination with crosslink prior to TAP also for transient interactions	pathogen has to be genetically manipulable tag might interfere with function and PPI detects only stable PPIs, unless crosslinking is performed prior to purification	*H. pylori*	[39, 40]

detectable 'bait' protein is used to probe the PPI with a target protein on the membrane. For *Legionella pneumophila*, PPIs between the proteins of a T4SS were detected by far-Western analysis on crude extracts of wild-type and the respective PPI partner-deficient mutant strains [21]. Another study investigated the PPIs between secreted Esat-6 proteins of *M. tuberculosis* [22], and detected among others the known Esat-6/Cfp-10 complex already found with other methods [17].

Surface Plasmon Resonance (SPR). Several PPI studies using SPR as detection method for in vitro PPIs of recombinant purified proteins have been applied to pathogenic bacteria. The method measures the changes of the refractive indices at the interface of two substrates under conditions of total internal reflection of polarized light. Thus, SPR can be used to detect PPIs between a surface-immobilized bait protein and a soluble interaction partner. Additionally, association and dissociation rates as well as the binding affinity can be determined. The binding kinetics of the PPI between two T3SS proteins from *Pseudomonas aeruginosa* were analyzed [23] and PPIs among proteins of the T3SS of *Yersinia pestis* were detected and subsequently validated by mass spectrometry [24].

Targeted Pull-Down via Immunoprecipitation (IP) and Single-Tag Affinity Purification. Only stable PPIs can be identified by biochemical isolation of bait proteins (pull-down), unless crosslinking is performed. The pull-down is targeted when a distinct prey protein is identified, e.g. by antibody detection. IP is a very common pull-down approach. Typically, cell lysates are incubated with an antibody that specifically recognizes one protein of interest. Subsequently, the antibody-antigen complexes are precipitated using antibody-binding beads and analyzed for PPI partners. This method has extensively been used in pathogenic bacteria, mostly for the validation of a defined subset of Y2H data. Examples are analyses of PPIs of virulence factors [25, 26], bacterial secretion machineries (e.g. Type III, Type IV [27, 28]), as well as PPIs involved in biosynthetic pathways (e.g. [29, 30]) (table 1). IP experiments strongly depend on the specificity of the antibody and of the beads used, frequently leading to the pull-down of unspecific proteins.

If specific antibodies for the proteins of interest are not available, the protein can be tagged by a generic, commercially availably polypeptide (e.g. His-, Myc-, Strep-, MBP-, GST-tag). The respective proteins are either tagged directly in the original organism or in model organisms (if the pathogen is genetically not manipulable or raises biosafety concerns). Many examples of PPI studies in pathogens using a single tag are found in the literature (see selected examples in table 1).

Complex Identification

Complex Pull-Down via Immunoprecipitation (IP) and Single-Tag Affinity Purification. In contrast to targeted pull-down, complex pull-down implicates the identification of protein complexes, which are copurified with the bait protein. As mentioned above, the specificity of the antibody for the endogenous protein or protein tags is decisive whether large amounts of the target protein at sufficient purification grade can be isolated. De novo identification of PPI partners is performed in combination with mass spectrometry (MALDI or SELDI [31, 32]). For example, Zenk et al. [33] used His-tagging of the needle complex of the *Shigella* T3SS and identified needle components that had not been found in previous studies.

Complex Pull-Down via Tandem-Affinity Purification (TAP). Since IP and single-tag affinity purification are usually hampered by non-specific pull-down, the use of two tags in tandem revolutionized the biochemical isolation of protein complexes. The TAP technique was originally developed for yeast [34] but has been applied to a variety of eukaryotes [35–37] and recently to *E. coli* and *H. pylori* [38–40]. Usually protein A of *Staphylococcus aureus* and a calmodulin-binding domain, which are separated by a specific protease cleavage site, are fused to a bait protein on the chromosome of the original organism. The bait protein in complex with its interaction partners is purified via two successive affinity columns under native conditions. Subsequently, the co-purified proteins are separated by one-dimensional PAGE and individually identified by mass spectrometry. TAP has been proven to be an efficient means to access multi-partner protein complexes with much reduced false-positive versus true-positive ratio than for Y2H [41]. In a pilot study, we have used this technique to decipher the interaction partners of the urease complex in *H. pylori* [39]. To capture transient protein complexes that are easily lost during pull-down, we additionally applied a crosslink procedure in vivo prior to TAP. The feasibility of the method was validated by the identification of the entire set of the well-characterized urease accessory proteins with the structural subunits of urease. Several novel interaction partners have been identified providing new clues about the maturation of iron-sulfur clusters in *H. pylori* and the coupling of ammonium production and assimilation.

Two-Dimensional Blue-Native/SDS Gel Electrophoresis. The 2D blue-native/SDS gel electrophoresis is based on the binding of coomassie brilliant blue to protein complexes, enabling their migration in a first dimension electrophoresis under native conditions [42]. The protein components of these multi-complexes are then separated under denaturating conditions in a second SDS gel electrophoresis. The method was used for the identification of PPIs in crude or partially purified extracts of *H. pylori* [43]. Several multi-subunit complexes were identified, among them known membrane complexes. However, due to the large molecular weights of the migrating multi-complexes, size separation is limited by the low resolution of 2D gels. In addition, it is indistinguishable whether protein components identified by co-migration stem from the same complex or belong to different complexes of similar molecular weight. Hence, the interpretation of the results is relatively subjective.

What is the Impact of PPI Studies on Subsequent Research?

H. pylori represents a unique case, for which PPI data are available from almost all methods and, therefore, it is an excellent example to access the impact of PPI studies on future research. When published in 2001, the first bacterial Y2H large-scale interaction map of *H. pylori* [7] served as a starting point for multiple subsequent studies (fig. 1).

T4SS

To estimate the reliability of the large-scale Y2H data, systematic biochemical validation experiments were performed for 17 PPIs using affinity purification [44]. This study affirmed nearly 80% of the interactions, including six PPIs of T4SS components. Because of this validation, a potential role in type IV secretion was proposed for proteins of previously unknown functions, among them HP1451. In a further study, the VirB11 homologue, HP0525, was co-crystallized with a fragment of HP1451 [45]. It was proposed that HP1451 regulates Cag-dependent secretion, which was in agreement with an HP1451-concentration dependent inhibition of HP0525 ATPase activity. The study by Rain et al. [7], however, also showed limitations. Primarily, the interactome is incomplete. Although nearly half of the proteome was connected, only a fraction of the entire *H. pylori* proteome was used as bait. Indeed, most of the T4SS

PPIs are missing in the large-scale study, since only four T4SS proteins were analyzed as bait proteins, giving rise to only four reciprocal T4SS PPIs, including two oligomerizations.

In the case of the missed PPIs, small-scale studies have advanced our knowledge of the T4SS. One of the two independent T4SS of *H. pylori*, the Cag system, is involved in protein and peptidoglycane translocation into host cells [46–48]. Using FLAG-tagging combined with co-immunoprecipitation, the translocated effector protein CagA was shown to interact with CagF in *H. pylori* [49]. Because *cagF* deficient mutants showed a lack of CagA translocation, a putative role as chaperone was attributed to this so far unknown protein. Using GST-CagF in pull-down experiments with truncated CagA derivatives, the interaction domain of CagA was established [50]. Information about PPIs between the Cag proteins was profoundly extended by comprehensive Y2H for exclusively Cag proteins [28] as well as by a previous study [51] using 19 or 14 Cag proteins as baits, respectively. Importantly, several PPIs identified by Y2H were verified by pull-down experiments. Thus, the identified PPIs combined with immuno-based localization [28] provided valuable data allowing the proposition of a low-resolution model for the Cag system, which will serve as a basis for future research.

Flagellar Proteins

The dataset of Rain et al. [7] identified an interaction between the σ^{28} factor and a protein of unknown function, HP1122. Homologues of the anti-σ^{28} factor, FlgM, which regulates timing of late flagellar synthesis in other bacteria, are absent from the genome of *H. pylori*. Since HP1122 inhibited PPI of σ^{28} with the β-region of RNA polymerase in a three-hybrid system and overproduction of HP1122 in *H. pylori* led to truncated flagella, a function as an anti-σ^{28} factor was attributed to HP1122 [52]. Two other studies [53, 54] further explored the interactions of HP0958 with either σ^{54} or FliH, a flagellar ATPase regulator [7]. Both studies observed an aflagellate phenotype of a mutant deficient in HP0958 and reduced levels of flagellin and hook protein production [53]. More work is needed to further decipher the molecular function of HP0958 in flagella biogenesis. Furthermore, GST pull-down experiments with truncated FliH proteins defined its interaction domain with FliI, a highly conserved flagellum-specific ATPase [55]. Finally, Y2H data were integrated with phenotypes of mutant strains deficient in putative motility proteins comparing *E. coli*, *C. jejuni*, *H. pylori* and *T. pallidum* in a comprehensive study [56]. This led to the identification of a core set of motility proteins, with an unexpected large number of species-specific components.

Other Proteins with Unknown Functions

There are further examples showing that PPI studies serve as a 'creative director' for the attribution of new functions to unknown proteins, for which homology search presented a 'dead end' due to the existence of evolutionary analogues.

A starting point for the identification of a novel protein implicated in chromosomal replication was the PPI between the main replication initiator protein DnaA and HP1230 [7]. Subsequent studies using in vitro and in vivo methods corroborated the PPI and suggested that HP1230 stabilizes the orisome (DnaA-oriC complex) [40]. Functional analysis of the essential HP1230 in *H. pylori* identified this protein, termed HobA, as a new replication initiation factor in ε-proteobacteria. Consistently, the crystal structure of HobA was solved, showing a striking structural homology to the analogous protein DiaA, which ensures timely initiation of chromosomal replication in *E. coli* [57].

The study of Rain et al. [7] also detected a PPI between a principal oxidative stress protein, catalase, and HP0874 of unknown function. Strains deficient in HP0874 exhibited wild-type catalase activity [58, 59], whereas resistance to hydrogen peroxide as well as the capability to persist at the gastric mucosa were significantly affected [59, 60], suggesting a role of HP0874 in oxidative stress response.

Urease

The dataset of Rain et al. [7] contained PPIs between the structural subunits and the accessory proteins of urease that is essential for acid resistance of this gastric pathogen. The incorporation of nickel ions into this metallo-enzyme requires several accessory proteins. Confirmation of the biological significance of the observed subset of PPIs stems, first, from genetic and biochemical data of the homologous system in *Klebsiella pneumoniae* [61–63]. Second, *H. pylori* mutants deficient in urease accessory proteins showed phenotypes that were consistent with their essential role for urease activity [64]. Third, several PPIs were confirmed by an independent Y2H analysis on a subset of urease proteins, by co-immunoprecipitation [25] and recently by TAP [39]. However, the large-scale study also suggested that urease baits physically interacted with several other proteins, not encoded by the urease gene cluster. None of these PPIs were identified by TAP [39] that additionally revealed other interaction partners. Whereas binary PPI approaches like Y2H fail to detect multi-component complexes, TAP, pull-down and two-dimensional native gel approaches can overcome this problem by identification of multi-protein complexes. However, the latter methods do not detect transient PPIs like most of the binary methods, unless crosslinking is performed prior to biochemical isolation of the protein complex. Thus, the use of multiple PPI approaches for the characterization of the same PPI network is required to achieve a comprehensive understanding of bacterial interactomes. The example of *H. pylori* nicely demonstrates that different PPI methods reveal distinct information and are, thus, complementary rather than opposed.

Perspectives

Homology searches across species, genomic context analyses as well as transcriptional and translational profiling are potent tools for functional annotations to unknown proteins. PPI studies add up with predictions for proteins that show functional analogy to known proteins of the classical model organism, *E. coli*. We have presented different PPI methods that gave insight into distinct aspects of the interactome of bacterial pathogens. Still, most PPI studies are not performed in the original pathogenic bacterium, since genetic tools for manipulation are often missing. Therefore, there is an exigent need to establish new methods that render pathogens accessible for advanced PPI studies, like e.g. the TAP technology. Furthermore, the integration of an increasing amount of PPI data from different experimental approaches and in different organisms is one of the future challenges. An example of PPI data integration from a variety of sources is the STRING (search tool for the retrieval of interacting proteins) database that is available online (http://string.embl.de/, [65]) and that enables to interconnect PPI information of currently 373 completely sequenced bacterial genomes.

The example of *H. pylori* conclusively demonstrates the complementarity of different PPI approaches and their immense impact on future research. PPI studies are powerful to deliver information about never anticipated functional connections, which will contribute to the global understanding of bacterial pathogenesis as well as its combat.

Acknowledgement

We thank H. de Reuse for helpful discussion and critical reading of the manuscript. K.Sch. was supported by a postdoctoral fellowship of the Fondation pour la Recherche Médicale (FRM).

References

1 Devos D, Russell RB: A more complete, complexed and structured interactome. Curr Opin Struct Biol 2007;17:370–377.

2 Moran NA: Microbial minimalism: genome reduction in bacterial pathogens. Cell 2002;108:583–586.

3 Dyer MD, Murali TM, Sobral BW: The landscape of human proteins interacting with viruses and other pathogens. PLoS Pathog 2008;4:e32.

4 Berggard T, Linse S, James P: Methods for the detection and analysis of protein-protein interactions. Proteomics 2007;7:2833–2842.

5 Tomb JF, White O, Kerlavage AR, Clayton RA, Sutton GG, et al: The complete genome sequence of the gastric pathogen *Helicobacter pylori*. Nature 1997;388: 539–547.

6 Fields S, Song O: A novel genetic system to detect protein-protein interactions. Nature 1989;340:245–246.

7 Rain JC, Selig L, De Reuse H, Battaglia V, Reverdy C, et al: The protein-protein interaction map of *Helicobacter pylori*. Nature 2001;409:211–215.

8 Parrish JR, Yu J, Liu G, Hines JA, Chan JE, et al: A proteome-wide protein interaction map for *Campylobacter jejuni*. Genome Biol 2007;8:R130.

9 Titz B, Rajagopala SV, Goll J, Hauser R, McKevitt MT, et al: The binary protein interactome of *Treponema pallidum*-the syphilis spirochete. PLoS ONE 2008;3:e2292.

10 Hart GT, Ramani AK, Marcotte EM: How complete are current yeast and human protein-interaction networks? Genome Biol 2006;7:120.

11 Huang H, Jedynak BM, Bader JS: Where have all the interactions gone? Estimating the coverage of two-hybrid protein interaction maps. PLoS Comput Biol 2007;3:e214.

12 Ladant D, Karimova G: Genetic systems for analyzing protein-protein interactions in bacteria. Res Microbiol 2000;151:711–720.

13 Hu JC, Kornacker MG, Hochschild A: *Escherichia coli* one- and two-hybrid systems for the analysis and identification of protein-protein interactions. Methods 2000;20:80–94.

14 Pelletier JN, Campbell-Valois FX, Michnick SW: Oligomerization domain-directed reassembly of active dihydrofolate reductase from rationally designed fragments. Proc Natl Acad Sci USA 1998; 95:12141–12146.
15 Malek JA, Wierzbowski JM, Tao W, Bosak SA, Saranga DJ, et al: Protein interaction mapping on a functional shotgun sequence of *Rickettsia sibirica*. Nucleic Acids Res 2004;32:1059–1064.
16 Singh A, Mai D, Kumar A, Steyn AJ: Dissecting virulence pathways of *Mycobacterium tuberculosis* through protein-protein association. Proc Natl Acad Sci USA 2006;103:11346–11351.
17 Renshaw PS, Panagiotidou P, Whelan A, Gordon SV, Hewinson RG, et al: Conclusive evidence that the major T-cell antigens of the *Mycobacterium tuberculosis* complex ESAT-6 and CFP-10 form a tight, 1:1 complex and characterization of the structural properties of ESAT-6, CFP-10, and the ESAT-6*CFP-10 complex. Implications for pathogenesis and virulence. J Biol Chem 2002;277:21598–21603.
18 McLaughlin B, Chon JS, MacGurn JA, Carlsson F, Cheng TL, et al: A mycobacterium ESX-1-secreted virulence factor with unique requirements for export. PLoS Pathog 2007;3:e105.
19 Tafelmeyer P, Johnsson N, Johnsson K: Transforming a (beta/alpha)8-barrel enzyme into a split-protein sensor through directed evolution. Chem Biol 2004; 11:681–689.
20 O'Hare H, Juillerat A, Dianiskova P, Johnsson K: A split-protein sensor for studying protein-protein interaction in mycobacteria. J Microbiol Methods 2008;73:79–84.
21 Coers J, Kagan JC, Matthews M, Nagai H, Zuckman DM, Roy CR: Identification of Icm protein complexes that play distinct roles in the biogenesis of an organelle permissive for *Legionella pneumophila* intracellular growth. Mol Microbiol 2000;38:719–736.
22 Okkels LM, Andersen P: Protein-protein interactions of proteins from the ESAT-6 family of *Mycobacterium tuberculosis*. J Bacteriol 2004;186: 2487–2491.
23 Nanao M, Ricard-Blum S, Di Guilmi AM, Lemaire D, Lascoux D, et al: Type III secretion proteins PcrV and PcrG from *Pseudomonas aeruginosa* form a 1:1 complex through high affinity interactions. BMC Microbiol 2003;3:21.
24 Swietnicki W, O'Brien S, Holman K, Cherry S, Brueggemann E, et al: Novel protein-protein interactions of the *Yersinia pestis* type III secretion system elucidated with a matrix analysis by surface plasmon resonance and mass spectrometry. J Biol Chem 2004;279:38693–38700.
25 Voland P, Weeks DL, Marcus EA, Prinz C, Sachs G, Scott D: Interactions among the seven *Helicobacter pylori* proteins encoded by the urease gene cluster. Am J Physiol Gastrointest Liver Physiol 2003;284: G96–G106.
26 Hett EC, Chao MC, Steyn AJ, Fortune SM, Deng LL, Rubin EJ: A partner for the resuscitation-promoting factors of *Mycobacterium tuberculosis*. Mol Microbiol 2007;66:658–668.
27 Jouihri N, Sory MP, Page AL, Gounon P, Parsot C, Allaoui A: MxiK and MxiN interact with the Spa47 ATPase and are required for transit of the needle components MxiH and MxiI, but not of Ipa proteins, through the type III secretion apparatus of *Shigella flexneri*. Mol Microbiol 2003;49:755–767.
28 Kutter S, Buhrdorf R, Haas J, Schneider-Brachert W, Haas R, Fischer W: Protein subassemblies of the *Helicobacter pylori* Cag type IV secretion system revealed by localization and interaction studies. J Bacteriol 2008;190:2161–2171.
29 Veyron-Churlet R, Guerrini O, Mourey L, Daffe M, Zerbib D: Protein-protein interactions within the Fatty Acid Synthase-II system of *Mycobacterium tuberculosis* are essential for mycobacterial viability. Mol Microbiol 2004;54:1161–1172.
30 Veyron-Churlet R, Bigot S, Guerrini O, Verdoux S, Malaga W, et al: The biosynthesis of mycolic acids in *Mycobacterium tuberculosis* relies on multiple specialized elongation complexes interconnected by specific protein-protein interactions. J Mol Biol 2005; 353:847–858.
31 Steyn AJ, Collins DM, Hondalus MK, Jacobs WR Jr, Kawakami RP, Bloom BR: *Mycobacterium tuberculosis* WhiB3 interacts with RpoV to affect host survival but is dispensable for in vivo growth. Proc Natl Acad Sci USA 2002;99:3147–3152.
32 Steyn AJ, Joseph J, Bloom BR: Interaction of the sensor module of *Mycobacterium tuberculosis* H37Rv KdpD with members of the Lpr family. Mol Microbiol 2003;47:1075–1089.
33 Zenk SF, Stabat D, Hodgkinson JL, Veenendaal AK, Johnson S, Blocker AJ: Identification of minor inner-membrane components of the *Shigella* type III secretion system 'needle complex'. Microbiology 2007;153:2405–2415.
34 Rigaut G, Shevchenko A, Rutz B, Wilm M, Mann M, Seraphin B: A generic protein purification method for protein complex characterization and proteome exploration. Nat Biotechnol 1999;17:1030–1032.
35 Gavin AC, Bosche M, Krause R, Grandi P, Marzioch M, et al: Functional organization of the yeast proteome by systematic analysis of protein complexes. Nature 2002;415:141–147.

36 Van Leene J, Stals H, Eeckhout D, Persiau G, Van De Slijke E, et al: A tandem affinity purification-based technology platform to study the cell cycle interactome in *Arabidopsis thaliana*. Mol Cell Proteomics 2007;6:1226–1238.

37 Koch HB, Zhang R, Verdoodt B, Bailey A, Zhang CD, et al: Large-scale identification of c-MYC-associated proteins using a combined TAP/MudPIT approach. Cell Cycle 2007;6:205–217.

38 Butland G, Peregrin-Alvarez JM, Li J, Yang W, Yang X, et al: Interaction network containing conserved and essential protein complexes in *Escherichia coli*. Nature 2005;433:531–537.

39 Stingl K, Schauer K, Ecobichon C, Labigne A, Lenormand P, et al: In vivo interactome of *Helicobacter pylori* urease revealed by tandem affinity purification. Mol Cell Proteomics 2008;7:2429–2441.

40 Zawilak-Pawlik A, Kois A, Stingl K, Boneca IG, Skrobuk P, et al: HobA-a novel protein involved in initiation of chromosomal replication in *Helicobacter pylori*. Mol Microbiol 2007;65:979–994.

41 Deng M, Sun F, Chen T: Assessment of the reliability of protein-protein interactions and protein function prediction. Pac Symp Biocomput 2003;8: 140–151.

42 Schägger H, von Jagow G: Blue native electrophoresis for isolation of membrane protein complexes in enzymatically active form. Anal Biochem 1991;199: 223–231.

43 Pyndiah S, Lasserre JP, Menard A, Claverol S, Prouzet-Mauleon V, et al: Two-dimensional blue native/SDS gel electrophoresis of multiprotein complexes from *Helicobacter pylori*. Mol Cell Proteomics 2007;6:193–206.

44 Terradot L, Durnell N, Li M, Ory J, Labigne A, et al: Biochemical characterization of protein complexes from the *Helicobacter pylori* protein interaction map: strategies for complex formation and evidence for novel interactions within type IV secretion systems. Mol Cell Proteomics 2004;3:809–819.

45 Hare S, Fischer W, Williams R, Terradot L, Bayliss R, et al: Identification, structure and mode of action of a new regulator of the *Helicobacter pylori* HP0525 ATPase. EMBO J 2007;26:4926–4934.

46 Stein M, Rappuoli R, Covacci A: Tyrosine phosphorylation of the *Helicobacter pylori* CagA antigen after cag-driven host cell translocation. Proc Natl Acad Sci USA 2000;97:1263–1268.

47 Odenbreit S, Puls J, Sedlmaier B, Gerland E, Fischer W, Haas R: Translocation of *Helicobacter pylori* CagA into gastric epithelial cells by type IV secretion. Science 2000;287:1497–1500.

48 Viala J, Chaput C, Boneca IG, Cardona A, Girardin SE, et al: Nod1 responds to peptidoglycan delivered by the *Helicobacter pylori* cag pathogenicity island. Nat Immunol 2004;5:1166–1174.

49 Couturier MR, Tasca E, Montecucco C, Stein M: Interaction with CagF is required for translocation of CagA into the host *via* the *Helicobacter pylori* type IV secretion system. Infect Immun 2006;74: 273–281.

50 Pattis I, Weiss E, Laugks R, Haas R, Fischer W: The *Helicobacter pylori* CagF protein is a type IV secretion chaperone-like molecule that binds close to the C-terminal secretion signal of the CagA effector protein. Microbiology 2007;153:2896–2909.

51 Busler VJ, Torres VJ, McClain MS, Tirado O, Friedman DB, Cover TL: Protein-protein interactions among *Helicobacter pylori* Cag proteins. J Bacteriol 2006;188:4787–4800.

52 Colland F, Rain JC, Gounon P, Labigne A, Legrain P, De Reuse H: Identification of the *Helicobacter pylori* anti-sigma28 factor. Mol Microbiol 2001;41:477–487.

53 Ryan KA, Karim N, Worku M, Moore SA, Penn CW, O'Toole PW: HP0958 is an essential motility gene in *Helicobacter pylori*. FEMS Microbiol Lett 2005;248: 47–55.

54 Pereira L, Hoover TR: Stable accumulation of sigma54 in *Helicobacter pylori* requires the novel protein HP0958. J Bacteriol 2005;187:4463–4469.

55 Lane MC, O'Toole PW, Moore SA: Molecular basis of the interaction between the flagellar export proteins FliI and FliH from *Helicobacter pylori*. J Biol Chem 2006;281:508–517.

56 Rajagopala SV, Titz B, Goll J, Parrish JR, Wohlbold K, et al: The protein network of bacterial motility. Mol Syst Biol 2007;3:128.

57 Natrajan G, Hall DR, Thompson AC, Gutsche I, Terradot L: Structural similarity between the DnaA-binding proteins HobA (HP1230) from *Helicobacter pylori* and DiaA from *Escherichia coli*. Mol Microbiol 2007;65:995–1005.

58 Odenbreit S, Wieland B, Haas R: Cloning and genetic characterization of *Helicobacter pylori* catalase and construction of a catalase-deficient mutant strain. J Bacteriol 1996;178:6960–6967.

59 Harris AG, Hinds FE, Beckhouse AG, Kolesnikow T, Hazell SL: Resistance to hydrogen peroxide in *Helicobacter pylori*: role of catalase (KatA) and Fur, and functional analysis of a novel gene product designated 'KatA-associated protein', KapA (HP0874). Microbiology 2002;148:3813–3825.

60 Harris AG, Wilson JE, Danon SJ, Dixon MF, Donegan K, Hazell SL: Catalase (KatA) and KatA-associated protein (KapA) are essential to persistent colonization in the *Helicobacter pylori* SS1 mouse model. Microbiology 2003;149:665–672.

61 Colpas GJ, Hausinger RP: In vivo and in vitro kinetics of metal transfer by the *Klebsiella aerogenes* urease nickel metallochaperone, UreE. J Biol Chem 2000;275:10731–10737.

62 Lee MH, Mulrooney SB, Renner MJ, Markowicz Y, Hausinger RP: *Klebsiella aerogenes* urease gene cluster: sequence of ureD and demonstration that four accessory genes (ureD, ureE, ureF, and ureG) are involved in nickel metallocenter biosynthesis. J Bacteriol 1992;174:4324–4330.

63 Soriano A, Hausinger RP: GTP-dependent activation of urease apoprotein in complex with the UreD, UreF, and UreG accessory proteins. Proc Natl Acad Sci USA 1999;96:11140–11144.

64 Ferrero RL, Cussac V, Courcoux P, Labigne A: Construction of isogenic urease-negative mutants of *Helicobacter pylori* by allelic exchange. J Bacteriol 1992;174:4212–4217.

65 von Mering C, Jensen LJ, Kuhn M, Chaffron S, Doerks T, et al: STRING 7-recent developments in the integration and prediction of protein interactions. Nucleic Acids Res 2007;35:D358–D362.

66 Ninio S, Zuckman-Cholon DM, Cambronne ED, Roy CR: The *Legionella* IcmS-IcmW protein complex is important for Dot/Icm-mediated protein translocation. Mol Microbiol 2005;55:912–926.

67 MacGurn JA, Raghavan S, Stanley SA, Cox JS: A non-RD1 gene cluster is required for Snm secretion in *Mycobacterium tuberculosis*. Mol Microbiol 2005; 57:1653–1663.

68 Lightbody KL, Renshaw PS, Collins ML, Wright RL, Hunt DM, et al: Characterisation of complex formation between members of the *Mycobacterium tuberculosis* complex CFP-10/ESAT-6 protein family: towards an understanding of the rules governing complex formation and thereby functional flexibility. FEMS Microbiol Lett 2004;238:255–262.

69 Sinha KM, Stephanou NC, Gao F, Glickman MS, Shuman S: Mycobacterial UvrD1 is a Ku-dependent DNA helicase that plays a role in multiple DNA repair events, including double-strand break repair. J Biol Chem 2007;282:15114–15125.

70 Deighan P, Beloin C, Dorman CJ: Three-way interactions among the Sfh, StpA and H-NS nucleoid-structuring proteins of *Shigella flexneri* 2a strain 2457T. Mol Microbiol 2003;48:1401–1416.

71 Page AL, Fromont-Racine M, Sansonetti P, Legrain P, Parsot C: Characterization of the interaction partners of secreted proteins and chaperones of *Shigella flexneri*. Mol Microbiol 2001;42:1133–1145.

72 Montagna LG, Ivanov MI, Bliska JB: Identification of residues in the N-terminal domain of the *Yersinia* tyrosine phosphatase that are critical for substrate recognition. J Biol Chem 2001;276:5005–5011.

73 Day JB, Plano GV: A complex composed of SycN and YscB functions as a specific chaperone for YopN in *Yersinia pestis*. Mol Microbiol 1998;30:777–788.

74 Jackson MW, Plano GV: Interactions between type III secretion apparatus components from *Yersinia pestis* detected using the yeast two-hybrid system. FEMS Microbiol Lett 2000;186:85–90.

75 Francis MS, Aili M, Wiklund ML, Wolf-Watz H: A study of the YopD-lcrH interaction from *Yersinia pseudotuberculosis* reveals a role for hydrophobic residues within the amphipathic domain of YopD. Mol Microbiol 2000;38:85–102.

76 Paschos A, Patey G, Sivanesan D, Gao C, Bayliss R, et al: Dimerization and interactions of *Brucella suis* VirB8 with VirB4 and VirB10 are required for its biological activity. Proc Natl Acad Sci USA 2006;103: 7252–7257.

77 Höppner C, Carle A, Sivanesan D, Hoeppner S, Baron C: The putative lytic transglycosylase VirB1 from *Brucella suis* interacts with the type IV secretion system core components VirB8, VirB9 and VirB11. Microbiology 2005;151:3469–3482.

Kerstin Stingl
Westfälische Wilhelms-Universität Münster, Institut für Allgemeine Zoologie und Genetik
Schlossplatz 5
DE–48149 Münster (Germany)
Tel. +49 251 83 23 926, Fax +49 251 83 24 723, E-Mail k.stingl@web.de

de Reuse H, Bereswill S (eds): Microbial Pathogenomics.
Genome Dyn. Basel, Karger, 2009, vol 6, pp 62–74

Helicobacter pylori Sequences Reflect Past Human Migrations

Y. Moodley · B. Linz

Department of Molecular Biology, Max-Plank Institute for Infection Biology, Berlin, Germany

Abstract

The long association between the stomach bacterium *Helicobacter pylori* and humans, in combination with its predominantly within-family transmission route and its exceptionally high DNA sequence diversity, make this bacterium a reliable marker for discerning both recent and ancient human population movements. As much of the diversity in *H. pylori* sequences is generated by recombination and mutation on a local scale, the partitioning of *H. pylori* sequences from a large globally distributed data set into six geographic populations enabled the detection of recent (<500 years) human population movements including the European colonial expansion and the slave trade. The further separation of bacterial populations into distinct sub-populations traced prehistoric population movements like the settlement of the Americas by Asians across the Bering Strait and the Bantu migrations in Africa. The ability to deduce ancestral population structure from modern sequences was a key development that allowed the detection of zones of admixture, such as Europe, and the inference of multiple migration waves into these zones. The significantly similar global population structure of both *H. pylori* and humans confirmed not only an evolutionary timescale association between host and parasite, but also that humans had carried *H. pylori* in their stomachs on their migrations out of Africa.

In the last decade, sequence differences in microbes from different geographical areas have increasingly been interpreted with regard to population movements of their human hosts. The spiral shaped stomach bacterium *Helicobacter pylori,* discovered by Barry Marshall and Robin Warren as the causative agent of gastritis and gastric ulcers [1, 2], was also found to be an attractive candidate for reconstructing ancient human migrations. *H. pylori* is usually acquired early in childhood, and once acquired, bacterial colonization often endures through most of the host's life. Initial analyses revealed predominantly intra-familiar transmission from parents to children [3, 4], however more recently, localized frequent horizontal transmission between unrelated people inhabiting the same area has also been observed [5]. Recombination between unrelated strains occurs during mixed colonization [6–9] resulting in numerous changes in the bacterium's genome. An unusually high mutation rate [10–12] and recombination rate [8, 13] generate a sequence diversity within *H. pylori* that is much greater than that of other bacteria. As a consequence, *H. pylori* is highly diverse and almost every isolate possesses a unique sequence type (ST) in a multi-locus sequence typing scheme [14, 15], unlike other bacteria where strains with identical STs are frequently found [16]. This unusually high sequence diversity led to the initial suggestion that the population structure in *H. pylori* was panmictic [17, 18].

Geographical Distribution of *H. pylori* Populations

Despite this exceptional sequence diversity, several bacterial populations were identifiable firstly by sequence similarity [19–21] and then later using model-based cluster and assignment analyses [14, 15]. These populations correlated with their continent of origin which argued against worldwide panmixia and suggested admixture on a regional or local scale only. Polymorphism in seven housekeeping gene fragments from a global collection of 769 *H. pylori* isolates was as high as 47%, and the six defined major bacterial populations (fig. 1a) were designated after the geographic location in which they were found most frequently [15]. Of these, five populations were found to be very closely related to each other and these included hpEurope, isolated from Europeans, from countries in the middle East and from India [14, 15, 22, 23]; hpAfrica1 from Morocco, Senegal, Burkina Faso and South Africa; hpNEAfrica, isolated in Ethiopia, Somalia, Sudan and from Nilo-Saharan speakers in northern Nigeria; hpAsia2, predominantly in Northern India and also among isolates from Bangladesh, Thailand and the Philippines; and hpEastAsia from continental East Asia, Oceania and the Americas (fig. 1b). Further within-regional clustering split the populations hpAfrica1 into western (hspWAfrica) and southern (hspSAfrica) subpopulations, and hpEastAsia into mainland East Asian (hspEAsia), Oceanic (hspMaori) and Native American (hspAmerind) subpopulations [14, 15]. The large proportion of diversity at the individual level appears to result in a phylogenetic continuum across these five populations (fig. 1a), however, a sixth and more distantly related population, hpAfrica2, has also been defined. hpAfrica2 is not only very divergent to all other *H. pylori* populations, but was only isolated in South Africa, among people with both African and European ancestry. The origin of hpAfrica2 is still unclear.

Recent Human Movements and Population Level Inconsistencies

The consistent partitioning of strains into geographically separated populations immediately yielded conclusive evidence for obvious recent human population movements, and much more readily than with human DNA. The European colonial expansion began approximately 500 years before present (BP) and led to the colonization

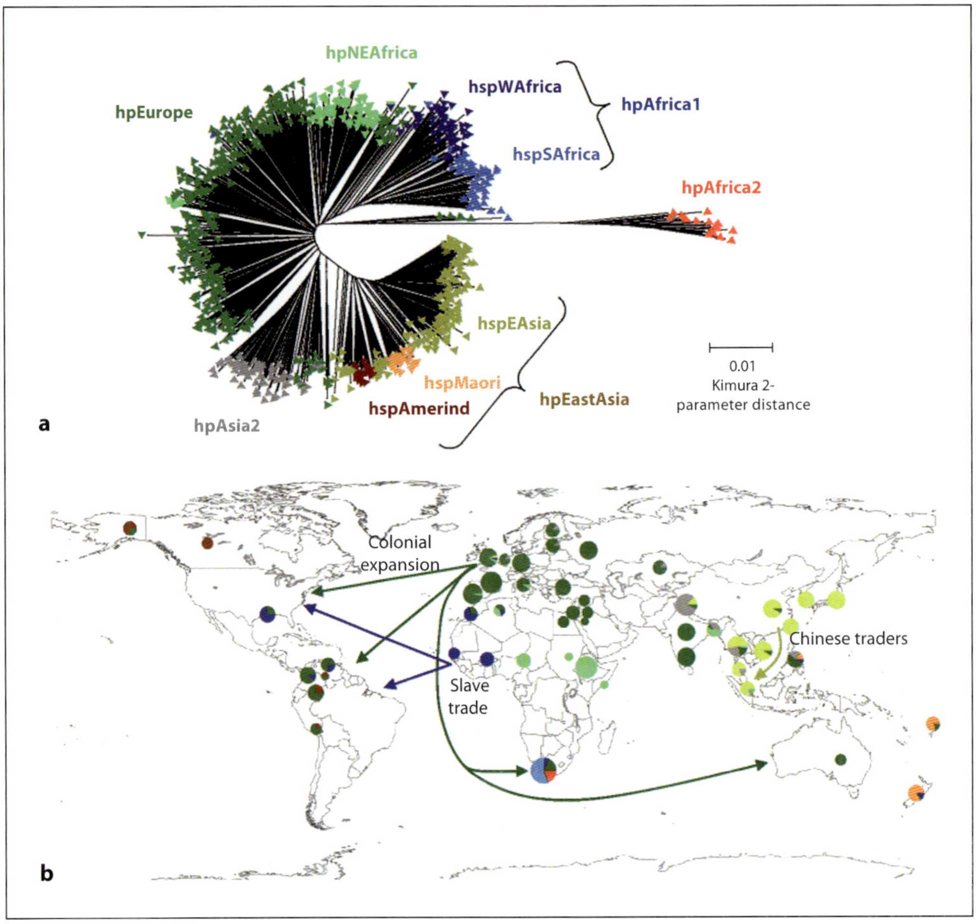

Fig. 1. Neighbour-joining tree (**a**) of 769 concatenated housekeeping gene sequences of *H. pylori* color-coded according to assignment into the populations hpEurope, hpAsia2, hpNEAfrica, hpAfrica2, hpAfrica1 (subpopulations: hspWAfrica, hspSAfrica) and hpEastAsia (subpopulations hspAmerind, hspMaori, hspEAsia). (**b**) The global distribution of extant *H. pylori* populations reflects recent (<500 years) human migrations.

of numerous regions in the world, a process that was often associated with the (near) extinction of indigenous human populations. It is now clear that the colonizers brought more than just their own genes to the lands they colonized. European *H. pylori* (hpEurope) was isolated in both North and South America, and in the former British [14, 15] and Russian Empires [22]. However, the high frequency of hpEurope strains among isolates from India was possibly associated with ancient migrations of Indo-Aryan speakers into the subcontinent, rather than with recent British introduction [23]. A direct result of the European colonial expansion was the slave trade, which saw the forced migration of West Africans to the Americas. As a result, bacteria of the West African subpopulation hspWAfrica can be found in the USA, Colombia and

Venezuela. Similar to the extensive spread of Europeans, recent movements of traders of Chinese origin across Southeast Asia into Thailand, Malaysia and Singapore are also reflected in the distribution of hspEAsia (fig. 1b) [15].

Prehistoric Human Migrations: The Sub-Population Level

Large scale human movements occurring since the recession of the most recent glacial maximum produced more subtle variations in *H. pylori* genetic structure. Using both a model-based approach to determine fine-scale structuring within populations as well as a phylogenetic approach to determine ancestral from derived populations revealed the indelible signatures for a number of prehistoric migrations.

Across the Bering Strait

The Americas were initially populated by people of Asian origin when a land-bridge connected Asia with North America during the last glacial maximum approximately 18,000 BP [24]. If humans carried *H. pylori* bacteria in their stomachs during migration across the Bering Strait, one would expect these bacteria to be of Asian origin, and thus to be related to strains from modern East Asians. However, initial analyses of *H. pylori* from Peruvian Amerindians showed higher sequence similarity to Spanish and not to East Asian isolates [21]. Given no other evidence, this was interpreted as an absence of *H. pylori* in the Pre-Colombian Americas and the introduction of *H. pylori* to the Americas by European conquerors. A direct consequence of this hypothesis was the first attempt to fix a time for the association between humans and *H. pylori*. The absence of East Asian *H. pylori* in the Americas was taken as evidence that humans had acquired *H. pylori* after Native Americans had already diverged from East Asians. *H. pylori* infection of humans was therefore thought to begin in early agricultural societies, e.g. the Fertile Crescent, as with numerous other pathogens, after a host jump from domesticated animals to humans in the last 10,000 years [21]. More recent studies have identified East Asian *H. pylori* among strains from Native Americans from North and South America [14, 25, 26]. These *H. pylori* sequences formed a separate sub-population, which argues against a recent introduction of the East Asian bacteria into the Americas from Chinese or Japanese immigrants. Instead, these differentiated East Asian strains provided the first evidence that *H. pylori* accompanied humans on their migrations across the Bering Strait and into the Americas (fig. 2a–c).

China and the Polynesian Expansion

During the last 3000 years, Chinese, a subfamily of the Sino-Tibetan language family, was spread south and eastwards across all of China, mainly by the expansion of Zhou Dynasty (1100 to 221 BC) [27]. This resulted in the fragmentation of the three major language families (Hmong-Mien, Tai-Kadai, and Austroasiatic) originally spoken in south China. The population structure of *H. pylori* across East Asia, including the

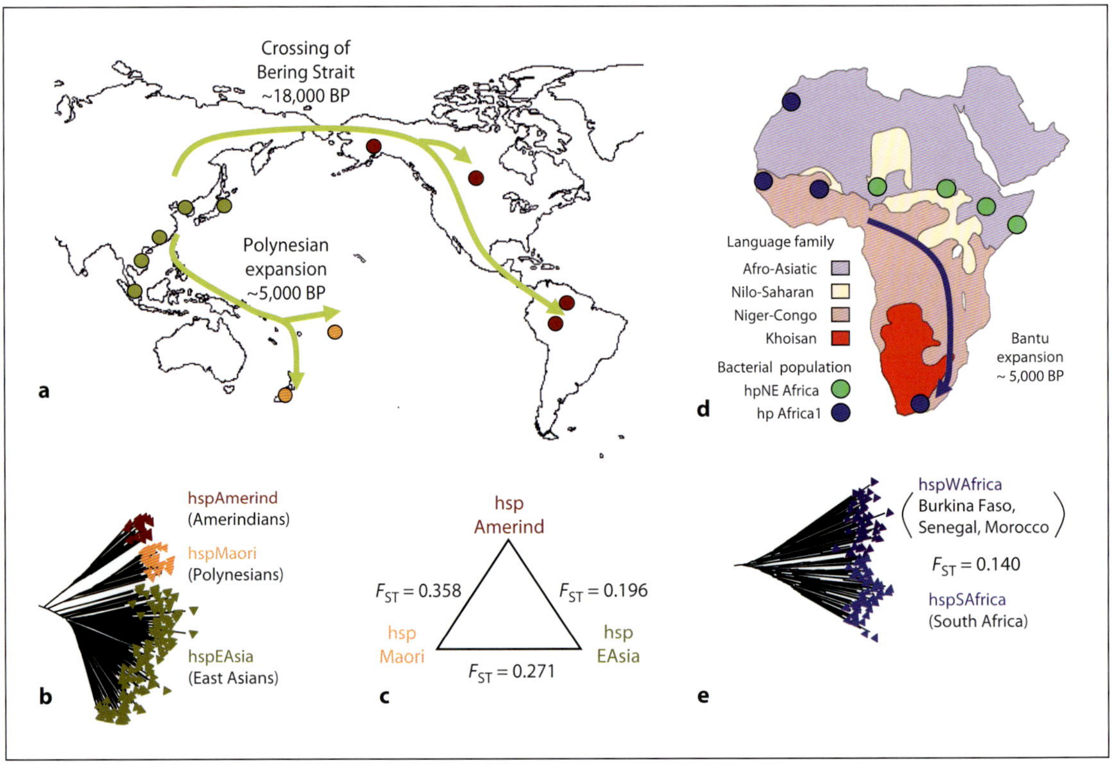

Fig. 2. Signals of prehistoric human migrations in *H. pylori* sequences. (**a**) Asian people carried *H. pylori* in their stomachs when they crossed the Bering Strait and populated the Americas. (**b**) The Polynesian expansion is traced by *H. pylori* of East Asian ancestry from Polynesians and Maoris from New Zealand. (**c**) The strong genetic drift in the hspMaori sub-population is presumably the result of human population bottlenecks that are due to the sequential island-hopping during the colonization of the Pacific. (**d**) The expansion of Niger-Congo speaking Bantu peoples across southern Africa brought hpAfrica1 bacteria to South Africa. (**e**) This expansion resulted in the formation of the two hpAfrica1 sub-populations hspWAfrica and hspSAfrica that are still so similar that the genetic distance between them is very low.

Korean peninsula and Japan, is therefore almost uniform, with most strains belonging to the population hpEastAsia. Within this continuum, the only exceptions are the Thai, whose bacteria are predominantly assigned to hpAsia2, and hence appear to have resisted the southern expansion of hpEastAsia [15]. The very close relationship between Korean and Japanese isolates is not surprising since the ancestors of modern Japanese migrated to Japan from the Korean peninsula [27] introducing rice agriculture to the Japanese archipelago.

The Austronesian language family was probably also present in East Asia, but has since disappeared from the mainland entirely, surviving only on Taiwan (Formosa). From Taiwan, Austronesian-speaking seafarers succeeded in settling the huge geographical expanse of Oceania in what is now referred to as the Polynesian expansion

[28–30]. Accordingly, hpEastAsia bacteria that were carried in their stomachs were disseminated across the Pacific from the Philippines, to Polynesia and to the Maori of New Zealand. This process of 'island hopping' where each succeeding population is seeded by a few founders led to an accelerated genetic drift, which differentiated these bacteria from those on the East Asian mainland, making them easily distinguishable as a distinct sub-population, hspMaori (fig. 2a–c) [14, 15].

Bantu Expansions in Africa

Tropical West Africa is the homeland of the Bantu, a group of people speaking a collection of closely related languages that constitute a single, low-order subfamily of the Niger-Congo language family. This group consists of approximately 500 languages [31] and is spoken in most of sub-Saharan Africa (fig. 2d). This reflects two major prehistoric events – the development of agriculture in tropical West Africa and the subsequent expansion of Bantu societies into the summer-rainfall regions of sub-equatorial Africa that were climatically suitable for their crops [27, 32]. The Bantu expansion began around 5,000 years ago and either replaced or absorbed most of the original hunter-gatherer societies in its wake. By 700 AD it had reached its southern limit in eastern South Africa. The stomachs of Niger-Congo speakers (including Bantu) were found to be infected by *H. pylori* from the population hpAfrica1 [14, 15]. Consistent with a rapid expansion, hpAfrica1 bacteria from South Africa differ only slightly from those found in Senegal, Gambia and Burkina Faso (fig. 2e). The short time period (5,000 years) since the beginning of the Bantu expansion only allowed the development of two closely related subpopulations, hspWAfrica and hspSAfrica [14]. The presence of hspWAfrica in the North African countries Morocco and Algeria [15] provides evidence for gene flow across the Sahara. However hpAfrica1 is completely absent in the Sahel in northeast Nigeria. This region is populated exclusively by *H. pylori* more closely related to strains isolated in Sudan, Ethiopia and Somalia which belong to the population hpNEAfrica (fig. 2d) [15]. The eastward spread of Bantu farmers across the Sahel was therefore limited, most likely due to the incompatibility of tropical crops in more arid climates and possibly to the presence of another, older society of Nilo-Saharan speaking farmers equipped with their own arid-adapted array of crops.

Looking further into a Complicated Past: Identifying Ancestral Populations

Until the recent past, admixture between strains occurred mainly within populations. Therefore, despite very high within-population genetic diversity, signals for more ancient events that occurred between populations still persisted. One very informative method for inferring ancient population structure was the linkage model [33], which assigned individual nucleotides to groups on the basis of their linkage to neighboring nucleotides. This method identified five ancestral populations: ancestral

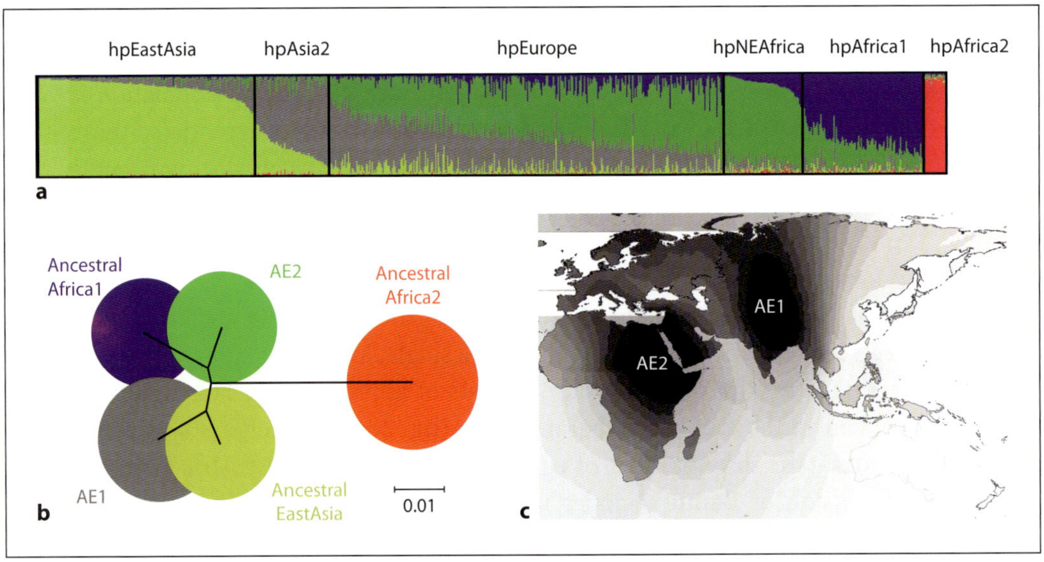

Fig. 3. Five ancestral populations in *H. pylori.* (**a**) The proportion of ancestry from each of the five ancestral sources varies in individual isolates and appears as a continuum across the modern bacterial populations, with the exception of hpAfrica2. (**b**) Neighbor-joining tree of the five ancestral populations. (**c**) Declining dark-light gradients in the proportion of ancestral nucleotides by distance from a geographic centre revealed Central Asian (AE1) and Northeast African (AE2) origins of the two predominant ancestral sources of extant European *H. pylori.*

Europe1 (AE1), ancestral Europe2 (AE2), ancestral EastAsia, ancestral Africa1 and ancestral Africa2 (fig. 3b) [14]. The terms 'ancestral Europe 1' and 'ancestral Europe2' were invented to designate the significant proportions of ancestry in European *H. pylori* isolates, but maybe misleading as the spatial distribution of ancestral nucleotides indicated that AE1 originated in Central Asia and AE2 in Northeast Africa (fig. 3c) [15]. Ancestral EastAsia originated in East Asia, ancestral Africa1 in West Africa and ancestral Africa2 in South Africa. The proportion of ancestry from each of the five ancestral sources varies in individual isolates and when grouped by modern population, appears as a continuum, suggesting clinal variation between ancestral populations from a common ancestor (fig. 3a). The concept of ancestral proportions then allowed the detection of the more complicated population events that occur in hybrid zones – areas where isolates from two ancestral sources meet. Ladakh in North India is one such zone inhabited by people of two major human groups, Muslims and Buddhists, who have coexisted for almost 1,000 years but remained largely isolated due to cultural and religious differences. Human microsatellites and mtDNA were only marginally informative in detecting differences between the two groups. However, an analysis of *H. pylori* housekeeping gene sequences using the linkage model showed that isolates from Muslims were quite uniform in AE1 ancestry indicating that the Islamic religion was introduced by few missionaries rather than extensive population

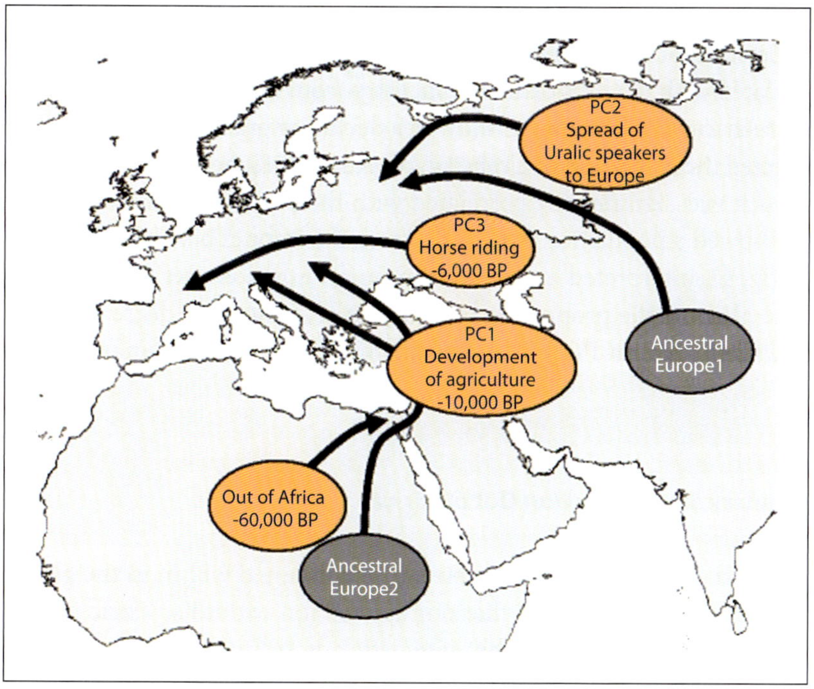

Fig. 4. The out of Africa event and human migrations to Europe as inferred from *H. pylori* sequences.

movements. In contrast, isolates from Buddhists showed a cline of introgression from almost pure ancestral EastAsia to almost pure AE1 which was taken as a clear signal for the introduction of Buddhism (as well as hpEastAsia *H. pylori)* by Tibetan migrants into a pre-existing Ladakhi population [34].

The pattern of ancestry in modern European *H. pylori* is more complex. Isolates assigned to the hpEurope population were found to be recombinants of mainly AE1 that originated in Central Asia and AE2 from Northeast Africa which is probably associated with the re-colonization of Europe after the ice age (figs. 3, 4). Furthermore, numerous hpEurope isolates also contained polymorphisms acquired from ancestral Africa1 and ancestral EastAsia. Europe, therefore, was a complex hybrid zone. A multivariate technique known as Principal Component Analysis (PCA) was used to partition the ancestral data into its varying layers of complexity. Each layer, known as a Principal Component (PC), describes only a proportion of the total variation in the data. A previous PCA of allozymes in human European populations revealed the existence of gradients in allele frequencies across Europe that traced a series of prehistoric human migrations [35]. When the same technique was implemented on *H. pylori* sequences, it unraveled a very similar series of population movements into Europe [15]. The first PC described a cline from the Southeast to Northwest Europe that correlated with archeological data on the westward spread of domesticated crops from the Fertile Crescent across Europe by Neolithic farmers (fig. 4). This cline was significantly correlated with the proportion of ancestry from the population AE2 that arose in Northeast Africa. The second PC showed a declining gradient from North to South, which reflected the migration of Uralic language speaking peoples from Siberia into Scandinavia as it was significantly correlated with the proportion of AE1 ancestry. The third PC showed a population expansion from the steppes between the Volga and Don Rivers (fig. 4), interpreted as spread of pastoral nomads after the domestication of the horse [36]. As the proportion of variance accounted for decreases with each subsequent PC, the fourth PC was not consistent between the human and *H. pylori* data.

Detecting Evolutionary Signal: *H. pylori* Out of Africa

Given the modern and ancestral population structure contained within in the global sample of *H. pylori* DNA, it is now clear that our association with this gastric pathogen is very old. While the exact age of this association is still not known, startling comparisons between human and *H. pylori* DNA provide evidence for a relationship on an evolutionary time-scale. Pairwise *FST*, a measure of genetic differentiation between populations, obtained from human microsatellite data was strongly correlated ($R^2 = 0.73$) with pairwise *FST* from *H. pylori* housekeeping gene sequences from the same human populations [15]. This quantifiably confirmed a similar and directly comparable population structure in both the bacterium and its human host. Further evidence for similarities in evolutionary trajectories was obtained when both *H. pylori* and human data showed the same pattern of isolation-by-distance where the genetic distance increases with geographic distance between populations [15].

In human populations, genetic diversity is known to decrease with distance from East Africa, the likely cradle of modern humans, due to serial founder effects where only a proportion of an original population migrates further to form a new population [37, 38]. Thus, the overall genetic diversity in the population's gene pool decreases in a stepwise nature with distance from the origin. When diversity for each *H. pylori* sampling locality was plotted against distance from East Africa, a similarly significant trend was observed, indicating an African origin for both host and parasite [15]. Computer simulations on human DNA data, using a stepping-stone model of migration, indicated that anatomically modern humans migrated from Africa around 56,000 BP [39]. The same simulation for *H. pylori* data resulted in an estimate of 58,000 years [15]. Taken together, these data strongly imply that, *H. pylori* arose in Africa and our forefathers carried this pathogen in their stomachs on their migrations out of Africa. However, the actual age of association between humans and *H. pylori* must be much older as the out-of-Africa computer simulations were calculated without the distant population hpAfrica2 and because of a host jump of *H. pylori* from early humans to large felines that gave rise to *H. pylori*'s closest relative, *Helicobacter acinonychis.* The timing of this host jump was originally estimated to have occurred 200,000 years BP [11], but this was recently adjusted to 100,000 years BP [12]. If the great apes, the closest relatives of humans, also carried gastric helicobacters phylogenetically related to *H. pylori,* then the human stomach could possibly have been colonized by *H. pylori* for several million years.

Other Microbes as Markers for Human Migrations

Besides *H. pylori,* several other human pathogens possess a global phylogeographical structure, however, each of those is associated with specific problems. One of the major drawbacks using microbial sequences is the microbes' transmission mode from one host generation to the next, because frequent horizontal transmission between unrelated hosts dramatically hampers, if not abolishes, any attempts to elucidate human population history. This problem appears peculiar to viruses but also affects the mycobacteria. Six major geographically associated lineages have been described for *Mycobacterium tuberculosis* [40], however, the phylogeny is not rooted in Africa arguing against an evolutionary association between humans and this bacterium. *M. tuberculosis* possesses a clonal genetic population structure which results in a further problem, the lack of resolution due to the limited amount of polymorphism, which is even more pronounced in its close relative *Mycobacterium leprae* [41]. Yet, three informative SNPs have been identified in *M. leprae,* the combination of which resulted in four SNP-types. The geographical distribution of these SNP-types traced a number of recent human migrations including the European colonial expansion, the slave trade and the spread of Asian traders to Caribbean islands. Moreover, these data were consistent with an African origin and subsequent, independent spread of leprosy to Asia and Europe [41].

Soon after the publication of geographical differences [42, 43] in sequences of the human polyomavirus JC (JCV), genetic structuring in this virus was extensively interpreted in the light of human population history, including an African origin and prehistoric migrations [44]. Virus infection is mostly acquired during adolescence, and due to its predominantly vertical transmission from parents to children as well as easy virus sample collection from urine, JCV became a very popular tool to unravel the population movements on a local scale (reviewed in [45, 46]). However, an analysis of the JC virus evolution and its association with human populations revealed no evidence for codivergence between JCV and human phylogenies [47]. Hence, this virus should not be used as a marker for human population history. Likewise, the analyses of other viruses like human papillomavirus, human T-cell lymphotropic virus or the hepatitis G virus that are usually useful at a local scale, are problematic on a global scale, as the geographical distribution of several virus types is difficult or even impossible to explain by known past human migrations [46, 48].

Thus, to date *H. pylori* is probably the most promising candidate although it is also associated with specific problems. Bacteria are usually grown from gastric biopsies which are taken during gastroendoscopy which is an invasive procedure, and the bacterial transmission is also not strictly vertically within families.

Concluding Remarks

H. pylori, a highly diverse bacterium at the sequence level, possesses strong phylogeographic structure that is wholly interpretable in the light of human population movements. Human genetic, archaeological and linguistic data have been used to explain the observed patterns in *H. pylori* sequences. *H. pylori* already accompanied our forefathers on their migration(s) out of Africa around 60,000 years BP, and the intimate association between host and parasite that has been maintained ever since enabled the reconstruction of numerous ancient and modern human population movements. These findings set the stage for the use of *H. pylori* sequences to unravel fiercely debated topics in human population history, and in unprecedented detail. A study about the source and trajectory of spread of two distinct waves of migrations into the Pacific [49] was published while this book was in press. A first migration reached New Guinea and Australia 31,000–37,000 years ago and a second, much later dispersal originated in Taiwan and spread hspMaori through the Pacific.

References

1 Marshall BJ, Warren JR: Unidentified curved bacilli in the stomach of patients with gastritis and peptic ulceration. Lancet 1984;1:1311–1315.

2 Marshall BJ, Armstrong JA, McGechie DB, Glancy RJ: Attempt to fulfil Koch's postulates for pyloric *Campylobacter.* Med J Aust 1985;142:436–439.

3 Tindberg Y, Bengtsson C, Granath F, Blennow M, Nyren O, Granstrom M: *Helicobacter pylori* infection in Swedish school children: lack of evidence of child-to-child transmission outside the family. Gastroenterology 2001;121:310–316.

4 Kivi M, Tindberg Y, Sorberg M, Casswall TH, Befrits R, et al: Concordance of *Helicobacter pylori* strains within families. J Clin Microbiol 2003;41: 5604–5608.

5 Delport W, Cunningham M, Olivier B, Preisig O, Van Der Merwe SW: A population genetics pedigree perspective on the transmission of *Helicobacter pylori.* Genetics 2006;174:2107–2118.

6 Taylor NS, Fox JG, Akopyants NS, Berg DE, Thompson N, et al: Long-term colonization with single and multiple strains of *Helicobacter pylori* assessed by DNA fingerprinting. J Clin Microbiol 1995;33:918–923.

7 Kersulyte D, Chalkauskas H, Berg DE: Emergence of recombinant strains of *Helicobacter pylori* during human infection. Mol Microbiol 1999;31:31–43.

8 Falush D, Kraft C, Taylor NS, Correa P, Fox JG, et al: Recombination and mutation during long-term gastric colonization by *Helicobacter pylori:* Estimates of clock rates, recombination size and minimal age. Proc Natl Acad Sci USA 2001;98:15056–15061.

9 Raymond J, Thiberg JM, Chevalier C, Kalach N, Bergeret M, et al: Genetic and transmission analysis of *Helicobacter pylori* strains within a family. Emerg Infect Dis 2004;10:1816–1821.

10 Björkholm B, Sjölund M, Falk PG, Berg OG, Engstrand L, Andersson DI: Mutation frequency and biological cost of antibiotic resistance in *Helicobacter pylori.* Proc Natl Acad Sci USA 2001; 98:14607–14612.

11 Eppinger M, Baar C, Linz B, Raddatz G, Lanz C, et al: Who ate whom? Adaptive *Helicobacter* genomic changes that accompanied a host jump from early humans to large felines. PLoS Genet 2006;2:e120.

12 Schuster SC, Wittekindt NE, Linz B: Molecular mechanisms of host-adaptation in *Helicobacter*; in Yamaoka Y (ed): *Helicobacter pylori:* Molecular Genetics and Cellular Biology. Wymondham, UK, Horizon Scientific Press, 2008, pp 193–204.
13 Suerbaum S, Smith JM, Bapumia K, Morelli G, Smith NH, et al: Free recombination within *Helicobacter pylori.* Proc Natl Acad Sci USA 1998; 95:12619–12624.
14 Falush D, Wirth T, Linz B, Pritchard JK, Stephens M, et al: Traces of human migrations in *Helicobacter pylori* populations. Science 2003;299:1582–1585.
15 Linz B, Balloux F, Moodley Y, Manica A, Liu H, et al: An African origin for the intimate association between humans and *Helicobacter pylori.* Nature 2007;445:915–918.
16 Urwin R, Maiden MC: Multi-locus sequence typing: a tool for global epidemiology. Trends Microbiol 2003;11:479–487.
17 Salaun L, Audibert C, Le Lay G, Burucoa C, Fauchere JL, Picard B: Panmictic structure of *Helicobacter pylori* demonstrated by the comparative study of six genetic markers. FEMS Microbiol Lett 1998;161:231–239.
18 Go MF, Kapur V, Graham DY, Musser JM: Population genetic analysis of *Helicobacter pylori* by multilocus enzyme electrophoresis: extensive allelic diversity and recombinational population structure. J Bacteriol 1996;178:3934–3938.
19 Achtman M, Azuma T, Berg DE, Ito Y, Morelli G, et al: Recombination and clonal groupings within *Helicobacter pylori* from different geographical regions. Mol Microbiol 1999;32:459–470.
20 Mukhopadhyay AK, Kersulyte D, Jeong JY, Datta S, Ito Y, et al: Distinctiveness of genotypes of *Helicobacter pylori* in Calcutta, India. J Bacteriol 2000; 182:3219–3227.
21 Kersulyte D, Mukhopadhyay AK, Velapatino B, Su W, Pan Z, et al: Differences in genotypes of *Helicobacter pylori* from different human populations. J Bacteriol 2000;182:3210–3218.
22 Momynaliev KT, Chelysheva VV, Akopian TA, Selezneva OV, Linz B, et al: Population identification of *Helicobacter pylori* isolates from Russia. Genetika 2005;41:1434–1437.
23 Devi SM, Ahmed I, Francalacci P, Hussain MA, Akhter Y, et al: Ancestral European roots of *Helicobacter pylori* in India. BMC Genomics 2007;8:184.
24 Fagundes NJ, Kanitz R, Eckert R, Valls AC, Bogo MR, et al: Mitochondrial population genomics supports a single pre-Clovis origin with a coastal route for the peopling of the Americas. Am J Hum Genet 2008;82:583–592.
25 Yamaoka Y, Orito E, Mizokami M, Gutierrez O, Saitou N, et al: *Helicobacter pylori* in North and South America before Columbus. FEBS Letters 2002;517: 180–184.
26 Ghose C, Perez-Perez GI, Dominguez-Bello MG, Pride DT, Bravi CM, Blaser MJ: East Asian genotypes of *Helicobacter pylori* strains in Amerindians provide evidence for its ancient human carriage. Proc Natl Acad Sci USA 2002;99:15107–15111.
27 Diamond J: Guns, Germs and Steel. London, Jonathan Cape, 1997.
28 Diamond JM: Express train to Polynesia. Nature 1988;336:307–308.
29 Diamond JM: Taiwan's gift to the world. Nature 2000;403:709–710.
30 Trejaut JA, Kivisild T, Loo JH, Lee CL, He CL, et al: Traces of archaic mitochondrial lineages persist in Austronesian-speaking Formosan populations. PLoS Biol 2005;3:e247.
31 Ruhlen M: The Origin of Language. New York, John Wiley & Sons, Inc., 1994.
32 Diamond J, Bellwood P: Farmers and their languages: the first expansions. Science 2003;300:597–603.
33 Falush D, Stephens M, Pritchard JK: Inference of population structure using multilocus genotype data: linked loci and correlated allele frequencies. Genetics 2003;164:1567–1587.
34 Wirth T, Wang X, Linz B, Novick RP, Lum JK, et al: Distinguishing human ethnic groups by means of sequences from *Helicobacter pylori:* Lessons from Ladakh. Proc Natl Acad Sci USA 2004;101:4746–4751.
35 Cavalli-Sforza LL, Menozzi P, Piazza A: The History and Geography of Human Genes. Princeton, NJ, Princeton University Press, 1994.
36 Piazza A, Rendine S, Minch E, Menozzi P, Mountain J, Cavalli-Sforza LL: Genetics and the origin of European languages. Proc Natl Acad Sci USA 1995; 92:5836–5840.
37 Prugnolle F, Manica A, Balloux F: Geography predicts neutral genetic diversity of human populations. Curr Biol 2005;15:R159–R160.
38 Ramachandran S, Deshpande O, Roseman CC, Rosenberg NA, Feldman MW, Cavalli-Sforza LL: Support from the relationship of genetic and geographic distance in human populations for a serial founder effect originating in Africa. Proc Natl Acad Sci USA 2005;102:15942–15947.
39 Liu H, Prugnolle F, Manica A, Balloux F: A geographically explicit genetic model of worldwide human-settlement history. Am J Hum Genet 2006; 79:230–237.

40 Gagneux S, DeRiemer K, Van T, Kato-Maeda M, de Jong BC, et al: Variable host-pathogen compatibility in *Mycobacterium tuberculosis*. Proc Natl Acad Sci USA 2006;103:2869–2873.
41 Monot M, Honore N, Garnier T, Araoz R, Coppee JY, et al: On the origin of leprosy. Science 2005;308: 1040–1042.
42 Agostini HT, Yanagihara R, Davis V, Ryschkewitsch CF, Stoner GL: Asian genotypes of JC virus in Native Americans and in a Pacific Island population: markers of viral evolution and human migration. Proc Natl Acad Sci USA 1997;94:14542–14546.
43 Sugimoto C, Kitamura T, Guo J, Al Ahdal MN, Shchelkunov SN, et al: Typing of urinary JC virus DNA offers a novel means of tracing human migrations. Proc Natl Acad Sci USA 1997;94:9191–9196.
44 Pavesi A: African origin of polyomavirus JC and implications for prehistoric human migrations. J Mol Evol 2003;56:564–572.
45 Yogo Y, Sugimoto C, Zheng HY, Ikegaya H, Takasaka T, Kitamura T: JC virus genotyping offers a new paradigm in the study of human populations. Rev Med Virol 2004;14:179–191.
46 Wirth T, Meyer A, Achtman M: Deciphering host migrations and origins by means of their microbes. Mol Ecol 2005;14:3289–3306.
47 Shackelton LA, Rambaut A, Pybus OG, Holmes EC: JC virus evolution and its association with human populations. J Virol 2006;80:9928–9933.
48 Holmes EC: The phylogeography of human viruses. Mol Ecol 2004;13:745–756.
49 Moodley Y, Linz B, Yamaoka Y, Windsor HM, Breurec S, et al: The peopling of the Pacific from a bacterial perspective. Science 2009;323:527–530.

Bodo Linz
Department of Molecular Biology, Max-Planck Institute for Infection Biology
Charitéplatz 1
DE 10117 Berlin (Germany)
Tel. +49 30 28460 169, Fax +49 30 28460 111, E-Mail bodo.linz@googlemail.com

de Reuse H, Bereswill S (eds): Microbial Pathogenomics.
Genome Dyn. Basel, Karger, 2009, vol 6, pp 75–90

Helicobacter pylori Genome Plasticity

D.A. Baltrus[a] · M.J. Blaser[b] · K. Guillemin[c]

[a]Department of Biology, University of North Carolina at Chapel Hill, Chapel Hill, N.C., [b]Department of Medicine, New York University School of Medicine, New York, N.Y., [c]Institute of Molecular Biology, University of Oregon, Eugene, Oreg., USA

Abstract

Helicobacter pylori, a Gram-negative pathogen associated with ulcers, chronic gastritis, and gastric cancers, has been a resident of the human stomach since early human history [1]. This association has only recently begun to erode with the advent of antibiotics and modern lifestyles, but even today *H. pylori* colonizes approximately half the world's population. To have remained a successful colonizer of humans during thousands of years of association, populations of *H. pylori* must have been able to survive and adapt to countless evolutionary challenges within and between hosts. As a species, *H. pylori* possesses one of the most fluid genomes within the prokaryotic kingdom [2], a characteristic that has likely aided its continued success. *H. pylori* exhibits exceptionally high rates of DNA point mutations, intragenomic recombination (facilitated by repetitive elements common in *H. pylori* genomes), and intergenomic recombination (mediated by natural transformation), all of which contribute to the high genomic variability between isolates. Previous reviews have focused on these processes as agents of evolutionary change within *H. pylori* [2–8]. The mechanisms of both mutation and natural transformation, and the evolutionary processes that retain genetic variation generated by these mechanisms, dictate the extent to which each contributes to genomic diversity in the context of different bacterial population structures [9–13]. Unlike well-studied evolutionary systems, such as *Salmonella* and *Escherichia coli*, *H. pylori* is notable in its lack of an environmental reservoir outside of human and other primate stomachs, suggesting that between-host survival is a relatively weak determinant of selection pressures [14, 15]. Given that *H. pylori* exist largely as distinct host-associated populations, it is possible to begin to model the evolutionary mechanisms that affect the long-term persistence of this species. In this chapter, we consider how the attributes of *H. pylori*'s natural history as a long-term resident of the human stomach and the specific mechanisms of mutation and genetic exchange in this organism have shaped the *H. pylori* genome. We begin with a survey of genome plasticity in *H. pylori*. We then discuss mechanisms of mutation and natural transformation in *H. pylori* and examine experimental evidence for the generation of genomic changes within populations. Finally, we consider how different models of *H. pylori* population structure affect the relative contributions of mutation and recombination to the evolutionary success of this organism. By bridging evolutionary studies with investigations of pathogenesis from a molecular perspective, we hope to shed new light on how *H. pylori* has and continues to evolve with its human hosts.

The Genomic Landscape of *H. pylori*

Early *H. pylori* researchers using random genotyping methods, such as random amplification of polymorphic DNA (RAPD) and multilocus sequence typing (MLST), were astonished at the level of diversity found within *H. pylori* populations [16]. These studies led to the conclusion that genotypic diversity was so high within *H. pylori* that each individual host harbored his or her own strain. This viewpoint was tempered slightly with the publication of the first two complete genomes from isolates of *H. pylori*, which demonstrated on a genome-wide scale that the high level of nucleotide variation consisted mainly of changes silent at the protein level and thus largely neutral to selection [17, 18]. These complete genome sequences enabled the creation of microarrays to investigate gene content differences among many isolates, which led to further observations of the level of genetic content differences within the species as a whole. Specifically, strains were found to vary in gene content by ~12–18% within pair-wise comparisons with ~22–32% of the 'pan-genome' (or collective set of genes contained in the genomes of all *H. pylori* strains) classified as variable because the gene was absent in at least one isolate [19–21]. Within a single individual host, strains from the same clonal lineage were found to vary in the presence of between 24 to 67 genes [22, 23], although another report did not identify such variation within family clusters [21]. Likewise, although there is usually only one dominant strain within any individual, simultaneous co-colonization by multiple strains has been documented and thus strains within the same stomach may vary in genomic content by more than ~18% of their genes [21]. However, because the microarrays were designed only with the genomes of J99 and 26695, these analyses were minimal measurements of strains' gene content diversity and could not estimate the full size of the *H. pylori* pan-genome.

The complete genome sequences for two additional *H. pylori* strains [24, 25], and draft sequences for additional strains [26] argue against the existence of an expansive, undiscovered *H. pylori* pan-genome. For several enteric bacteria, such as *E. coli*, sequential genome sequence determination of different strains has been characterized by a high rate of novel gene discovery, resulting in a collective pan-genome size for these species far larger than any individual strain's genome [27]. In contrast, the additional *H. pylori* genome sequences have uncovered few new genes within these commonly studied strains, with around 10% within any pair-wise comparison defined as strain-specific. Most of these strain specific features are predicted to encode genes of unknown function, restriction endonucleases and methylases that are elements of restriction modification (RM) systems, and outer membrane proteins. Thus, based on gene content, *H. pylori* strains appear to possess relatively similar consensus genomes, as opposed to being subsets of a vast pan-genome.

In addition, the genome sizes of *H. pylori* strains have proven to be quite constant, with between 1485 and 1600 genes, in line with the relatively small genome sizes of the ε-proteobacteria as compared to other known free-living prokaryotes, which may reflect their specialist lifestyles [28]. The genomes are also generally syntenic with one another, but with evidence of large intragenomic rearrangements such as inversions. Many of the strain-specific genes are located in large tracts within regions of the genome referred to as plasticity regions (1 region in J99, HPAG, G27; 2 regions in 26695) or found as singlets or doublets scattered throughout the chromosome.

Acquisition of novel genetic information by horizontal gene transfer has played an important role in the evolutionary history of the *H. pylori* genome. One of the most important virulence determinants of this species, the *cytotoxin associated gene (cag)* pathogenicity island (PAI), which encodes a Type Four Secretion System (TFSS), is located on a genomic island characterized by a lower percent GC than the rest of the chromosome, indicative of having been acquired from a different bacterial species. The acquisition of the *cag* PAI likely occurred after one of the African subpopulations branched off from the rest of the strains [19]. The plasticity regions of the *H. pylori* genome also are characterized by a lower than normal percent GC. However, although there is evidence that genes have been horizontally transferred into the genome from other species [29], many of the variable genes appear to be species-specific with identifiable homologs found only within other *H. pylori* strains. [30]. In contrast, so much genetic exchange has occurred between two different *Campylobacter* species that it appears as though the two separate species are collapsing into one [31].

As discussed below, *H. pylori* is naturally competent to take up free DNA, which provides an important route for genome diversification. The strong species-specific bias in genetic exchange for *H. pylori* may be due to the fact that this species is the dominant member of the human stomach microbiome, contributing approximately three quarters of the bacterial 16S ribosomal RNA clones isolated from this tissue [32]. Therefore, it is likely that a large percentage of the free DNA available for transformation in stomachs colonized by *H. pylori* consists of fragmented *H. pylori* genomes. In addition, abundant RM systems and an apparent ability of the transformation machinery to discriminate between species-specific and foreign DNA [33] likely limit incorporation of DNA from other species. With these restrictions on inter-species genetic exchange, natural transformation may even be a force in maintaining *H. pylori* as a cohesive species by ensuring exchange of genetic information that promotes similarities between strains [34, 35].

The *H. pylori* genome sequences do not suggest a major role for extrachromosomal elements in generating genome diversity for this species. Plasmids were present within two of the sequenced strains, HPAG and G27, but neither appeared to encode many genes other than those required for plasmid transfer [24, 25], similar to other reports of *H. pylori* plasmids [36–41]. Two cryptic plasmids have been shown to include genes similar to loci located within the plasticity zones [37], suggesting that recombination between plasmids and the chromosome does occur. Evidence has also been provided for conjugation among cells within laboratory populations [42, 43], but RM systems provide a barrier against plasmid transfer between strains [44, 45]. Although phage may be associated with *H. pylori* [46], the genomes do not contain sequences of lysogenic phage, arguing against a major role for virally-mediated genetic diversity in this species.

Thus, although *H. pylori* strains' genomes are highly variable at the sequence level, they are remarkably similar to each other in their gene content, size, and paucity of extrachromosomal elements or horizontal transmission of genes from other species. Unlike strain comparisons in *E. coli*, in which phenotypic differences such as tissue tropisms can often be attributed to the presence of gene cassettes, frequently encoded by plasmids or prophages [47], with the exception of the *cag* PAI, the complement of strain-specific genes present in the different *H. pylori* strains whose genome sequences have been determined does not readily explain these strains' phenotypic differences. Instead, it appears that more subtle genetic variation, such as combinations of strain-specific alleles of genes, generated through mutation and reshuffled through genetic exchange, contribute to the phenotypic diversity of *H. pylori* isolates.

Mechanisms that Generate Genomic Variation in *H. pylori*

Mutation

We broadly define mutation as those genetic changes that occur intra-genomically and consist of single nucleotide changes, gene conversion, rearrangements, or deletions mediated by intragenomic recombination. Novel mutations arise within populations because the cellular machinery for DNA replication is not completely faithful or because repair mechanisms are not completely capable of reversing damage due to mutagenic insults [13]. *H. pylori* strains have been found to possess abnormally high rates of nucleotide mutation compared to other representative members of the prokaryotic kingdom [6, 48–51], although other researchers have reported mutation rates in *H. pylori* more similar to those in *E. coli* [52]. Laboratory mutation rate measurements are only a crude approximation of genetic changes occurring across genomes of bacterial populations growing in vivo. For example, within other bacterial systems, mutation rates are known to change with growth phase [53], and vary over different portions of the chromosome as well as with transcriptional level [54, 55]. For *H. pylori*, the environment of the inflamed human stomach would be expected to be rich in mutation-generating chemicals such as reactive oxygen species, which may result in higher mutation rates than those observed during growth in a test tube [56].

H. pylori appears to lack many DNA repair genes including most of the methyl-directed mismatch repair system and the SOS-repair triggered mutagenesis (reviewed in [2]). Variation within these pathways plays a major role in explaining mutation rate diversity for many well-studied bacterial systems [57–59]. However, absence of identifiable sequences of DNA repair genes does not prove the absence of these functions, as demonstrated by the recent description of an *addAB* recombination repair system in *H. pylori* [60].

The *H. pylori* genome is poised to generate phenotypic variation through frame shift mutations at sites throughout the genome that can be referred to as contingency loci. These sites consist of repeated tracts of single or oligo nucleotides that promote slipped-strand mispairing during DNA replication, thereby promoting inactivation of the genes in which they are located by introducing premature stop codons [61–63]. Mutations at contingency loci often occur at rates substantially higher than the normal single nucleotide mutation rate and thus can lead to dramatic shift (phase variation) in the phenotypic characteristics of *H. pylori* populations over very short periods of time [10, 64–66]. At least 46 genes within the 26695 and J99 genomes contain tracts that could act as contingency loci [61]. These predicted contingency loci are enriched in *H. pylori* genes involved in synthesis of cell wall components, which function in host cell-binding but are also targeted by the host immune system [3]. Importantly, compared to single base pair changes resulting in missense mutations, frameshift mutations at contingency loci are more readily reversed. Therefore, the presence of contingency loci within genes important for survival within a host allow bacterial populations to substantially change antigenic profile in a way that can be reversed relatively easily if the selective conditions change, such as introductions into a new host.

The *H. pylori* genome is also organized to generate genetic variation through intragenomic recombination mediated by repeated nucleotide sequences that are distributed non-randomly throughout the genome and oriented to mediate deletion or duplication of intervening sequences [67–71]. Recombination rates increase with the size of the repeated sequences [67]. In a few instances, deletion of sequences flanked by direct repeats has been shown to alter *H. pylori*-host interactions [68, 70, 72]. For example the *cagY* gene, which encodes a surface protein of the *cag* PAI TFSS important for *H. pylori* pathogenesis, contains multiple repeated sequences. Recombination between any of these sequences results in deletions or duplications that maintain an intact open reading frame but would be expected to alter the antigenic properties of this surface protein [68]. Thus, recombination-promoting repeat sequences, like contingency loci, may contribute to rapid adaptation of *H. pylori* populations subject to selective pressure imposed by the host immune system. Importantly, unlike contingency loci, deletions mediated by intra-genomic recombination are not easily revertible through mutation alone. However, deleted sequences can readily be restored by natural transformation with DNA provided by related organisms that have not undergone intra-genomic recombination, as discussed below.

Natural Transformation

Natural transformation is a process by which bacterial cells take up DNA fragments from the extracellular environment, and incorporate these fragments into their own chromosome using conserved recombination machinery [73, 74]. *H. pylori* possesses a unique transformation apparatus that is derived from a TFSS, as opposed to the type IV pili-like structure used by most naturally competent bacteria [4]. The *H. pylori* transformation system is saturable with increasing amounts of DNA and, through an unknown mechanism, can discriminate between its own DNA and that of other species [33].

The process of natural transformation is regulated at multiple steps. Competence, or the ability to take up DNA by natural transformation, differs among cells within bacterial populations. *H. pylori* strains exhibit multiple peaks of competence within both logarithmic and stationary growth phases, with the timing and number of competence peaks differing between strains [75, 76]. The rate of transformation of a cell by any particular allele is dependent on the frequency of fragments containing that allele within the free DNA pool. The nature of the DNA pool available to *H. pylori* is not known, but it could arise through active processes (as with *Neisseria gonorrhoeae*) [77], or through random cell death and lysis [75]. Once a given fragment of DNA is internalized in the cell, its ability to be incorporated into the genome will depend on the extent of sequence homology with the chromosome, to allow for recombination, and its methylation pattern, to resist attack from the cells' complement of RM-systems.

Despite these barriers, natural transformation is thought to be responsible for generating extraordinarily high frequencies of recombination between *H. pylori* strains [5, 78]. Indeed, due to the extent of genetic exchange in *H. pylori*, it can be difficult to establish phylogenetic relationships between strains at a local level because each gene fragment within a genome can have very different evolutionary histories. However, at the global level, sequences from housekeeping genes have been used to show that there are 7 major subpopulations of *H. pylori* that are strongly associated with different populations of humans [1, 79]. Given the intimate association of these strains with their human hosts, and isolation between human populations until recently, phylogenetic information from the bacteria has been used successfully to trace human migration patterns [79] as well as clarify ethnic relationships [80].

Experimental Evidence for Adaptive Benefits of Mutation and Natural Transformation within *H. pylori*

A number of experiments provide evidence that both mutation and natural transformation generate genetic variation that is selected during *H. pylori* adaptation to different growth conditions. These conditions range from growth in liquid culture medium, which offers a controlled experimental design, to experimental colonization of mice and primates, to clinical data from colonized humans, which provide more relevant environmental conditions but greater challenges for interpretation.

Laboratory Culture

Studies have used growth in laboratory medium to show that genotypic [81] and phenotypic properties of *H. pylori* strains [62] are stable in vitro as opposed to their plasticity in vivo. Such studies provide evidence that mutation and transformation alone, within the context of laboratory culture, do not explain inherent diversity within the species. The simplest and most direct test of the importance of natural transformation for adaptation was the measurement of fitness in parallel competent and non-competent nearly isogenic lines derived from strain G27 that were subjected to daily liquid passage in the laboratory [82]. Within this system, all mutations arose de novo during the course of passage, and competence was indeed demonstrated to provide an evolutionary advantage. However, this advantage did not arise until after at least 360 generations of growth in vitro, presumably after linkage disequilibrium had arisen within the populations.

Experimental Animal Infections

Adaptation to growth in mice has also been used as an experimental selective pressure for *H. pylori*. Mice are not natural hosts for *H. pylori* and therefore bacterial adaptation to growth within the mouse stomach would be expected to require genetic changes. This host barrier is highlighted by the demonstration that mouse-adapted strains reproducibly incur mutations in the *cag* PAI or otherwise become attenuated in their capacity to induce inflammatory responses in cell culture assays [83, 84]. Furthermore, genetic variation arising at contingency loci was demonstrated in experimentally infected mice after 360 days of infection [61]. Although it is difficult to determine whether these genetic variants underwent positive selection, it is noteworthy that many genes whose transcriptional patterns were altered were outer membrane proteins or involved in acid resistance. Intriguingly, recombination has been shown to significantly affect the ability of *H. pylori* to successfully colonize mice, although it is not possible to distinguish whether this is due to a requirement for recombination-mediated DNA repair or for gene exchange [60, 85].

In contrast to mice, rhesus monkeys are a natural host for *H. pylori*. Experimental challenge of rhesus monkeys with *H. pylori* has demonstrated the capacity of this bacterium to rapidly adapt its cell surface to complement the glycoproteins of its host [64, 86]. In one study, strains adapted to this environment by eliminating production of the BabA outer membrane protein, which is used for adherence to host Lewis B epithelial antigens [86]. There were multiple routes towards disruption of the *babA* locus, including *babA* replacement with a copy of an alternative outer membrane protein encoding locus *babB* found at another region of the genome. This change presumably occurred by either natural transformation or gene conversion. Similar mechanisms for generating antigenic variation through recombination of homologous regions have been reported in other bacteria [87]. Alternatively, *babA* expression was eliminated through alteration of dinucleotide (CT) repeats in the 5′ coding region of the gene. Selection within this system occurred quickly, with Lewis B adherence by these strains disappearing between 4 and 8 weeks after inoculation. Another study using experimental infection of rhesus monkeys also showed rapid adaptation of the bacteria through modification of Lewis antigens to better match those of the host [64]. In two animals, strain J166, which originally possessed Lewis^{Y} as the dominant antigen, phenotypes switched such that Lewis^{x} was dominant. The phenotypic switch was caused by a single base pair frameshift in a 9 cytosine tract in the fucosyltransferase *futC* required for difucosylated Lewis antigen biosynthesis. In both of these studies it was impossible to determine whether the genetic variants that came to dominate in the stomach were present before inoculation or generated de novo during infection. Still, these results illustrate the importance of both transformation and frameshift mutations in generating genetic variation that can confer adaptive benefits within a host.

Human Studies

In humans, *H. pylori* host adaptation has been studied by sampling bacterial isolates at different times during the course of persistent colonization. However, because of sampling limitations, it is not certain whether strains containing genetic changes identified after months or years of growth within the same individual were present during the first sampling period but were simply not isolated. Therefore, while these studies provide insight into the types of changes that can occur within individual stomachs, it becomes difficult to understand accurately the time scales over which these changes took place. Collectively, these studies have shown that substantial genotypic and phenotypic changes can occur during the course of colonization within a single individual [23, 26, 72, 81, 88–91]. By sampling paired isolates (mean interval of 1.8 years) from 26 individuals, Falush et al. demonstrated that extensive recombination can take place within a single individual, and that natural transformation is likely a dominant contributor to evolutionary dynamics in vivo [88]. Surprisingly, the average size of successfully recombined fragments was 417 bp, very small relative to other bacteria. However, within this same study, 13 of the paired isolates displayed no signs of recombination at all, while 3 displayed only single nucleotide differences indicative of de novo mutation or transformation by small fragments. In another study, a patient from whom the completely sequenced strain J99 was isolated continued to carry *H. pylori* for another 6 years, after which *H. pylori* strains were reisolated [89]. Not only had some of these later isolates developed resistance to the antibiotic clarithromycin (through point mutation in the 23S rRNA gene), but ~2.3% of known J99 ORFs had been lost in at least one isolate. Furthermore, novel DNA sequences were found to have been inserted into the genome, including multiple genes that were similar to loci found in strain 26695. A separate study, using molecular evolutionary methods to compare bacterial populations from different hosts, also showed evidence for recombination of genes that encode enzymes involved in biosynthesis of lipopolysaccharide [92], again pointing to the importance of host selective pressures on shaping the bacterial cell wall.

In cases in which temporal sampling from patients is not possible, multiple isolates from within a single stomach have been collected and analyzed to attempt to infer their evolutionary histories and relationships. For instance, metronidazole has been used as a standard component in the multiple agent therapy for *H. pylori* eradication, and resistance to this drug can occur through mutational inactivation of the oxygen-insensitive NADPH nitroreductase *(rdxA)* gene [93]. In one study, a mixed population of closely related isolates showed phenotypic diversity in metronidazole resistance, with this diversity presumably arising through de novo mutation [94]. However, another study also investigating mixed populations of metronidazole-resistant and sensitive cells demonstrated that resistant *rdxA* alleles were being exchanged between strain backgrounds [95]. Using the same type of logic, the emergence of *cag* PAI negative strains can be followed over the course of persistent colonization. In one case, an individual harbored two distinct *H. pylori* strains, one of which was composed entirely of *cag* PAI$^-$ isolates [96]. The second lineage contained mostly *cag* PAI$^+$ cells, but also some isolates in which the empty site allele of the *cag* PAI had been recombined into the genome from the alternative lineage. This recombination event could be identified by the presence of nucleotide polymorphisms on each side of the empty site allele. Although recombination by an empty site allele cannot be ruled out, a second study showed that *cag* PAI$^-$ isolates could originate by deletion due to homologous recombination between 31 bp repeat regions on each side of the pathogenicity island [72]. Finally, differences in cellular interaction due to the *vacA* gene product between two very closely related strains within one stomach were shown to have arisen through recombination [97]. Collectively, these studies demonstrate that both mutation and natural transformation contribute to adaptation of *H. pylori* populations within human hosts.

Host and Bacterial Population Structures that Promote Adaptive Changes

In the above sections, we have discussed mutation and natural transformation as the dominant forces for generating novel genetic variation within *H. pylori*. Importantly, the generation of novel alleles within populations is completely distinct from the processes that act to drive alleles to measurable frequencies [98]. According to population genetics theory, the extent to which randomly generated mutations will become fixed within bacterial populations is dependent on the selective pressures exerted on these mutations as well as the effective population size [99]. Although the equilibrium frequencies of neutral alleles should be proportional to mutation rate, it is unlikely that even extremely high rates of mutation and natural transformation alone can explain the substantial levels of genetic diversity between strains of *H. pylori*. Two models that incorporate population dynamics could help explain this high diversity. In one model, recurrent selection on *H. pylori* creates high levels of neutral genetic diversity between populations. This would occur when selection for beneficial mutations resulted in retention of linked neutral alleles within the same genome. Different suites of neutral alleles would be expected to arise within separate populations, thus increasing between population divergence. It is currently unclear whether populations of *H. pylori* undergo sufficiently frequent bouts of selection to explain the measured levels of diversity.

Alternatively, genomic diversity in this species could be maintained if *H. pylori* were sequestered in anatomically or nutritionally distinct subpopulations within a single stomach and were not subject to extensive mixing. Population subdivision could allow different *H. pylori* clones to 'explore' their own evolutionary space [100] and fix different beneficial mutations. Also, if the sizes of these isolated subpopulations were sufficiently small, this would reduce the efficiency by which mildly deleterious mutations were eliminated by selection, thereby creating a much larger role for genetic drift in the generation of genotypic diversity. The effective population size of *H. pylori* populations within the human stomach is not known, but it may be quite small if subpopulations are able to occupy distinct niches or experience large bottlenecks during transmission.

A non-homogeneous or subdivided *H. pylori* population would also increase the contributions of natural transformation to evolutionary adaptation. Transformation can counteract the tendency of high mutation rates to generate deleterious mutations by reintroducing functioning copies of genes [71, 101]. Furthermore, as a form of natural transformation can reshuffle genes to bring together beneficial combinations of alleles, but such advantages are only manifest under certain population parameters. Natural transformation will be beneficial in this regard only in populations in linkage disequilibrium in which there are non-random associations between genotypes [9, 11]. Due to this limitation, it is difficult to imagine scenarios within single large populations in which competence provides an extensive evolutionary benefit [12]. However, co-colonization within a single stomach by multiple divergent strains generates high levels of linkage disequilibrium and potentially provides evolutionary advantages for competent strains. Additionally, small divergent subpopulations inhabiting different niches within a colonized host would be predicted to benefit from exchanging DNA via natural transformation. If globally beneficial mutations were to arise within different subpopulations, there would be significant linkage disequilibrium and thus transformation could have a large evolutionary effect within the global metapopulation. Individual subpopulations could also act as reservoirs for each other to replace genes lost through deletion. On the other hand, the presence and expression of different RM systems may prevent the population from collapsing to a single dominant genotype in linkage equilibrium [44].

The human stomach differs dramatically between different anatomic sites, which could easily accommodate differently adapted subpopulations of *H. pylori*. Support for the presence of bacterial subpopulations within single infected individuals comes from multiple studies that have identified genotypically different strains from distinct parts of the stomach [22, 97, 102–105], although the absence of strains from certain anatomic locations could simply be due to sampling limitations. Increasing evidence from animal models also demonstrates that different strains of *H. pylori* preferentially colonize different parts of the stomach [22, 106].

H. pylori population structure within individual stomachs is thought to be dominated by single strains that undergo extensive diversification, even to the point where they begin to resemble viral quasi-species [48, 91]. However, initial inoculation by one strain does not appear to provide strong immunity against super-infection by additional strains [107], thereby providing the potential for access to novel *H. pylori* DNA which can then be incorporated through natural transformation. In fact, one potential explanation for the dramatic difference in rates of genetic divergence found for *H. pylori* populations within two different studies [81, 88] is that sampling for these studies concentrated on areas with marked differences in the prevalence of *H. pylori* and thus potential for co-colonization by multiple strains.

Transmission of *H. pylori* has primarily been thought to occur by passage between close family members, but recent epidemiological studies have shown the situation to be more complex [34, 108]. In developed countries that possess relatively high levels of sanitation, there is a high level of similarity among strains isolated within families, suggesting that transmission occurs predominantly among family members. However, in rural communities there appears to be a much higher prevalence of horizontal transmission between unrelated hosts and there is no significant correlation between kinship and *H. pylori* genotype. Additionally, these rural communities appear more likely to harbor multiple genomic subpopulations of *H. pylori* [109]. Since the presence of multiple divergent strains within a single stomach generates linkage disequilibrium, this significantly increases the evolutionary potential of natural transformation systems. Therefore, the extent to which *H. pylori* strains can generate novel genetic diversity for adaptation may differ significantly between the developing and the developed worlds.

Conclusion

The unique biology of *H. pylori* provides an exceptional opportunity to study the evolutionary importance of mutation and transformation in the context of evolving bacterial populations. With the advent of next generation sequencing technologies, we are on the verge of being able to combine physiological characterization of these processes with extensive genotypic information from population-level sequencing studies. We propose that *H. pylori* microdiversification within subpopulations inhabiting distinct niches of the human stomach maximizes the adaptive benefits of high rates of mutation and natural transformation in this species. The ease with which *H. pylori* genomes can generate antigenic variation through frame shift mutations, intra-genomic recombination, and natural transformation, may allow *H. pylori* populations to respond to the host immune system much like a rheostat, with different genotypes ramping up in response to continually fluctuating selective pressures. Despite the potential for generation of genetic variation within hosts, co-colonization with multiple *H. pylori* strains is likely the dominant factor for generating novel allelic combinations by natural transformation. Therefore, increased levels of sanitation could create a decline in the availability of genetic diversity of *H. pylori* within colonized individuals [109], which, along with increased antibiotic use, could account for the declining fitness of *H. pylori* in modern human societies.

References

1 Linz B, Balloux F, Moodley Y, Manica A, Liu H, et al: An African origin for the intimate association between humans and *Helicobacter pylori*. Nature 2007;445:915–918.

2 Kang J, Blaser MJ: Bacterial populations as perfect gases: genomic integrity and diversification tensions in *Helicobacter pylori*. Nat Rev Microbiol 2006;4: 826–836.

3 Cooke CL, Huff JL, Solnick JV: The role of genome diversity and immune evasion in persistent infection with *Helicobacter pylori*. FEMS Immunol Med Microbiol 2005;45:11–23.

4 Smeets LC, Kusters JG: Natural transformation in *Helicobacter pylori*: DNA transport in an unexpected way. Trends Microbiol 2002;10:159–162; discussion 162.

5 Suerbaum S, Achtman M: Evolution of *Helicobacter pylori*: the role of recombination. Trends Microbiol 1999;7:182–184.

6 Wang G, Humayun MZ, Taylor DE: Mutation as an origin of genetic variability in *Helicobacter pylori*. Trends Microbiol 1999;7:488–493.

7 Kraft C, Suerbaum S: Mutation and recombination in *Helicobacter pylori*: mechanisms and role in generating strain diversity. Int J Med Microbiol 2005; 295:299–305.

8 Suerbaum S, Josenhans C: *Helicobacter pylori* evolution and phenotypic diversification in a changing host. Nat Rev Microbiol 2007;5:441–452.

9 de Visser JA, Elena SF: The evolution of sex: empirical insights into the roles of epistasis and drift. Nat Rev Genet 2007;8:139–149.

10 Moxon R, Bayliss C, Hood D: Bacterial contingency loci: the role of simple sequence DNA repeats in bacterial adaptation. Annu Rev Genet 2006;40:307–333.

11 Otto SP, Gerstein AC: Why have sex? The population genetics of sex and recombination. Biochem Soc Trans 2006;34:519–522.

12 Redfield RJ: Do bacteria have sex? Nat Rev Genet 2001;2:634–639.

13 Sniegowski PD, Gerrish PJ, Johnson T, Shaver A: The evolution of mutation rates: separating causes from consequences. Bioessays 2000;22:1057–1066.

14 Go MF: Review article: natural history and epidemiology of *Helicobacter pylori* infection. Aliment Pharmacol Ther 2002;16(suppl 1):3–15.

15 Brown LM: *Helicobacter pylori*: epidemiology and routes of transmission. Epidemiol Rev 2000;22:283–297.

16 Marshall DG, Dundon WG, Beesley SM, Smyth CJ: *Helicobacter pylori* – a conundrum of genetic diversity. Microbiology 1998;144:2925–2939.

17 Alm RA, Ling LS, Moir DT, King BL, Brown ED, et al: Genomic-sequence comparison of two unrelated isolates of the human gastric pathogen *Helicobacter pylori*. Nature 1999;397:176–180.

18 Tomb JF, White O, Kerlavage AR, Clayton RA, Sutton GG, et al: The complete genome sequence of the gastric pathogen *Helicobacter pylori*. Nature 1997;388:539–547.

19 Gressmann H, Linz B, Ghai R, Pleissner KP, Schlapbach R, et al: Gain and loss of multiple genes during the evolution of *Helicobacter pylori*. PLoS Genet 2005;1:e43.

20 Salama N, Guillemin K, McDaniel TK, Sherlock G, Tompkins L, Falkow S: A whole-genome microarray reveals genetic diversity among *Helicobacter pylori* strains. Proc Natl Acad Sci USA 2000;97:14668–14673.

21 Kivi M, Rodin S, Kupershmidt I, Lundin A, Tindberg Y, et al: *Helicobacter pylori* genome variability in a framework of familial transmission. BMC Microbiol 2007;7:54.

22 Salama NR, Gonzalez-Valencia G, Deatherage B, Aviles-Jimenez F, Atherton JC, et al: Genetic analysis of *Helicobacter pylori* strain populations colonizing the stomach at different times postinfection. J Bacteriol 2007;189:3834–3845.

23 Kraft C, Stack A, Josenhans C, Niehus E, Dietrich G, et al: Genomic changes during chronic *Helicobacter pylori* infection. J Bacteriol 2006;188:249–254.

24 Baltrus DA, Amieva MR, Covacci A, Lowe TM, Merrell DS, et al: The complete genome sequence of *Helicobacter pylori* strain G27. J Bacteriol 2009; 191:447–448.

25 Oh JD, Kling-Bäckhed H, Giannakis M, Xu J, Fulton RS, et al: The complete genome sequence of a chronic atrophic gastritis *Helicobacter pylori* strain: evolution during disease progression. Proc Natl Acad Sci USA 2006;103:9999–10004.
26 Giannakis M, Chen SL, Karam SM, Engstrand L, Gordon JI: *Helicobacter pylori* evolution during progression from chronic atrophic gastritis to gastric cancer and its impact on gastric stem cells. Proc Natl Acad Sci USA2008;105:4358–4363.
27 Medini D, Donati C, Tettelin H, Masignani V, Rappuoli R: The microbial pan-genome. Curr Opin Genet Dev 2005;15:589–594.
28 Linz B, Schuster SC: Genomic diversity in *Helicobacter* and related organisms. Res Microbiol 2007;158:737–744.
29 Saunders NJ, Boonmee P, Peden JF, Jarvis SA: Interspecies horizontal transfer resulting in core-genome and niche-adaptive variation within *Helicobacter pylori*. BMC Genomics 2005;6:9.
30 Janssen PJ, Audit B, Ouzounis CA: Strain-specific genes of *Helicobacter pylori*: distribution, function and dynamics. Nucleic Acids Res 2001;29:4395–4404.
31 Sheppard SK, McCarthy ND, Falush D, Maiden MC: Convergence of *Campylobacter* species: implications for bacterial evolution. Science 2008;320:237–239.
32 Bik EM, Eckburg PB, Gill SR, Nelson KE, Purdom EA, et al: Molecular analysis of the bacterial microbiota in the human stomach. Proc Natl Acad Sci USA 2006;103:732–737.
33 Levine SM, Lin EA, Emara W, Kang J, DiBenedetto M, et al: Plastic cells and populations: DNA substrate characteristics in *Helicobacter pylori* transformation define a flexible but conservative system for genomic variation. FASEB J 2007;21:3458–3467.
34 Delport W, Cunningham M, Olivier B, Preisig O, van der Merwe SW: A population genetics pedigree perspective on the transmission of *Helicobacter pylori*. Genetics 2006;174:2107–2118.
35 Fraser C, Hanage WP, Spratt BG: Recombination and the nature of bacterial speciation. Science 2007;315:476–480.
36 Höfler C, Fischer W, Hofreuter D, Haas R: Cryptic plasmids in *Helicobacter pylori*: putative functions in conjugative transfer and microcin production. Int J Med Microbiol 2004;294:141–148.
37 Hofreuter D, Haas R: Characterization of two cryptic *Helicobacter pylori* plasmids: a putative source for horizontal gene transfer and gene shuffling. J Bacteriol 2002;184:2755–2766.
38 Hosaka Y, Okamoto R, Irinoda K, Kaieda S, Koizumi W, et al: Characterization of pKU701, a 2.5-kb plasmid, in a Japanese *Helicobacter pylori* isolate. Plasmid 2002;47:193–200.
39 Minnis JA, Taylor TE, Knesek JE, Peterson WL, McIntire SA: Characterization of a 3.5-kbp plasmid from *Helicobacter pylori*. Plasmid 1995;34:22–36.
40 Heuermann D, Haas R: Genetic organization of a small cryptic plasmid of *Helicobacter pylori*. Gene 1995;165:17–24.
41 Kleanthous H, Clayton CL, Tabaqchali S: Characterization of a plasmid from *Helicobacter pylori* encoding a replication protein common to plasmids in gram-positive bacteria. Mol Microbiol 1991;5:2377–2389.
42 Kuipers EJ, Israel DA, Kusters JG, Blaser MJ: Evidence for a conjugation-like mechanism of DNA transfer in *Helicobacter pylori*. J Bacteriol 1998;180: 2901–2905.
43 Backert S, Kwok T, Konig W: Conjugative plasmid DNA transfer in *Helicobacter pylori* mediated by chromosomally encoded relaxase and TraG-like proteins. Microbiology 2005;151:3493–3503.
44 Ando T, Xu Q, Torres M, Kusugami K, Israel DA, Blaser MJ: Restriction-modification system differences in *Helicobacter pylori* are a barrier to interstrain plasmid transfer. Mol Microbiol 2000;37: 1052–1065.
45 Donahue JP, Israel DA, Peek RM, Blaser MJ, Miller GG: Overcoming the restriction barrier to plasmid transformation of *Helicobacter pylori*. Mol Microbiol 2000;37:1066–1074.
46 Heintschel von Heinegg E, Nalik HP, Schmid EN: Characterisation of a *Helicobacter pylori* phage (HP1). J Med Microbiol 1993;38:245–249.
47 Welch RA, Burland V, Plunkett G 3rd, Redford P, Roesch P, et al: Extensive mosaic structure revealed by the complete genome sequence of uropathogenic *Escherichia coli*. Proc Natl Acad Sci USA 2002;99: 17020–17024.
48 Björkholm B, Sjölund M, Falk PG, Berg OG, Engstrand L, Andersson DI: Mutation frequency and biological cost of antibiotic resistance in *Helicobacter pylori*. Proc Natl Acad Sci USA 2001, 98:14607–14612.
49 Huang S, Kang J, Blaser MJ: Antimutator role of the DNA glycosylase mutY gene in *Helicobacter pylori*. J Bacteriol 2006;188:6224–6234.
50 Kang J, Huang S, Blaser MJ: Structural and functional divergence of MutS2 from bacterial MutS1 and eukaryotic MSH4-MSH5 homologs. J Bacteriol 2005;187:3528–3537.
51 Kang J, Blaser MJ: Repair and antirepair DNA helicases in *Helicobacter pylori*. J Bacteriol 2008;190: 4218–4224.
52 O'Rourke EJ, Chevalier C, Pinto AV, Thiberge JM, Ielpi L, et al: Pathogen DNA as target for host-generated oxidative stress: role for repair of bacterial DNA damage in *Helicobacter pylori* colonization. Proc Natl Acad Sci USA 2003;100:2789–2794.

53 Loewe L, Textor V, Scherer S: High deleterious genomic mutation rate in stationary phase of *Escherichia coli*. Science 2003;302:1558–1560.
54 Hudson RE, Bergthorsson U, Ochman H: Transcription increases multiple spontaneous point mutations in *Salmonella enterica*. Nucleic Acids Res 2003;31:4517–4522.
55 Hudson RE, Bergthorsson U, Roth JR, Ochman H: Effect of chromosome location on bacterial mutation rates. Mol Biol Evol 2002;19:85–92.
56 Kang JM, Iovine NM, Blaser MJ: A paradigm for direct stress-induced mutation in prokaryotes. FASEB J 2006;20:2476–2485.
57 Gonzalez C, Hadany L, Ponder RG, Price M, Hastings PJ, Rosenberg SM: Mutability and importance of a hypermutable cell subpopulation that produces stress-induced mutants in *Escherichia coli*. PLoS Genet 2008;4:e1000208.
58 Foster PL: Stress-induced mutagenesis in bacteria. Crit Rev Biochem Mol Biol 2007;42:373–397.
59 Sundin GW, Weigand MR: The microbiology of mutability. FEMS Microbiol Lett 2007;277:11–20.
60 Amundsen SK, Fero J, Hansen LM, Cromie GA, Solnick JV, et al: *Helicobacter pylori* AddAB helicase-nuclease and RecA promote recombination-related DNA repair and survival during stomach colonization. Mol Microbiol 2008;69:994–1007.
61 Salaün L, Linz B, Suerbaum S, Saunders NJ: The diversity within an expanded and redefined repertoire of phase-variable genes in *Helicobacter pylori*. Microbiology 2004;150:817–830.
62 Sanabria-Valentin E, Colbert MT, Blaser MJ: Role of futC slipped strand mispairing in *Helicobacter pylori* Lewisy phase variation. Microbes Infect 2007;9: 1553–1560.
63 de Vries N, Duinsbergen D, Kuipers EJ, Pot RG, Wiesenekker P, et al: Transcriptional phase variation of a type III restriction-modification system in *Helicobacter pylori*. J Bacteriol 2002;184:6615–6623.
64 Wirth HP, Yang M, Sanabria-Valentín E, Berg DE, Dubois A, Blaser MJ: Host Lewis phenotype-dependent *Helicobacter pylori* Lewis antigen expression in rhesus monkeys. FASEB J 2006;20:1534–1536.
65 Tannaes T, Dekker N, Bukholm G, Bijlsma JJ, Appelmelk BJ: Phase variation in the *Helicobacter pylori* phospholipase A gene and its role in acid adaptation. Infect Immun 2001;69:7334–7340.
66 Kudo T, Nurgalieva ZZ, Conner ME, Crawford S, Odenbreit S, et al: Correlation between *Helicobacter pylori* OipA protein expression and oipA gene switch status. J Clin Microbiol 2004;42:2279–2281.
67 Aras RA, Kang J, Tschumi AI, Harasaki Y, Blaser MJ: Extensive repetitive DNA facilitates prokaryotic genome plasticity. Proc Natl Acad Sci USA 2003; 100:13579–13584.
68 Aras RA, Fischer W, Perez-Perez GI, Crosatti M, Ando T, et al: Plasticity of repetitive DNA sequences within a bacterial (Type IV) secretion system component. J Exp Med 2003;198:1349–1360.
69 Ayraud S, Janvier B, Salaun L, Fauchère JL: Modification in the *ppk* gene of *Helicobacter pylori* during single and multiple experimental murine infections. Infect Immun 2003;71:1733–1739.
70 Aras RA, Lee Y, Kim SK, Israel D, Peek RM Jr, Blaser MJ: Natural variation in populations of persistently colonizing bacteria affect human host cell phenotype. J Infect Dis 2003;188:486–496.
71 Aras RA, Takata T, Ando T, van der Ende A, Blaser MJ: Regulation of the *HpyII* restriction-modification system of *Helicobacter pylori* by gene deletion and horizontal reconstitution. Mol Microbiol 2001; 42:369–382.
72 Björkholm B, Lundin A, Sillén A, Guillemin K, Salama N: Comparison of genetic divergence and fitness between two subclones of *Helicobacter pylori*. Infect Immun 2001;69:7832–7838.
73 Lorenz MG, Wackernagel W: Bacterial gene transfer by natural genetic transformation in the environment. Microbiol Rev 1994;58:563–602.
74 Johnsborg O, Eldholm V, Havarstein LS: Natural genetic transformation: prevalence, mechanisms and function. Res Microbiol 2007;158:767–778.
75 Baltrus DA, Guillemin K: Multiple phases of competence occur during the *Helicobacter pylori* growth cycle. FEMS Microbiol Lett 2006;255:148–155.
76 Israel DA, Lou AS, Blaser MJ: Characteristics of *Helicobacter pylori* natural transformation. FEMS Microbiol Lett 2000;186:275–280.
77 Hamilton HL, Dillard JP: Natural transformation of *Neisseria gonorrhoeae*: from DNA donation to homologous recombination. Mol Microbiol 2006; 59:376–385.
78 Suerbaum S, Smith JM, Bapumia K, Morelli G, Smith NH: Free recombination within *Helicobacter pylori*. Proc Natl Acad Sci USA 1998;95:12619–12624.
79 Falush D, Wirth T, Linz B, Pritchard JK, Stephens M: Traces of human migrations in *Helicobacter pylori* populations. Science 2003;299:1582–1585.
80 Wirth T, Wang X, Linz B, Novick RP, Lum JK, et al: Distinguishing human ethnic groups by means of sequences from *Helicobacter pylori*: lessons from Ladakh. Proc Natl Acad Sci USA 2004;101:4746–4751.
81 Lundin A, Björkholm B, Kupershmidt I, Unemo M, Nilsson P, et al: Slow genetic divergence of *Helicobacter pylori* strains during long-term colonization. Infect Immun 2005;73:4818–4822.

82 Baltrus DA, Guillemin K, Phillips PC: Natural transformation increases the rate of adaptation in the human pathogen *Helicobacter pylori*. Evolution 2008;62:39–49.
83 Philpott DJ, Belaid D, Troubadour P, Thiberge JM, Tankovic J, et al: Reduced activation of inflammatory responses in host cells by mouse-adapted *Helicobacter pylori* isolates. Cell Microbiol 2002;4:285–296.
84 Sozzi M, Crosatti M, Kim SK, Romero J, Blaser MJ: Heterogeneity of *Helicobacter pylori* cag genotypes in experimentally infected mice. FEMS Microbiol Lett 2001;203:109–114.
85 Robinson K, Loughlin MF, Potter R, Jenks PJ: Host adaptation and immune modulation are mediated by homologous recombination in *Helicobacter pylori*. J Infect Dis 2005;191:579–587.
86 Solnick JV, Hansen LM, Salama NR, Boonjakuakul JK, Syvanen M: Modification of *Helicobacter pylori* outer membrane protein expression during experimental infection of rhesus macaques. Proc Natl Acad Sci USA 2004;101:2106–2111.
87 van der Woude MW, Bäumler AJ: Phase and antigenic variation in bacteria. Clin Microbiol Rev 2004;17:581–611.
88 Falush D, Kraft C, Taylor NS, Correa P, Fox JG, et al: Recombination and mutation during long-term gastric colonization by *Helicobacter pylori*: estimates of clock rates, recombination size, and minimal age. Proc Natl Acad Sci USA 2001;98:15056–15061.
89 Israel DA, Salama N, Krishna U, Rieger UM, Atherton JC, et al Jr: *Helicobacter pylori* genetic diversity within the gastric niche of a single human host. Proc Natl Acad Sci USA 2001;98:14625–14630.
90 Prouzet-Mauléon V, Hussain MA, Lamouliatte H, Kauser F, Mégraud F, Ahmed N: Pathogen evolution in vivo: genome dynamics of two isolates obtained 9 years apart from a duodenal ulcer patient infected with a single *Helicobacter pylori* strain. J Clin Microbiol 2005;43:4237–4241.
91 Kuipers EJ, Israel DA, Kusters JG, Gerrits MM, Weel J, et al: Quasispecies development of *Helicobacter pylori* observed in paired isolates obtained years apart from the same host. J Infect Dis 2000;181:273–282.
92 Salaun L, Saunders NJ: Population-associated differences between the phase variable LPS biosynthetic genes of *Helicobacter pylori*. BMC Microbiol 2006;6:79.
93 Albert TJ, Dailidiene D, Dailide G, Norton JE, Kalia A, et al: Mutation discovery in bacterial genomes: metronidazole resistance in *Helicobacter pylori*. Nat Methods 2005;2:951–953.
94 Goodwin A, Kersulyte D, Sisson G, Veldhuyzen van Zanten SJ, Berg DE, Hoffman PS: Metronidazole resistance in *Helicobacter pylori* is due to null mutations in a gene (*rdxA*) that encodes an oxygen-insensitive NADPH nitroreductase. Mol Microbiol 1998;28:383–393.
95 Smeets LC, Arents NL, van Zwet AA, Vandenbroucke-Grauls CM, Verboom T, et al: Molecular patchwork: Chromosomal recombination between two *Helicobacter pylori* strains during natural colonization. Infect Immun 2003;71:2907–2910.
96 Kersulyte D, Chalkauskas H, Berg DE: Emergence of recombinant strains of *Helicobacter pylori* during human infection. Mol Microbiol 1999;31:31–43.
97 Aviles-Jimenez F, Letley DP, Gonzalez-Valencia G, Salama N, Torres J, Atherton JC: Evolution of the *Helicobacter pylori* vacuolating cytotoxin in a human stomach. J Bacteriol 2004;186:5182–5185.
98 Webb GF, Blaser MJ: Dynamics of bacterial phenotype selection in a colonized host. Proc Natl Acad Sci USA 2002;99:3135–3140.
99 Hartl DL, Clark AG: Principles of Population Genetics. Sunderland, Sinauer Associates, Inc., 2007.
100 Ostrowski EA, Woods RJ, Lenski RE: The genetic basis of parallel and divergent phenotypic responses in evolving populations of *Escherichia coli*. Proc Biol Sci 2008;275:277–284.
101 Szollosi GJ, Derenyi I, Vellai T: The maintenance of sex in bacteria is ensured by its potential to reload genes. Genetics 2006;174:2173–2180.
102 Matteo MJ, Granados G, Pérez CV, Olmos M, Sanchez C, Catalano M: *Helicobacter pylori* cag pathogenicity island genotype diversity within the gastric niche of a single host. J Med Microbiol 2007;56:664–669.
103 Wirth HP, Yang M, Peek RM Jr, Höök-Nikanne J, Fried M, Blaser MJ: Phenotypic diversity in Lewis expression of *Helicobacter pylori* isolates from the same host. J Lab Clin Med 1999;133:488–500.
104 Carroll IM, Ahmed N, Beesley SM, Khan AA, Ghousunnissa S: Microevolution between paired antral and paired antrum and corpus *Helicobacter pylori* isolates recovered from individual patients. J Med Microbiol 2004;53:669–677.
105 Lee YC, Lee SY, Pyo JH, Kwon DH, Rhee JC, Kim JJ: Isogenic variation of *Helicobacter pylori* strain resulting in heteroresistant antibacterial phenotypes in a single host in vivo. Helicobacter 2005;10:240–248.
106 Akada JK, Ogura K, Dailidiene D, Dailide G, Cheverud JM, Berg DE: *Helicobacter pylori* tissue tropism: mouse-colonizing strains can target different gastric niches. Microbiology 2003;149:1901–1909.

107 Dubois A, Berg DE, Incecik ET, Fiala N, Heman-Ackah LM, et al: Host specificity of *Helicobacter pylori* strains and host responses in experimentally challenged nonhuman primates. Gastroenterology 1999;116:90–96.

108 Blaser MJ, Kirschner D: The equilibria that allow bacterial persistence in human hosts. Nature 2007;449:843–849.

109 Schwarz S, Morelli G, Kusecek B, Manica A, Balloux F, et al: Horizontal versus familial transmission of *Helicobacter pylori*. PLoS Pathog 2008;4:e1000180.

Karen Guillemin
Institute of Molecular Biology, University of Oregon
1370 Franklin Blvd
Eugene, OR 97403 (USA)
Tel. +1 541 346 5360, Fax +1 541 346 5891, E-Mail guillemin@molbio.uoregon.edu

de Reuse H, Bereswill S (eds): Microbial Pathogenomics.
Genome Dyn. Basel, Karger, 2009, vol 6, pp 91–109

Genomics of Thermophilic *Campylobacter* Species

D.J.H. Gaskin · M. Reuter · N. Shearer · F. Mulholland · B.M. Pearson · A.H.M. van Vliet

Institute of Food Research, Norwich Research Park, Norwich, UK

Abstract

The thermophilic *Campylobacter* species *C. jejuni* and *C. coli* are important human pathogens, which are major causes of bacterial gastroenteritis. The recent progress in genomics techniques has allowed for a rapid increase in our knowledge of the molecular biology of *Campylobacter* species, but needs to be matched by concurrent increases in our understanding of the unique biology of these organisms. *Campylobacter* species display significant levels of genomic variation via natural transformation, phase variation, plasmid transfer and infection with bacteriophages, and this poses a continuous challenge for studies on pathogenesis, physiology, epidemiology and evolution of *Campylobacter*. In this chapter we will review the current state of the art of the genomics of thermophilic *Campylobacter* species, and opportunities where genomics can further contribute to our understanding of the biology of these successful human pathogens.

Members of the genus *Campylobacter* colonise the gastrointestinal tract of a broad range of mammals and birds, where they can be either commensal or act as pathogens [1]. The best studied members of the genus *Campylobacter* are the thermophilic, foodborne pathogens *Campylobacter jejuni* and *Campylobacter coli*, which are considered to be commensal organisms in poultry and other avian hosts, but are important causes of human bacterial gastroenteritis in both industrialised and developing countries [1]. Although the gastroenteritis is usually self-limiting, sequelae of *Campylobacter* infection include the development of neurodegenerative diseases like Guillain-Barré syndrome and Miller-Fisher Syndrome [2].

Despite its importance as a human pathogen, our understanding of the mechanisms of *Campylobacter*-associated diseases is still relatively poor. The first complete *C. jejuni* genome sequence was published in 2000 [3], and coupled to the rapid developments in genomics in the last ten years this has contributed significantly to increasing our knowledge about the biology of *Campylobacter*. In this chapter we will discuss the different aspects of *Campylobacter* genomics in the light of the biology of the organism, and the current state of the art in technical developments. We will also discuss the contribution of genomics to a better understanding of *Campylobacter* physiology and virulence, and suggest areas where developments are still required in the coming years. Since most of the research on *Campylobacter* has been focused on *C. jejuni*, we will mostly discuss data on *C. jejuni*, and will specifically indicate it when we are discussing other *Campylobacter* species, including *C. coli*.

***Campylobacter* Biology**

The thermophilic campylobacters are small (~0.5 μm wide and ~3 μm long) Gram-negative rods, which have a spiral or curved shape. Cells have single, unsheathed polar flagella, which are commonly present on both poles. Both the 'corkscrew'-like morphology and flagella allow for rapid motility in viscous environments, like the gastro-intestinal mucosal layer. The thermophilic *Campylobacter* species are catalase- and oxidase-positive while being urease negative. *Campylobacter* is a microaerophile, and grows at oxygen concentration of 3–15%, but also displays capnophilic characteristics as it requires a carbon dioxide concentration of at least 3%. Most laboratories use an atmosphere of 85% N_2, 5% O_2 and 10% CO_2, although growth can also be observed at 85% N_2, 10% O_2 and 5% CO_2, or in a tissue culture incubator set at 10% CO_2 and 90% air. The temperature range for growth of the thermophilic *Campylobacter* species is 34–44°C, with an optimal growth at 42°C, which probably reflects an adaptation to the intestines of warm-blooded birds. However, many laboratories worldwide routinely grow *Campylobacter* at 37°C, as this most closely resembles the human body temperature.

Metabolism

Campylobacter species are fastidious organisms, which are unable to ferment carbohydrates, but are thought to primarily use amino acids as their carbon source [4]. This is reflected in the relative rarity of carbohydrate transporters in the *C. jejuni* genome, whereas there is a relative abundance of transporter systems for amino acids and organic acids [4, 5]. *C. jejuni* contains a complete citric-acid cycle, but several enzyme components of this cycle differ from that of respiratory aerobes, and resemble more their counterparts found in obligate anaerobic bacteria. Next to amino acids, *C. jejuni* can also use pyruvate as a carbon source, and is thought to be able to use hydrogen and formate as energy sources in vivo [4].

The respiratory chain in *C. jejuni* is quite complex, and this may well link in with its lifestyle, which includes exposure to atmospheric oxygen levels as well as potentially lack of oxygen in anaerobic niches. Organic acids like fumarate, formate, lactate, but also sulphite can act as electron donors [4], while oxygen and hydrogen peroxide act as primary electron acceptors. However, in oxygen-limited conditions, *C. jejuni* can use alternative electron acceptors like fumarate, nitrite and nitrate, and potentially also S- and N-oxides like DMSO and TMAO [4]. Many of the genes encoding the enzymes involved in *C. jejuni* respiration have now been identified and characterised, but their exact role in *C. jejuni* biology still requires investigation in how these systems interact during the different phases in the *C. jejuni* lifestyle.

An Overview of the *Campylobacter* Genome

To date, the Complete Microbial Genomes database at the National Center for Biotechnology Information (NCBI, http://www.ncbi.nlm.nih.gov/genomes/lproks.cgi) contains ongoing and completed genome sequence projects on nine different *Campylobacter* species (listed in table 1) [3, 6–11]. These include four complete genome sequences from different *C. jejuni* isolates (NCTC11168, RM1221, 81–176 and 81116), and several incomplete or unfinished *C. jejuni* sequences and sequences from other *Campylobacter* species. The fact that the genomes of so many different strains of *C. jejuni* have been sequenced reflects the fact that this species causes the majority of reported cases of *Campylobacter*-related food poisoning. The list in table 1 also contains comparative information from related genera from the epsilon subdivision of the Proteobacteria, like *Helicobacter pylori*, *Wolinella succinogenes* and *Arcobacter butzleri* [10, 12].

Compared to other enteric pathogens, *C. jejuni* has a relatively small genome, with a rather low G+C percentage of 30%, with the notable exception of *C. curvus* which has a G+C percentage of 44% (table 1) [3, 6–9, 11]. All *C. jejuni* genomes are between 1.6 and 1.8 megabase pairs, with other *Campylobacter* species like *C. concisus* having a slightly larger genome of 1.9 to 2.2 megabasepairs (table 1) [6]. However, these genome sizes should not be considered small, as obligate intracellular pathogens, insect endosymbionts and pathogens like *Mycoplasma pneumoniae* have genomes of less than 1 megabasepairs. The size of the *C. jejuni* genome most likely represents adaptation to its particular niche (the chicken caecum), with a minimum requirement for growth outside that niche, although the bacterium is clearly capable of environmental survival.

Several plasmids and integrated mobile elements similar to insertion sequences have been described for *C. jejuni* and *C. coli*, but relatively little is known for most of these elements.

Plasmids

Of the *C. jejuni* strains used in many laboratories, strains NCTC11168 and 81116 do not contain plasmids, whereas strain 81–176 can contain two plasmids, named pVir and pTet [13]. The pVir plasmid has been suggested to contain some genes of a

Table 1. Overview of available genome sequences of *Campylobacter* and related species

Genus/Species	Strain	Accession number	Length (nt)	(in)complete	Predicted ORFs	%GC
Campylobacter						
jejuni	NCTC11168	NC_002163	1,641,481	complete	1634	30
	81116	NC_009839	1,628,115	complete	1626	30
	RM1221	NC_003912	1,777,831	complete	1838	30
	81-176	NC_008787	1,616,554	complete	1653	30
	CG8486	NZ_AASY00000000	1,597,692	unfinished	1425	30
	84-25	NZ_AANT00000000	1,671,624	unfinished	1748	30
	HB93-13	NZ_AANQ00000000	1,694,788	unfinished	1710	30
	260.94	NZ_AANK00000000	1,657,846	unfinished	1716	30
	CF93-6	NZ_AANJ00000000	1,676,304	unfinished	1757	30
coli	RM2228	NZ_AAFL00000000	1,860,666	unfinished	1967	31
concisus	13826	NC_009802	2,052,007	unfinished	1929	39
curvus	525.92	NC_009715	1,971,264	unfinished	1931	44
doylei	269.97	NC_009707	1,845,106	unfinished	1731	30
fetus	82-40	NC_008599	1,773,615	unfinished	1719	33
hominis	ATCC BAA-381	NC_009714	1,711,273	unfinished	1682	31
lari	RM2100	NZ_AAFK00000000	1,562,926	unfinished	1599	29
upsaliensis	RM3195	NZ_AAFJ00000000	1,773,834	unfinished	1934	34
Helicobacter pylori	26695	NC_000915	1,667,867	complete	1576	38
acinonychis	Sheeba	NC_008229	1,553,927	complete	1612	38
hepaticus	ATCC 51449	NC_004917	1,799,146	complete	1875	35
Arcobacter butzleri	RM4018	NC_009850	2,341,251	complete	2259	27
Wolinella succinogenes	DSM 1740	NC_005090	2,110,355	complete	2042	48

potential type IV secretion system [13], but the contribution of the pVir plasmid to *C. jejuni* virulence is still under debate [14]. The pTet plasmid confers tetracycline resistance and contains genes encoding for a type IV secretion system, which is thought to function in conjugative transfer [15, 16]. A plasmid found in *C. coli*, pCC31, is >90% identical to pTet. A microarray screening for the presence of the pVir, pTet and pCC31 genes in 27 *C. jejuni* and 2 *C. coli* isolates indicated that 83% of these contained most of the genes present on pTet/pCC31, whereas pVir was not present in any strain except 81–176.

Integrated Elements

C. jejuni strain RM1221 contains 4 genomic islands which are absent in strain NCTC11168, and these were named *C. jejuni* integrated elements (CJIE) 1–4 [17]. These CJIEs contain many phage and plasmid related sequences, in particular CJIE1 is a *Campylobacter* Mu-like phage (CMLP1). Comparative genetic hybridization and PCR showed that these CJIEs are widely distributed among *C. jejuni* and *C. coli* strains [17, 18], and the *C. lari* genome sequence contains a homolog of CJIE4 [11]. The extent of genomic variation within *Campylobacter* species resulting from the vertical and horizontal transfer of plasmids and other phage like sequences is currently unclear, but given their importance in other organisms it is an area worthy of further investigation [18].

Gene Regulation

In addition to being a relatively small genome, the *C. jejuni* genome is also rather compact. There are few intergenic regions over 200 bp and the average intergenic region is ~50 bp (ranging from 1–2,070 bp). This lack of intergenic space may limit the organism's regulatory capacity, with little space for regulatory binding sites or transcription of regulatory small RNA species. This is consistent with the scarcity of regulatory proteins when compared to other enteric pathogens like *Salmonella*: the *C. jejuni* NCTC11168 genome encodes 5 complete two-component systems consisting of a histidine kinase sensor protein and cognate response regulator, plus an additional 4 orphan response regulators [3, 5]. At least two of the response regulators (Cj0355c, Cj1227c) are thought to be essential based on the observation that these genes could not be insertionally inactivated [19]. Two of the response regulators (Cj0285c – CheV, and Cj1118c – CheY) are known to be involved in chemotaxis. Analysis of the *C. jejuni* NCTC11168 genome sequence reveals that much of the archetypal chemotaxis system is intact [20]. However, there are some key differences: the CheB protein lacks a receiver domain while the CheA protein contains an appended C-terminal receiver domain. Originally, the NCTC11168 genome was thought not to contain a CheZ protein which stimulates dephosphorylation of CheY [20]. However, a recent study in *H. pylori* identified a putative CheZ protein (HP0170) which also has an ortholog (Cj0700) in *C. jejuni* [21]. The *C. jejuni* genome also encodes nine so-called one-component regulators, proteins containing a DNA-binding domain linked directly to signal-sensing domain. For example Cj0368c (CmeR) is a TetR-family transcriptional regulator [22] and Cj1042 is an AraC-family transcriptional regulator. To date, the physiological function of some of these systems, particularly the two-component systems, are beginning to be elucidated; however, the complete picture of signal transduction in *C. jejuni* is far from complete.

Sigma Factors

Similar published *Helicobacter* genomes, *C. jejuni* contains only 3 sigma factors: the 'house-keeping' sigma factor σ^{70}, and the alternative sigma-factors σ^{28} and σ^{54} [23].

Both σ^{28} and σ^{54} appear to be involved largely in flagellar biosynthesis, modification and regulation (see the transcriptomics section). Interestingly, a *Campylobacter* RpoS homolog has not been identified to date [23]. In light of physiological growth studies that show a lack of a typical 'stationary phase' growth period, this is perhaps not surprising. However, *Campylobacter* may exhibit a stringent response mediated via a bifunctional SpoT-RelA ortholog which has been linked to invasion, stress responses and environmental survival [24].

Riboregulation

The compact genome and short intergenic regions may also limit the capacity to encode small regulatory RNAs. Indeed, to date, a *Campylobacter* homolog of Hfq has not been identified [25]. However, studies comparing gene expression and protein profiles (proteomics) show a high level of discordance suggesting a role for post-transcriptional regulation. In one study where iron stress was investigated, an overlap of only 16 genes (10% of the total differentially-regulated genes) was observed when comparing gene expression and proteomics data [26]. It has been suggested that small regulatory RNAs may comprise around 2% of the genome, which means there may be as many as 30 small regulatory RNA molecules yet to be found in the genomes of *Campylobacter*.

Function Unknown (FUN) Genes

Despite the comparatively small size of the genome, and hence small number of genes (1,643 at the last count [5]), much still remains to be learned. The TIGR functional role classification of NCTC11168 annotates 600 genes as either 'hypothetical' or 'unknown function', with approximately 8% of these genes having no sequence database matches [3]. This number will have dropped following a recent re-annotation of this genome (18% of the annotations were revised [5]) and continuing biochemical and genetic analysis. However, the number of genes encoding proteins of unknown function remains high. A recent high-throughput study aiming to provide insight into protein function based on protein-protein interactions utilised the Yeast Two-Hybrid system [27]. This comprehensive study measured 11,687 interactions between 80% of the genome-encoding proteins. While such screens need validation, this initial screen identified proteins which may be involved in the chemotaxis system [27].

Phase Variation

Next to differences in plasmids, genes and mobile genetic elements, *C. jejuni* isolates also have the potential for phase variation using hypervariable homo- and heteropolymeric mononucleotide and dinucleotide stretches [3]. During replication, the number of mononucleotides or dinucleotides can change due to slipped-strand mispairing, which results in a change of reading frame and premature ending of translation [3]. This process is fully reversible, and hence populations can contain different combinations of genes switched ON and OFF at the translational level. In the original study on the *C. jejuni* NCTC11168 genome sequence, 29 hypervariable G-tracts were identified which contained more than 7 G-residues, and many of these are thought to be phase-variable [3]. These hypervariable sequences were mostly found in genes involved in biosynthesis of the capsular polysaccharide and lipooligosaccharide (LOS), genes involved in flagellar modification, and two autotransporters [3, 28, 29]. Interestingly, the number and location of the hypervariable sequences is only partially conserved between *C. jejuni* strains [7], and hence phase variation described in one strain may not be representative for the whole species. In addition to poly(G) tracts, it has also been reported that poly(A) or poly(T) tracts may function in phase variation of flagellar expression [30, 31].

Pseudogenes

One common feature of all sequenced *C. jejuni* strains is the presence of pseudogenes, regions of DNA that have homology to known genes but contain one or more in frame stop-codons [3, 5]. These regions represent ancestrally useful functions that are no longer required for growth and survival in the organism's current niche. Interestingly, a comparison of different *C. jejuni* genomes reveals that some pseudogenes in one genome are intact in other genomes (e.g. *cj0044* in strain NCTC11168), suggesting that the process of losing gene function may be a recent process and one that is still in flux. In support of this, there is evidence that some pseudogenes have retained their regulatory control; for example, the transcript levels of the *cj0444* pseudogene were increased in response to iron limitation [26].

Comparative Genomic Analysis of *Campylobacter*

One of the major issues when investigating *Campylobacter* biology and pathogenesis is the inherent variation between isolates and/or strains. The different genera of the Campylobacteraceae display high levels of genotypic and phenotypic variation, and this variation is thought to be reflected in the differences in colonisation potential, host specificity, environmental survival and other important aspects of *Campylobacter* biology. This genomic variation has important implications for evolutionary analyses as well as epidemiology of infections. *C. jejuni* and *C. coli* are capable of natural transformation, and lack several DNA repair mechanisms, which contribute to genome plasticity [5]. In addition, bacteriophages and plasmids are thought to contribute to genome plasticity and DNA rearrangements [16, 32].

Several techniques have been employed for the analysis of genetic variability and for typing of *C. jejuni* and *C. coli*. These include Amplified Fragment Length Polymorphism (AFLP) and Pulsed Field Gel Electrophoresis (PFGE), as well as Multi-Locus Sequence Typing (MLST). However, while very valuable for epidemiological and evolutionary purposes, these techniques do not give additional information about the genetic content of the genome of the tested isolates. In contrast, de novo genome sequencing [3, 6, 7, 9], subtractive hybridisation [33] or screening microarrayed clones [34] can reveal new genes.

Comparative Genomic Hybridization

Comparative genomic hybridization (CGH) is a technique suited to readily and rapidly compare the gene complement of a variety of strains. The CGH technique makes use of DNA microarrays for pairwise comparisons. Microarrays for *C. jejuni* for example, have been made from cloned genomic fragments [34], amplicons [35] and oligonucleotides [36]. The whole genomic DNA from two strains are fluorescently labelled with two different fluorophores, mixed and hybridised to the microarray. After washing, the microarrays are laser-scanned to determine the relative amounts of each fluorophore-labelled DNA bound to each gene feature. On a gene by gene basis, their presence and absence can then be scored. The technique is also sensitive enough to indicate regions of gene duplications and sequence divergence. Two minor issues should be remembered: that gene order cannot be directly determined using this technique, and perhaps more obviously, that microarrays can only ever give information about genes which are printed on the arrays. The latter issue is diminishing in significance as DNA microarray printing densities increase and new genome sequences become more readily available.

Studies using the CGH technique are beginning to reveal some of the aspects which account for strain diversity in *C. jejuni*. It appears that much of the variability is accounted for in plasticity regions [35], regions of the genome that change as a unit, probably by natural transformation and recombination. An example of such genetic variation that has a direct impact on *C. jejuni* biology is the apparent variation in iron acquisition systems, which is displayed in figure 1. While all *C. jejuni* strains contain the Chu heme uptake system, there is considerable variation in the number and distribution of *tonB* genes, as well as the *cfrA* gene which is either absent or present. Since this gene is responsible for enterochelin uptake [37], this may directly influence the potential of *C. jejuni* strains in utilisation of different iron sources. Other differences in *C. jejuni* gene content are found in regions containing genes involved in LOS and capsule biosynthesis, as well as flagellar modification [35, 36, 38].

Multi-Locus Sequence Typing

Genomic comparison techniques like MLST have also been used to compare strains from different sources or against different clinical outcomes to identify markers or virulence genes, and have been compared with data obtained by CGH [6, 35, 36, 38, 39]. This demonstrated that *C. jejuni* strains show genetic similarities based on their host species rather than geographical or temporal location, implying a genetic adaptation to that host [39]. Other MLST studies have suggested convergent evolution of *C. jejuni* and *C. coli* into novel ecological niches, created by human farming activity [40].

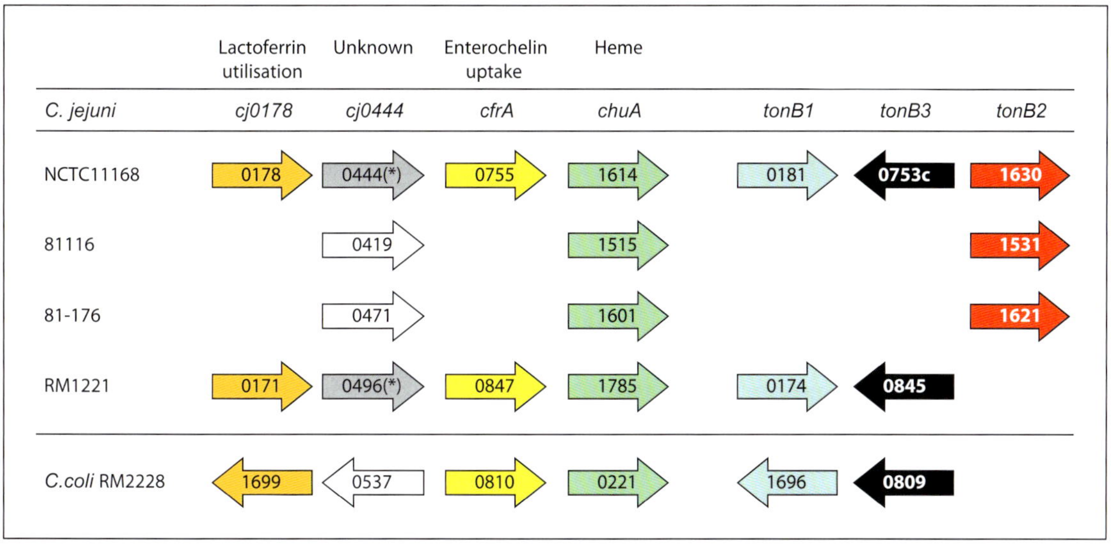

Fig. 1. Variation of *Campylobacter jejuni* and *C. coli* genes encoding TonB-dependent outer membrane proteins involved in iron acquisition, and the genes encoding the TonB energy transduction systems required for this transport. The genomes of *C. jejuni* strains NCTC11168, 81116, 81–176 and RM1221 were compared by using the ACT Program, while the *C. coli* RM2228 genome sequence was searched using the BLAST program on the xBase website (http://www.xbase2.bham.ac.uk). Numbers included in the arrows represent the gene numbers in the respective genome annotation. Grey arrows indicate sequences interrupted by mutations, which are annotated as pseudogenes. Iron sources associated with specific outer membrane receptors are indicated.

Campylobacter Transcriptomics and Proteomics

Microarray and proteomic analysis are amongst two of the most powerful methods for assigning putative functions to unknown genes, often using the 'guilt by association' hypothesis to link genes displaying similar changes. They involve the comparison of samples to identify differences. They are heavily reliant on the information available from genome sequences, and have thus benefitted from the ever increasing number of *Campylobacter* sequences available.

Transcriptomics

In addition to the comparison of genomic DNA samples mentioned above, DNA microarrays allow the comparison of transcript levels of all genes between two or more samples. These transcriptomic approaches allow a genomewide view of the response to a variety of environmental conditions and of the responses to specific gene mutations. For *Campylobacter*, only *C. jejuni*-specific microarrays have been reported and these are available for all the major strains studied including NCTC11168, 81–176 (including the virulence plasmid pVir), RM1221 and 81116. Commercially available microarrays are commonly based on the NCTC11168 genome sequence, but the

Table 2. Overview of transcriptomic studies on *C. jejuni* gene expression

Growth condition / process / mutant	Accession number[a]	Reference
Immobilised growth	GSE3028	[54]
Acid shock, stomach transit	GSE9938, GSE9920, GSE9937	[69]
Bile acid exposure	GSE10110	[70]
Nitrosative stress	GSE5439, GSE5438, GSE7048	[46]
Temperature variation (37°C vs 42°C)	N/A	[71]
Cold shock	N/A	[72]
Intestinal lifestyle	N/A	[73]
Chick colonisation	N/A	[48]
Flagellar biosynthesis (*fliK, rpoN, flgR*)	E-BUGS-50	[42]
Flagellar biosynthesis (*fliA, flhA*)	GSE708	[47]
Bile tolerance (*cmeR*)	GSE5412	[22]
Colonisation (*dccRS*)	GSE3198	[43]
Stringent response (*spoT/relA*)	GSE3209	[24]
Phosphate starvation (*phosRS*)	N/A	[41]
Nitrosative stress (*nssR*)	N/A	[45]
Iron homeostasis (*fur*)	N/A	[26, 37]
Heat shock (*hspR*)	N/A	[49]
Quorum sensing (*luxS*)	N/A	[74]

[a] GSE-accession numbers refer to the GEO and Arrayexpress databases (http://www.ncbi.nlm.nih.gov/geo and http://www.ebi.ac.uk/microarray-as/aer). The BUGS accession numbers refer to the BUGS database (http://www.bugs.sgul.ac.uk). Further microarray data may be found at the Stanford Microarray Database (http://genome-www5.stanford.edu/). N/A: not available.

developments in on-chip oligonucleotide synthesis techniques now allow rapid production of strain-specific arrays if required.

Microarray analysis of *C. jejuni* strains containing mutations in regulatory genes (table 2) has provided key insights into the regulation of stress responses and virulence gene expression. Of the 5 two-component regulatory systems present in *C. jejuni*, the PhosRS (Cj0890-Cj0889), DccRS (Cj1223c-Cj1222c) and FlgRS (Cj1024-Cj0793) regulons have been characterised through transcriptomic analysis of response regulator mutants [41–43]. Alignment of the promoter sequences of genes regulated by DccR and PhosR allowed consensus binding motifs for these two transcriptional activators to be identified, with binding to selected promoters confirmed by gel shift assays [41, 43].

In addition to identifying the target promoters for these two-component systems, microarray studies have also revealed the targets of several other transcriptional regulators. As in many bacteria, iron homeostasis is regulated in *C. jejuni* by the *fur* repressor [44]. Microarray analysis of Fur mutants has shown that it regulates expression of all the known iron uptake systems, and has helped identify additional components of the iron homeostasis system [26, 37]. The single domain globin (Cgb) of *C. jejuni* plays a major role in nitric oxide (NO) scavenging and detoxification and its expression has been shown to be regulated by an Fnr-Crp superfamily member, NssR [45]. Microarray studies have shown that NssR regulates expression of three additional genes (Cj0465c, Cj0761 and Cj0830) and real time PCR experiments, comparing their expression in wild type and Δ*nss*R strains in response to the NO releaser GSNO, confirmed that NssR is a positive regulator of gene expression [45, 46].

One area of *Campylobacter* research that has especially benefited from transcriptomic studies is that of the regulation of flagella synthesis. In Gram-negative bacteria flagella biosynthesis is tightly controlled in a hierarchical manner, such that genes are expressed in the order in which they are required for flagella assembly. Microarrays have confirmed the roles of σ^{54} (RpoN) and σ^{28} (FliA) in activating expression of middle and late flagellum biosynthesis genes respectively, and identified additional genes whose expression is regulated by these two alternative sigma factors of *C. jejuni* [42, 47]. Additionally, σ^{54} promoters were seen to be upregulated in a *fliA* mutant, indicating that σ^{28} or a gene controlled by it represses σ^{54} activity or σ^{54}-regulated genes [47]. Importantly, this study by Carrillo and colleagues showed that mutation of *flhA* (a key component of the flagella export apparatus) results in global changes in virulence gene expression including decreased expression of both σ^{54} and σ^{28} regulated genes. This suggests that FlhA may be a master regulator of flagellum expression and virulence, as has been suggested in other bacteria [47].

Transcriptomic studies of wild type strains under various conditions provide much information about stress responses and their regulatory control mechanisms. Gaynor and colleagues [24] identified a number of genes that were strongly and rapidly induced during *C. jejuni* 81–176 infection of epithelial cells. These included *spoT* (Cj1272c), which in other bacteria has been shown to regulate the global response to amino acid starvation. Whilst the previous example clearly identified the role of *spoT* in regulating the *C. jejuni* stringent response other microarray studies have only suggested the presence of as yet unidentified regulatory networks. An example of this is the *C. jejuni* response to variations in oxygen tension. It has been found that during colonisation of the chick caecum, *C. jejuni* upregulates genes that, in other bacteria, are induced at low oxygen concentrations [48]. *C. jejuni* does not contain the classic FNR or Arc systems responsible for regulating the transcriptional response to anaerobiosis in other bacteria and so a key focus for future research is to identify the genes(s) responsible for regulating the *C. jejuni* oxygen stress response.

There is now a substantial list of *C. jejuni* microarray data deposited in the online, MIAME-compliant databases at NCBI (GEO), EBI (ArrayExpress), Stanford and BUGS. An overview of such studies is presented in table 2. The recent increase in the availability of commercial *C. jejuni* microarrays is likely to result in further additions to these depositories. The available data are derived from analysis of several strains of *C. jejuni* and has been analysed in a variety of different ways. Consequently, this potentially hampers a global meta-analysis of *C. jejuni* gene expression. However, as more data emerge, coordinated gene expression patterns and regulons will be easier to identify.

Proteomics

Proteomics is the examination of the protein complement of an organism. Proteomic techniques include both one-dimensional (1D) SDS-PAGE and two-dimensional (2D) gel electrophoresis, and chromatography methods to separate the complex mixtures and mass spectrometry (MS) techniques to identify and in some cases quantify proteins. Relative quantification by 2D gel analysis software is also widely used to determine differences between protein complements.

The relatively recent expansion in genome sequence databases plus advances in the speed and sensitivity of MS as a tool for identification of proteins, coupled with the advances in computing have allowed proteomics to develop as an effective technology. An early proteomic analysis carried out on *Campylobacter* used 2D gels to observe changes in protein expression under high and low iron conditions [44]. At that time the only available technology for protein identification was N-terminal sequencing via Edman degradation, and *Campylobacter* protein sequences were relatively scarce prior to the release in 2000 of the genome sequence [3]. This limited the ability to identify all the proteins changing but effectively demonstrated the role of *fur* as a global regulator of iron metabolism in *C. jejuni*. A later study with access to the now-available genome sequence was able to identify further the proteins involved in the iron response [26]. This latter study used genome-wide transcriptomic analysis to provide supporting evidence at the transcriptional level for the proteins changing on a 2D gel, plus evidence of increased expression of integral membrane proteins such as the ABC-transporter permease, ChuB (Cj1615), that could not be observed in the gel based system. The iron transport and storage proteins observed on the 2D gel corresponded to the highest changes on the microarray. This was also the case in the combined analysis of the heat shock protein regulator (HspR) in *C. jejuni* where the comparison of the wild-type strain with an *hpsR* mutant showed that the chaperonin proteins, GroEL/ES, DnaK, GrpE, and ClpB were negatively regulated by HspR [49].

Most studies using proteomic techniques are limited to a number of specific proteins. Examples of this are the recent study on the temperature dependence of gluconate dehydrogenase (Cj0414 and Cj0415) which used a 2D gel approach to show increased levels of these proteins at 42°C compared to 37°C [50] and the effect of oxygen limitation causing increased expression and activity of aspartase (Cj0087) [51].

One clear limitation of proteomic technologies to date is the inability to cover the whole protein complement. Several important classes of proteins are still very difficult to detect due to their low abundance. Trans-membrane spanning proteins are also problematic. A study using both a 2D gel/matrix-assisted laser desorption/

ionization (MALDI) MS and Multi-Dimensional Protein Identification Technology (MuDPIT) approach coupled to MS-MS identification of the peptides of a tryptic digest of *C. jejuni* proteins resulted in the largest identification of *C. jejuni* proteins to date with 453 unique proteins being detected by the combination of techniques but still corresponding to 27.4% of the theoretical proteome giving an indication of the challenge still facing analysts [52].

2D gel electrophoresis has been used to examine the protein complements of a robust and poor chicken gastrointestinal colonizing isolates of *C. jejuni* [53]. Isolates were grown up in broth culture to produce the protein extracts and the specific expression of an outer membrane-fibronectin binding protein (CadF). A serine protease (HrtA), and a putative aminopeptidase (Cj0653c) were found in the soluble portion of the robust colonizer. Several proteins including a cysteine synthase (CysM) and aconitate hydratase (AcnB) were detected specifically in the poor colonizer protein extract. Several of the proteins observed in the robust colonising strain were also identified as significant in immobilised growth, where the use of 2D gels also showed increased expression of motility and chemotaxis proteins [54]. While these studies are of interest with regard to comparison of strains, the connection with colonisation properties is mostly circumstantial, could be influenced by in vitro passaging, and these data require further experimental validation in vivo.

An important reason for performing proteomics is the ability to examine post-translational modification of proteins. Typically these would not be detected by other ‘omic techniques. In analysis of 2D gels it is apparent that many proteins identify as a series of spots rather than just one suggesting some form of post-translational change. A common modification of proteins that has biological significance is phosphorylation and this was examined in *C. jejuni* using an SDS-PAGE and 2D gel approach following enrichment of the phospho-proteins using an Immobilised Metal Affinity Chromatography (IMAC) [55]. Fifty-eight phosphopeptides derived from tryptic digests of 1D SDS-PAGE bands were sequenced corresponding to 36 proteins. The major phosphoproteins following IMAC enrichment and separation on 2D gels were bacterioferritin (Cj1534c), superoxide dismutase (Cj0169) and a thiol peroxidase (Cj0779). Sequence analysis of the phosphopeptides showed threonine to be the most commonly phosphorylated amino acid with tyrosine modifications rarely found [55].

All studies to date that have been published are either qualitative, identifying the proteins present in *C. jejuni*, or comparative where the relative change in a protein in an experimental condition is measured against a suitable control. In some cases a suitable control is not always obvious which limits the interpretation of the observations. Although not yet used in *Campylobacter* the advent of absolute quantitative proteomic technologies such as Absolute QUAntification (AQUA) and the related Quantification conCATamers (QconCAT) techniques [56] for preparation of tryptic peptide standards will allow interesting high throughput techniques to be applied in *Campylobacter* proteomics.

Development of Genetic Tools for *Campylobacter*

Research on *Campylobacter* species has suffered from a lack of tractable genetic tools such as those available for other bacteria like *Salmonella* and *E. coli*. The construction of specific gene knockouts is relatively straightforward, but other tools such as reporter constructs, conditional or unmarked mutations and gene complementation are still not widely employed. The available techniques are often cumbersome and usually rely on the use of specific strains and plasmids, thus limiting their general application to *Campylobacter* research.

Random Mutant Libraries

Libraries of random insertional mutants in an organism can allow the identification of genes involved in processes without prior knowledge or hypotheses. All that is required is a suitable selection assay mimicking some aspect of the area of interest to apply to a library of mutants. However, the application of such methods to *Campylobacter* research has not been as widespread or successful as for other species. In the main this has been due to the genetically less tractable nature of *Campylobacter* limiting the creation of suitable libraries. Initial attempts to create libraries of mutants in *C. jejuni* relied on the non-random insertion of antibiotic resistance marker genes into libraries of chromosomal DNA via traditional restriction enzyme sites [57]. These produced small libraries that were used to identify genes involved in motility. Later several groups reported the use of transposon based methods for constructing libraries. These relied on both the in vivo and in vitro activity of different transposases [58–60]. Although these methods produced more complex libraries of essentially random mutants, they were still initially used to identify motility associated genes. As with the earlier studies, these studies relied on screening individual colonies from the libraries.

Signature Tagged Mutagenesis

The application of methods developed to screen whole complex libraries such as Signature Tagged Mutagenesis (STM), DNA microarray based library comparison methods or Genomic Analysis and Mapping by In Vitro Transposition (GAMBIT) to *C. jejuni* has also been limited with only 2 reports of the use of STM [61, 62]. In both studies, STM was applied to try and identify genes involved in chicken colonisation. Their results differed due to many factors such as using different *C. jejuni* strains and different colonisation models. In one study it was reported that it was not possible to recover reproducibly the same surviving and non-surviving mutants [61], while the other study failed to identify genes known to be essential for colonisation, such as *cadF* and *racRS* [62]. This may be due to many factors including the population dynamics of *Campylobacter* strains during chicken colonisation as well as possible limitations with the STM method in such a complex process.

The application of random mutant library screening in *C. jejuni* has not resulted in the same level of information as for other species. Although the availability of several annotated genome sequences may negate some of the usefulness of library screening, it is the ability to identify genes by functional screening rather than inferred function that is the principle advantage of random library methods. As more refined relevant screening methods are developed and applied to library screening, it is likely that useful insights into the roles of many previously unknown genes will be gained.

Mutant Complementation

Conveniently, pseudogenes provide a means of genetically modifying the organism without affecting any gene function. In particular, complementation constructs can be inserted into pseudogenes without adversely affecting functioning genes. This approach was first described in *C. jejuni* where an *nssR* mutant was successfully complemented by insertion of a functional copy of *nssR* with its own promoter into the pseudogene *cj0752* [45]. Given the variable success with introducing and maintaining plasmids in many *C. jejuni* strains, it is likely that the use of pseudogenes as targets for genetic tools such as complementation and reporter genes will become commonplace. An alternative target for such insertions has been described which utilizes ribosomal RNA gene clusters [63]. However, this system suffers from variability due to the varying number of rRNA gene duplications and transfer of the inserted sequence between them, resulting in varying numbers of copies of the genes within a population.

Reporter Genes

The application of predictive computational algorithms to genomic sequences results in many suggested functions that need validating. Such validation can be obtained by a variety of methods. In the case of promoter prediction, the use of libraries of short genomic DNA fragments fused to a reporter gene has allowed the identification of functional promoters in vivo as shown in *C. jejuni* [64]. Several reporter genes are now available for use in *C. jejuni*, and include systems based on β-galactosidase [64] and green, yellow and cyan fluorescent protein [65]. Unfortunately these systems are still relatively cumbersome in their use, and often limited to specific *C. jejuni* strains. Further development of these techniques is warranted to realise their full potential in *Campylobacter* research.

Conclusions

During the last three decades the role of *Campylobacter* as a human pathogen has become more apparent, and the organism is now recognised as the major cause of bacterial gastroenteritis worldwide. Despite the rapid development of genomic techniques in recent years, there are still gaps in our understanding of some of the basic aspects of the biology and pathogenicity of *Campylobacter*. Targets of future *Campylobacter* research will include further elucidation of its pathogenic mechanisms, including its interaction with the intestinal microbiota, the identification of invasion and translocation factors [66], the role and regulation of chemotactic motility [31, 67], and the elucidation of the roles of inflammation and toxin production by *Campylobacter* species [68]. These investigations will be aided by the rapid developments in high-throughput genome sequencing techniques, and hence we can predict an increase in the understanding of *Campylobacter* physiology and virulence, and this will subsequently aid the identification of novel targets for prevention and intervention strategies. It will however also have to be matched by the development of other high-throughput phenotypic and molecular approaches to test the hypotheses generated from genomics approaches, and this will be a major challenge in the coming years. The availability of several *Campylobacter* genome sequences should be coupled to the further development and improvement of (semi-)random mutagenesis strategies, to allow further insight in the role of specific genes in *Campylobacter* virulence. However, to complement the chicken colonisation model there is a need to improve the animal model of diarrhoeal disease, in order to be able to investigate the role of host immune pathways in *Campylobacter*-associated diseases.

Acknowledgements

The *Campylobacter* research in the Institute of Food Research is supported by the Core Strategic Grant from the Biotechnology and Biological Sciences Research Council (BBSRC), and N.S. and B.M.P. are supported by BBSRC grant BBD0131351. We apologise for not being able to cite many publications due to space limitations.

References

1 Young KT, Davis LM, Dirita VJ: *Campylobacter jejuni*: molecular biology and pathogenesis. Nat Rev Microbiol 2007;5:665–679.

2 Hughes R: *Campylobacter jejuni* in Guillain-Barre syndrome. Lancet Neurol 2004;3:644.

3 Parkhill J, Wren BW, Mungall K, Ketley JM, Churcher C, et al: The genome sequence of the food-borne pathogen *Campylobacter jejuni* reveals hypervariable sequences. Nature 2000;403:665–668.

4 Kelly DJ: Complexity and versatility in the physiology and metabolism of *Campylobacter jejuni*; in Nachamkin I, Szymanski CM, Blaser MJ (eds): *Campylobacter*, ed 3. Washington, DC, ASM Press, 2008, pp 41–61.

5 Gundogdu O, Bentley SD, Holden MT, Parkhill J, Dorrell N, Wren BW: Re-annotation and re-analysis of the *Campylobacter jejuni* NCTC11168 genome sequence. BMC Genomics 2007;8:162.

6 Fouts DE, Mongodin EF, Mandrell RE, Miller WG, Rasko DA, et al: Major structural differences and novel potential virulence mechanisms from the genomes of multiple *Campylobacter* species. PLoS Biol 2005;3:e15.

7 Pearson BM, Gaskin DJ, Segers RP, Wells JM, Nuijten PJ, van Vliet AHM: The complete genome sequence of *Campylobacter jejuni* strain 81116 (NCTC11828). J Bacteriol 2007;189:8402–8403.

8 Poly F, Read T, Tribble DR, Baqar S, Lorenzo M, Guerry P: Genome sequence of a clinical isolate of *Campylobacter jejuni* from Thailand. Infect Immun 2007;75:3425–3433.

9 Hofreuter D, Tsai J, Watson RO, Novik V, Altman B, et al: Unique features of a highly pathogenic *Campylobacter jejuni* strain. Infect Immun 2006;74: 4694–4707.

10 Miller WG, Parker CT, Rubenfield M, Mendz GL, Wosten MM, et al: The complete genome sequence and analysis of the epsilonproteobacterium *Arcobacter butzleri*. PLoS ONE 2007;2:e1358.

11 Miller WG, Wang G, Binnewies TT, Parker CT: The complete genome sequence and analysis of the human pathogen *Campylobacter lari*. Foodborne Pathog Dis 2008;5:371–386.

12 Eppinger M, Baar C, Raddatz G, Huson DH, Schuster SC: Comparative analysis of four Campylobacterales. Nat Rev Microbiol 2004;2:872–885.

13 Bacon DJ, Alm RA, Burr DH, Hu L, Kopecko DJ, et al: Involvement of a plasmid in virulence of *Campylobacter jejuni* 81–176. Infect Immun 2000; 68:4384–4390.

14 Louwen RP, van Belkum A, Wagenaar JA, Doorduyn Y, Achterberg R, Endtz HP: Lack of association between the presence of the pVir plasmid and bloody diarrhea in *Campylobacter jejuni* enteritis. J Clin Microbiol 2006;44:1867–1868.

15 Batchelor RA, Pearson BM, Friis LM, Guerry P, Wells JM: Nucleotide sequences and comparison of two large conjugative plasmids from different *Campylobacter* species. Microbiology 2004;150: 3507–3517.

16 Friis LM, Pin C, Taylor DE, Pearson BM, Wells JM: A role for the tet(O) plasmid in maintaining *Campylobacter* plasticity. Plasmid 2007;57:18–28.

17 Parker CT, Quinones B, Miller WG, Horn ST, Mandrell RE: Comparative genomic analysis of *Campylobacter jejuni* strains reveals diversity due to genomic elements similar to those present in *C. jejuni* strain RM1221. J Clin Microbiol 2006;44: 4125–4135.

18 Clark CG, Ng LK: Sequence variability of *Campylobacter* temperate bacteriophages. BMC Microbiol 2008;8:49.

19 Raphael BH, Pereira S, Flom GA, Zhang Q, Ketley JM, Konkel ME: The *Campylobacter jejuni* response regulator, CbrR, modulates sodium deoxycholate resistance and chicken colonization. J Bacteriol 2005;187:3662–3670.

20 Marchant J, Wren B, Ketley J: Exploiting genome sequence: predictions for mechanisms of *Campylobacter* chemotaxis. Trends Microbiol 2002;10:155–159.

21 Terry K, Go AC, Ottemann KM: Proteomic mapping of a suppressor of non-chemotactic cheW mutants reveals that *Helicobacter pylori* contains a new chemotaxis protein. Mol Microbiol 2006;61: 871–882.

22 Guo B, Wang Y, Shi F, Barton Y-W, Plummer P, et al: CmeR functions as a pleiotropic regulator and is required for optimal colonization of *Campylobacter jejuni* in vivo. J Bacteriol 2008;190:1879–1890.

23 Wosten MM: Eubacterial sigma-factors. FEMS Microbiol Rev 1998;22:127–150.

24 Gaynor EC, Wells DH, MacKichan JK, Falkow S: The *Campylobacter jejuni* stringent response controls specific stress survival and virulence-associated phenotypes. Mol Microbiol 2005;56:8–27.

25 Valentin-Hansen P, Eriksen M, Udesen C: The bacterial Sm-like protein Hfq: a key player in RNA transactions. Mol Microbiol 2004;51:1525–1533.

26 Holmes K, Mulholland F, Pearson BM, Pin C, McNicholl-Kennedy J, et al: *Campylobacter jejuni* gene expression in response to iron limitation and the role of Fur. Microbiology 2005;151:243–257.

27 Parrish JR, Yu J, Liu G, Hines JA, Chan JE, et al: A proteome-wide protein interaction map for *Campylobacter jejuni*. Genome Biol 2007;8:R130.

28 Linton D, Gilbert M, Hitchen PG, Dell A, Morris HR, et al: Phase variation of a beta-1,3 galactosyltransferase involved in generation of the ganglioside GM1-like lipo-oligosaccharide of *Campylobacter jejuni*. Mol Microbiol 2000;37:501–514.

29 Ashgar SS, Oldfield NJ, Wooldridge KG, Jones MA, Irving GJ, et al: CapA, an autotransporter protein of *Campylobacter jejuni*, mediates association with human epithelial cells and colonization of the chicken gut. J Bacteriol 2007;189:1856–1865.

30 Hendrixson DR: A phase-variable mechanism controlling the *Campylobacter jejuni* FlgR response regulator influences commensalism. Mol Microbiol 2006;61:1646–1659.

31 Hendrixson DR: Restoration of flagellar biosynthesis by varied mutational events in *Campylobacter jejuni*. Mol Microbiol 2008;70:519–536.

32 Scott AE, Timms AR, Connerton PL, Loc Carrillo C, Adzfa Radzum K, Connerton IF: Genome dynamics of *Campylobacter jejuni* in response to bacteriophage predation. PLoS Pathog 2007;3:e119.

33 Ahmed IH, Manning G, Wassenaar TM, Cawthraw S, Newell DG: Identification of genetic differences between two *Campylobacter jejuni* strains with different colonization potentials. Microbiology 2002; 148:1203–1212.

34 Dorrell N, Mangan JA, Laing KG, Hinds J, Linton D, et al: Whole genome comparison of *Campylobacter jejuni* human isolates using a low-cost microarray reveals extensive genetic diversity. Genome Res 2001;11:1706–1715.

35 Pearson BM, Pin C, Wright J, I'Anson K, Humphrey T, Wells JM: Comparative genome analysis of *Campylobacter jejuni* using whole genome DNA microarrays. FEBS Lett 2003;554:224–230.

36 Rodin S, Andersson AF, Wirta V, Eriksson L, Ljungstrom M, et al: Performance of a 70-mer oligonucleotide microarray for genotyping of *Campylobacter jejuni*. BMC Microbiol 2008;8:73.

37 Palyada K, Threadgill D, Stintzi A: Iron acquisition and regulation in *Campylobacter jejuni*. J Bacteriol 2004;186:4714–4729.

38 Champion OL, Gaunt MW, Gundogdu O, Elmi A, Witney AA, et al: Comparative phylogenomics of the food-borne pathogen *Campylobacter jejuni* reveals genetic markers predictive of infection source. Proc Natl Acad Sci USA 2005;102:16043–16048.
39 McCarthy ND, Colles FM, Dingle KE, Bagnall MC, Manning G, et al: Host-associated genetic import in *Campylobacter jejuni*. Emerg Infect Dis 2007;13:267–272.
40 Sheppard SK, McCarthy ND, Falush D, Maiden MC: Convergence of *Campylobacter* species: implications for bacterial evolution. Science 2008;320:237–239.
41 Wosten MM, Parker CT, van Mourik A, Guilhabert MR, van Dijk L, van Putten JP: The *Campylobacter jejuni* PhosS/PhosR operon represents a non-classical phosphate-sensitive two-component system. Mol Microbiol 2006;62:278–291.
42 Kamal N, Dorrell N, Jagannathan A, Turner SM, Constantinidou C, et al: Deletion of a previously uncharacterized flagellar-hook-length control gene fliK modulates the sigma54-dependent regulon in *Campylobacter jejuni*. Microbiology 2007;153:3099–3111.
43 MacKichan JK, Gaynor EC, Chang C, Cawthraw S, Newell DG, et al: The *Campylobacter jejuni* dccRS two-component system is required for optimal in vivo colonization but is dispensable for in vitro growth. Mol Microbiol 2004;54:1269–1286.
44 van Vliet AHM, Wooldridge KG, Ketley JM: Iron-responsive gene regulation in a *Campylobacter jejuni* fur mutant. J Bacteriol 1998;180:5291–5298.
45 Elvers KT, Turner SM, Wainwright LM, Marsden G, Hinds J, et al: NssR, a member of the Crp-Fnr superfamily from *Campylobacter jejuni*, regulates a nitrosative stress-responsive regulon that includes both a single-domain and a truncated haemoglobin. Mol Microbiol 2005;57:735–750.
46 Monk CE, Pearson BM, Mulholland F, Smith HK, Poole RK: Oxygen- and NssR-dependent globin expression and enhanced iron acquisition in the response of *Campylobacter* to nitrosative stress. J Biol Chem 2008;283:28413–28425.
47 Carrillo CD, Taboada E, Nash JHE, Lanthier P, Kelly J, et al: Genome-wide expression analyses of *Campylobacter jejuni* NCTC11168 reveals coordinate regulation of motility and virulence by flhA. J Biol Chem 2004;279:20327–20338.
48 Woodall CA, Jones MA, Barrow PA, Hinds J, Marsden GL, et al: *Campylobacter jejuni* gene expression in the chick cecum: evidence for adaptation to a low-oxygen environment. Infect Immun 2005;73:5278–5285.
49 Andersen MT, Brondsted L, Pearson BM, Mulholland F, Parker M, et al: Diverse roles for HspR in *Campylobacter jejuni* revealed by the proteome, transcriptome and phenotypic characterization of an hspR mutant. Microbiology 2005;151: 905–915.
50 Pajaniappan M, Hall JE, Cawthraw SA, Newell DG, Gaynor EC, et al: A temperature-regulated *Campylobacter jejuni* gluconate dehydrogenase is involved in respiration-dependent energy conservation and chicken colonization. Mol Microbiol 2008; 68:474–491.
51 Guccione E, Leon-Kempis MdelR, Pearson BM, Hitchin E, Mulholland F, et al: Amino-acid dependent growth of *Campylobacter jejuni*: Key roles for aspartase (AspA) under microaerobic and oxygen-limited conditions and identification of AspB (Cj0762), essential for growth on glutamate. Mol Microbiol 2008;69:77–93.
52 Cordwell SJ, Len AC, Touma RG, Scott NE, Falconer L, et al: Identification of membrane-associated proteins from *Campylobacter jejuni* strains using complementary proteomics technologies. Proteomics 2008;8:122–139.
53 Seal BS, Hiett KL, Kuntz RL, Woolsey R, Schegg KM, et al: Proteomic analyses of a robust versus a poor chicken gastrointestinal colonizing isolate of *Campylobacter jejuni*. J Proteome Res 2007;6:4582–4591.
54 Sampathkumar B, Napper S, Carrillo CD, Willson P, Taboada E, et al: Transcriptional and translational expression patterns associated with immobilized growth of *Campylobacter jejuni*. Microbiology 2006; 152:567–577.
55 Voisin S, Watson DC, Tessier L, Ding W, Foote S, et al: The cytoplasmic phosphoproteome of the Gram-negative bacterium *Campylobacter jejuni*: evidence for modification by unidentified protein kinases. Proteomics 2007;7:4338–4348.
56 Gerber SA, Rush J, Stemman O, Kirschner MW, Gygi SP: Absolute quantification of proteins and phosphoproteins from cell lysates by tandem MS. Proc Natl Acad Sci USA 2003;100:6940–6945.
57 Bleumink-Pluym NM, Verschoor F, Gaastra W, van der Zeijst BA, Fry BN: A novel approach for the construction of a *Campylobacter* mutant library. Microbiology 1999;145:2145–2151.
58 Colegio OR, Griffin TJ 4th, Grindley ND, Galan JE: In vitro transposition system for efficient generation of random mutants of *Campylobacter jejuni*. J Bacteriol 2001;183:2384–2388.
59 Golden NJ, Camilli A, Acheson DW: Random transposon mutagenesis of *Campylobacter jejuni*. Infect Immun 2000;68:5450–5453.

60 Hendrixson DR, Akerley BJ, DiRita VJ: Transposon mutagenesis of *Campylobacter jejuni* identifies a bipartite energy taxis system required for motility. Mol Microbiol 2001;40:214–224.
61 Grant AJ, Coward C, Jones MA, Woodall CA, Barrow PA, Maskell DJ: Signature-tagged transposon mutagenesis studies demonstrate the dynamic nature of cecal colonization of 2-week-old chickens by *Campylobacter jejuni*. Appl Environ Microbiol 2005;71:8031–8041.
62 Hendrixson DR, DiRita VJ: Identification of *Campylobacter jejuni* genes involved in commensal colonization of the chick gastrointestinal tract. Mol Microbiol 2004;52:471–484.
63 Karlyshev AV, Wren BW: Development and application of an insertional system for gene delivery and expression in *Campylobacter jejuni*. Appl Environ Microbiol 2005;71:4004–4013.
64 Wosten MM, Boeve M, Koot MG, van Nuenen AC, van der Zeijst BA: Identification of *Campylobacter jejuni* promoter sequences. J Bacteriol 1998;180:594–599.
65 Miller WG, Bates AH, Horn ST, Brandl MT, Wachtel MR, Mandrell RE: Detection on surfaces and in Caco-2 cells of *Campylobacter jejuni* cells transformed with new gfp, yfp, and cfp marker plasmids. Appl Environ Microbiol 2000;66:5426–5436.
66 Hu L, Tall BD, Curtis SK, Kopecko DJ: Enhanced microscopic definition of *Campylobacter jejuni* 81–176 adherence to, invasion into, translocation across, and exocytosis from polarized human intestinal Caco-2 cells. Infect Immun 2008;76:5294–5304.
67 Joslin SN, Hendrixson DR: Analysis of the *Campylobacter jejuni* FlgR response regulator suggests integration of diverse mechanisms to activate an NtrC-like protein. J Bacteriol 2008;190:2422–2433.
68 Istivan TS, Smith SC, Fry BN, Coloe PJ: Characterization of *Campylobacter concisus* hemolysins. FEMS Immunol Med Microbiol 2008;54:224–235.
69 Reid AN, Pandey R, Palyada K, Naikare H, Stintzi A: Identification of *Campylobacter jejuni* genes involved in the response to acidic pH and stomach transit. Appl Environ Microbiol 2008;74:1583–1597.
70 Malik-Kale P, Parker CT, Konkel ME: Culture of *Campylobacter jejuni* with sodium deoxycholate induces virulence gene expression. J Bacteriol 2008;190:2286–2297.
71 Stintzi A: Gene expression profile of *Campylobacter jejuni* in response to growth temperature variation. J Bacteriol 2003;185:2009–2016.
72 Stintzi A, Whitworth L: Investigation of the *Campylobacter jejuni* cold-shock response by global transcript profiling. Genome Letters 2003;2:18–27.
73 Stintzi A, Marlow D, Palyada K, Naikare H, Panciera R, et al: Use of genome-wide expression profiling and mutagenesis to study the intestinal lifestyle of *Campylobacter jejuni*. Infect Immun 2005;73:1797–1810.
74 He Y, Frye JG, Strobaugh TP, Chen CY: Analysis of AI-2/LuxS-dependent transcription in *Campylobacter jejuni* strain 81–176. Foodborne Pathog Dis 2008;5:399–415.

Arnoud H.M. van Vliet
Institute of Food Research, Norwich Research Park
Colney Lane
Norwich NR4 7UA (UK)
Tel. +44 1603 255250, Fax +44 1603 255288, E-Mail arnoud.vanvliet@bbsrc.ac.uk

de Reuse H, Bereswill S (eds): Microbial Pathogenomics.
Genome Dyn. Basel, Karger, 2009, vol 6, pp 110–125

Adaptation of Pathogenic *E. coli* to Various Niches: Genome Flexibility is the Key

E. Brzuszkiewicz[a,b] · G. Gottschalk[a] · E. Ron[c] · J. Hacker[b,d] · U. Dobrindt[d]

[a]Göttingen Genomics Laboratory, University of Göttingen, Göttingen, [b]Robert-Koch-Institute, Berlin, [d]University of Würzburg, Institute for Molecular Infection Biology, Würzburg, Germany; [c]Tel Aviv University, Department of Molecular Microbiology and Biotechnology, Ramat Aviv, Israel

Abstract

It is a well-known observation and a long-standing hypothesis that pathogen genome dynamics are important in infectious disease processes. Recent achievements in large-scale genome sequencing, comparative genomics and molecular epidemiology help to unravel current challenges of *E. coli* pathogenomics, i.e. to gain insights into the in vivo relevance of genome dynamics. Data from comparative genomics support the hypothesis of widespread involvement of horizontal gene transfer in the evolution of *E. coli*, leading to the presence of distinct and variable 'genomic islands' within the conserved 'chromosomal backbone' in several bacterial lineages. Extensive gene acquisition and loss provide different lineages with distinct metabolic, pathogenic and other capabilities. Not only mobile genetic modules but also point mutations facilitate rapid adaptation of *E. coli* to changing environmental conditions and hence extend the spectrum of sites that can be infected. We report on recent research efforts to analyze pathoadaptive and other genomic alterations of the *E. coli* genome that affect disease severity and may have consequences for diagnostics and treatment of *E. coli* infections.

Escherichia coli is a commensal member of the physiological gastrointestinal flora of man and warm-blooded animals. This facultative anaerobic species exhibits considerable physiologic and metabolic versatility that facilitates efficient colonization of the gut [1]. Additionally, several facultative and obligate pathogenic variants have been identified which cause various types of intestinal or extraintestinal infections in men and animals [2]. Facultative pathogenic variants belong to the normal intestinal flora and may mainly cause infections such as urinary tract infections, newborn meningitis and sepsis once they reach the corresponding sites of the body. In contrast, obligate pathogenic *E. coli* variants are not part of the physiological bacterial gut flora and cause different types of diarrhoea. The different pathogenic potential can be attributed to the presence of virulence- and fitness-associated gene sets coding for factors that are required for establishment of the infection. The importance of ordered gene acquisition events by horizontal gene transfer and loss of genetic information as well as of DNA rearrangements and point mutations for still ongoing evolution of different *E. coli* variants has been well documented during the last years [3, 4].

Genome Structure of Pathogenic and Non-Pathogenic *E. coli*

Whole genome sequencing projects have opened new possibilities in the bacterial genomics and evolution research field. In the wake of large-scale genome sequencing and comparative genomics, knowledge on genome content, the diversity between non-pathogenic and pathogenic *E. coli* in general, and between different *E. coli* pathotypes in particular, has started to accumulate. Eight complete *E. coli* genome sequences have been published so far and about 40 additional complete genome sequences of pathogenic and non-pathogenic *E. coli* isolates (sequenced by different institutions; http://www.genomesonline.org/) will be available in the near future [5–12]. The *E. coli* genome is composed of a conserved core of genes, providing the backbone of genetic information required for essential cellular processes [13], and of an additional, flexible gene pool. The latter one consists of strain-specific 'assortments' of genetic information, which provide additional metabolic and pathogenic properties enabling these strains to adapt to special environmental conditions (e.g., virulence-associated factors, antibiotic resistances). Accessory genetic elements like transposons, integrons, insertion elements and genomic or pathogenicity islands (GEIs, PAIs) represent major constituents of the flexible gene pool. The islands are seldom fixed but rather bear the potential for ongoing rearrangements, deletions and insertions. Accordingly, the stable chromosomal backbone and the flexible gene pool are constantly undergoing repeated insertions and deletions. Thus, the *E. coli* genome is composed of clonally evolving DNA regions that are periodically disrupted due to exchange of already existing gene blocks by homologous recombination, and insertion of horizontally acquired DNA segments. The majority of strain- or pathotype-specific regions accumulated over time by repeated horizontal gene transfer, frequently with successive transfers of different elements into identical loci of the core chromosome [4]. The existence of so many different horizontally acquired sequences in genomic islands differentiating closely related *E. coli* strains indicates that many of them are only temporarily present in the genome or provide a specific advantage to the individual lifestyle of particular strains.

Genome Size and Content

The pheno- and genotypic variability of pathogenic and commensal *E. coli* correlate with their genome content. *E. coli* genomes vary in size from 4.6 to 5.6 Mb

Table 1. General features of completely sequenced *E. coli* genomes

Strain	MG1655 (K-12)	W3110 (K-12)	CFT073 (UPEC) (O6:K2:H1)	536 (UPEC) (O6:K15:H31)	UTI89 (UPEC) (O18:K1:H7)	APEC O1 (APEC) (O1:K1:H7)	EDL933 (EHEC) (O157:H7)	Sakai (EHEC) (O157:H7)
Chromosome size (bp)	4,639,221	4,646,332	5,231,428	4,938,875	5,065,741	5,082,025	5,528,445	5,498,450
Plasmids (size in bp)	–		–	–	pUTI89 (114,230)	pAPEC-O1-ColBM (174,241) pAPEC-O1-R (241,387) pAPEC-O1-Cryp1 (105,834) pAPEC-O1-Cryp2 (46,870)	pO157 (92,721)	pO157 (92,721), pOSAK1 (3,308)
No. of prophage-like elements	10	n.d.	5	1	n.d.	10	16	24
DNA coding sequence (%)	89	86	89	88	88	86	88	88
No. of ORFs	4,411	4,226	5,533	4,747	5,021	4,467	5,361	5,981
G+C content (%)	50.8	50	50.5	50.5	50	50.6	50.5	50.5
No. of rRNA genes	22	22	22	22	22	22	22	22
No. of tRNA genes	87	135	88	81	98	93	100	103
No. of predicted misc. RNAs	47	n.d.	51	45	n.d.	n.d.	53	13
Backbone (%)	81	n.d.	71	77	n.d.	78.3	67	n.d.
No. of strain-specific ORFs (%)[a]	406 (9)	n.d.	867 (16)	374 (8)	n.d.	201 (4.5)	1270 (24)	n.d.

[a]For comparative analysis, each ORF was searched against all ORFs of the other *E. coli* strains using the BLAST tool. Orthologous proteins were defined with an amino acid identity of >90% over >90% of query and reference sequence.

[14]. These size differences among individual *E. coli* genomes indicate the presence of different amounts of strain-specific genetic information, which may represent up to 30% of the complete genome content (table 1). Comparison between different *E. coli* genomes revealed a mosaic-like genome structure in terms of the distribution of backbone genes conserved in *E. coli*, and 'foreign' genes, which presumably have been horizontally acquired [6, 12]. Genes for many virulence traits of intestinal pathogenic *E. coli* (IPEC) and extraintestinal pathogenic *E. coli* (ExPEC), especially those characteristic for one or another pathotype, may be encoded on mobile and accessory

genetic elements, e.g., GEIs and PAIs [15, 16], plasmids and bacteriophages, the latter of which contribute significantly to *E. coli* genome diversity [11, 17–20].

ExPEC are epidemiologically and phylogenetically distinct from many commensal strains as well as from IPEC. A variety of virulence factors directly contribute to pathotype-specific disease and their distribution is thus restricted to the corresponding pathotypes. For instance, the ETT-1 type III secretion system and its translocated effectors are usually indicative of enterohemorrhagic *E. coli* (EHEC) and enteropathogenic *E. coli* (EPEC). The heat-stable or heat-labile enterotoxins are characteristic of enterotoxigenic *E. coli* (ETEC) [2]. Certain invasion genes like *ibeA* as well as the K1 capsule determinant are frequently present in invasive ExPEC [21]. In many cases, however, ExPEC and commensal *E. coli* [22, 23] share a large fraction of their genome. There are also many so-called virulence-associated factors in ExPEC such as colicins, certain fimbriae, siderophore systems and toxins [22, 24–26] that have probably evolved to enhance survival in the gut and/or transmission between hosts, and therefore will be shared with at least some commensal strains and sometimes even with IPEC.

Genome Plasticity and its Impact on Evolution of Different Pathotypes/Variants

The Locus of Enterocyte Effacement (LEE) in EHEC, EPEC and atypical EPEC
Many PAI regions exhibit notable homology to fragments of mobile genetic elements such as bacteriophages and virulence plasmids. In addition, multiple copies of accessory DNA elements in one genome facilitate homologous recombination within one or between different islands or horizontally acquired DNA elements thus leading to rearrangements, deletions and acquisition of 'foreign' DNA. Consequently, many PAIs have a mosaic-like, modular structure. Although many of them superficially resemble each other with respect to the presence and/or genetic linkage of certain virulence determinants, PAI composition, structural organization and chromosomal localization can be highly variable even among strains of the same patho- or serotype [27, 28].

The 'locus of enterocyte effacement' (LEE) island, which encodes a type three secretion system (ETT-1) and its translocated effectors required for the attaching and effacing phenotype of EHEC and EPEC was considered for a long time to be a clonal unit inside a clonally evolving host. It was thus expected to evolve as a single unit, but has been recently shown to exhibit a mosaic-like composition and to be genetically divergent [29–32]. Comparative analysis of the evolutionary history of type three secretion systems indicated that horizontal gene transfer is a major driving force in evolution of corresponding determinants [33]. Based on the sequence polymorphism of the *eaeA* gene coding for the adhesin intimin, 28 alleles have been identified so far [34]. Although the core regions of each LEE type encode almost identical sets of genes, their DNA sequences are significantly divergent. Data based on comparative

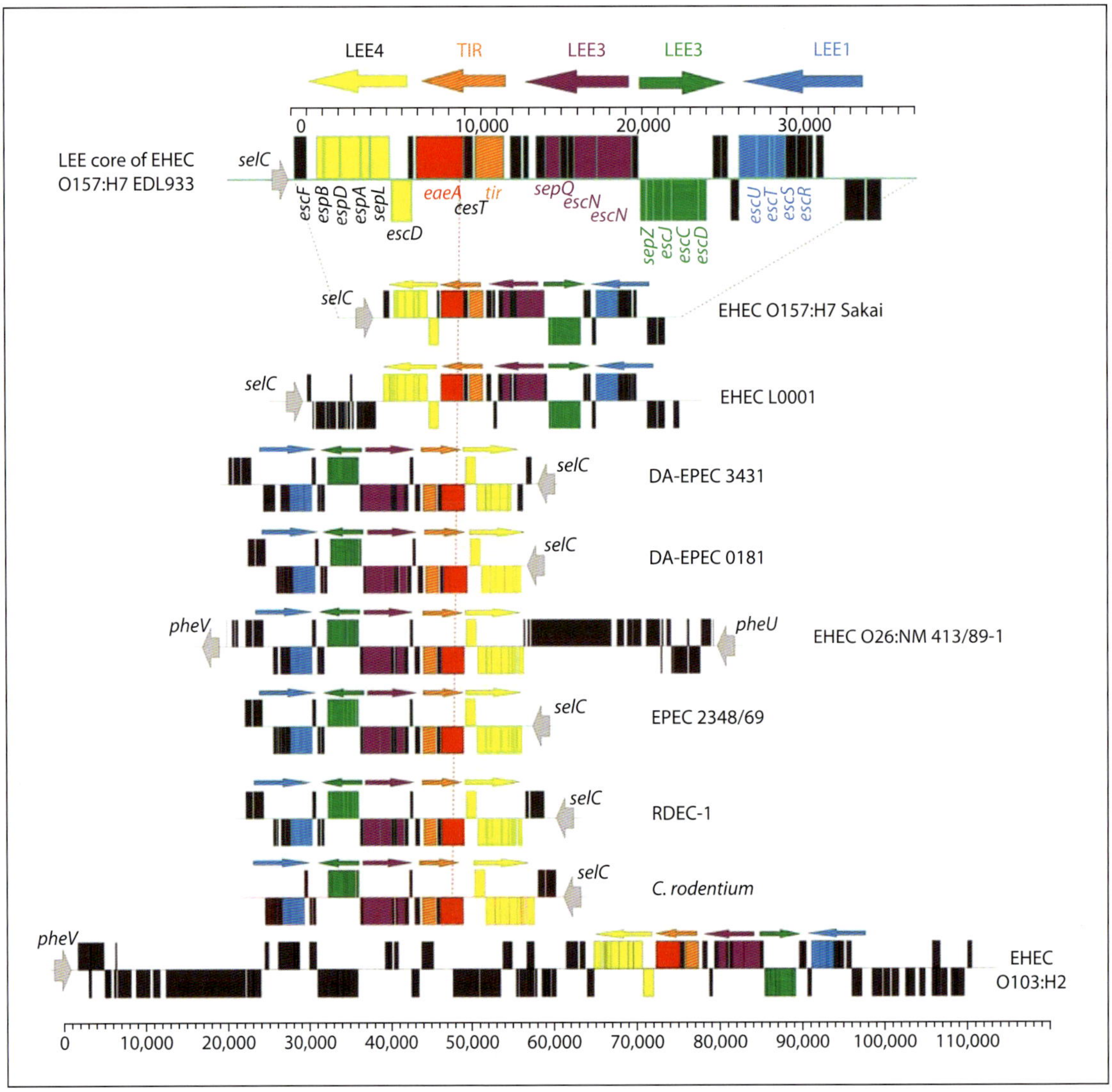

Fig. 1. Comparison of the genetic organization of the locus of enterocyte effacement (LEE) and its flanking chromosomal regions in intestinal pathogenic *E. coli* and *Citrobacter rodentium*. Identical regions of individual islands are highlighted by the same color and the orientation of the corresponding transcriptional units is indicated by an arrow above. tRNA genes in the vicinity of the LEE island serving as chromosomal insertion site of the LEE island are shown as light grey arrows. Additional ORFs within individual LEE islands which are not conserved are marked in dark grey. LEE, locus of enterocyte effacement; EHEC, enterohemorrhagic *E. coli*; EPEC, enteropathogenic *E. coli*; DA-EPEC, diffuse-adhering enteropathogenic *E. coli*; RDEC, rabbit diarrheagenic *E. coli*.

genome hybridization suggest that the core genes of non-O157 EHEC strains, which include seven LEE-encoded effector genes, also have significantly diverged nucleotide sequences [17].

Comparative genomics indicates that these LEE-PAIs contain a conserved 34 kb large core region. However, there are a number of alleles and size differences of individual LEE-containing PAIs as the LEE-core region flanking sequences are very different [30, 31, 35–37]. Furthermore, these LEE-PAIs are chromosomally inserted in different chromosomal tRNA loci (fig. 1). The LEE of O157:H7 strain EDL933 is 43,359 bp in size. The core region contains 41 ORFs, which are 93.9% identical relative to those of EPEC strain 348/69 [36]. The size difference between these LEE variants originates from the presence/absence of a 7.5 kb long 933L prophage. The ends of these two LEE islands have a weak similarity to elements of the IS*600* family and contain a small ORF with similarity to a putative transposase [38]. This indicates that the LEE has been transferred by mobilizing elements and that this mechanism has been inactivated in the course of its evolution.

In EHEC strain EDL933 and EPEC strain E2348/69, the LEE island is inserted in the tRNA gene *selC* [11, 39]. The LEE of bovine EHEC strain 413/89–1, however, has a size of 59.4 kb and is composed of the LEE island as found in EPEC strain 2348/69 and of an O-island 122-homologue of EHEC strain EDL933. This mosaic island is located in the *pheU*-tRNA locus [40]. The 34-kb LEE core region of strain RDEC-1 comprises only 40 ORFs, which are only 89.3% similar to those of the LEE in EPEC strain E2348/69. The 36-kb core region of the *Citrobacter rodentium* LEE contains 41 ORFs which are 98% identical to the LEE of E2348/69 and EDL933 [35]. In bovine EHEC isolate RW1374, the LEE core is located on a large mosaic-like 111.5-kb PAI at *pheV* [30]. The presence of IS elements and homologous 23-bp 3′-ends of *pheV* and *pheU* adjacent to the LEE suggests that this island has been inserted into an already existing PAI [32]. Differences in the genetic structure of the LEE core and its flanking regions do not only mirror different phylogenetic backgrounds and different histories of LEE acquisition, but they also affect the set of effector proteins translocated by the ETT-1 [41] which are often encoded in the LEE boundary regions. This variation in ETT-1 effector genes probably mirrors a distinct role in infection. Interestingly, a second type of type three secretion system (ETT-2) has been described in pathogenic and non-pathogenic *E. coli* [42, 43]. However, the role of ETT-2 in *E. coli* pathogenicity is still unclear. It has been recently demonstrated that a degenerate ETT-2 system from a colibacillosis isolate contributed to virulence in an experimental chicken infection model [44].

The D-Serine Utilization Determinant in ExPEC, IPEC and Commensal E. coli

Comparative genomics demonstrates that *E. coli* pathotypes reveal extensive genetic variability in the *argW-dsdCXA* island. The *dsdCXA* genes for D-serine utilization are usually intact in ExPEC strains but missing in diarrheagenic pathogens, in part due to a substitution with the sucrose utilization genes *cscRAKB*. Interestingly, many ExPEC strains, especially *E. coli* K1 strains that are able to cause newborn meningitis, have two copies of the *dsdCXA* genes for D-serine utilization at the *argW* and *leuX* islands. In addition, diarrheagenic *E. coli* exhibit a reciprocal pattern of sucrose fermentation versus D-serine utilization. Diarrheagenic *E. coli* do not efficiently colonize body sites outside of the mammalian intestine, which provides many sugars including sucrose. This may have been a driving force for the replacement of the *dsdCXA* genes by the *cscRAKB* determinant in these intestinal pathotypes. The ability of ExPEC to use D-serine has probably been selected during adaptation to their nutritional opportunities. ExPEC can colonize a wide range of extraintestinal niches which are, compared to the intestine, relatively carbohydrate-poor but peptide- and amino acid-rich environments [45]. D-serine is mostly found in the host brain but also in human urine, and can be toxic to certain *E. coli* strains. Consequently the ability to efficiently utilize D-serine has a positive effect on fitness of ExPEC that are able to cause meningitis or urinary tract infection.

The Interplay between Chromosomal and Episomal Elements (Plasmids, Phages, Islands): Comparison of Colicin Plasmids and Pathogenicity Islands of ExPEC

Many *E. coli* virulence-associated genes may be encoded on transmissible genetic elements such as bacteriophages, plasmids or transposons and thus play an important role in the spreading of such genes. As a consequence, individual DNA regions can be exchanged between the chromosome and mobile genetic elements with the capacity to integrate into and excise from the bacterial chromosome. Accordingly, several identical or closely related virulence determinants can be found on the chromosome or on mobile DNA elements. So-called colicin plasmids represent an interesting example of such mobile elements which in large parts exhibit considerable sequence similarity to PAIs in *E. coli* and contribute to PAI evolution and the spread of virulence traits among individual strains.

Colicins are plasmid-encoded toxic proteins produced by *E. coli* and some related species of Enterobacteriaceae. They inhibit growth of closely related bacterial strains and thus reduce the number of competitors in their growth niche. Until now, more than 30 types of colicins have been described [46]. Large colicin plasmids are found primarily in virulent, mainly septicaemic *E. coli* strains and they seem to be a characteristic marker for avian pathogenic *E. coli* (APEC), causing systemic infections in poultry.

The 174,240-bp ColBM plasmid of APEC strain O1 can be subdivided into an F-like transfer region and a virulence-related part [47]. The genetic structure of pAPEC-O1-ColBM highly resembles that of other large colicin and related plasmids and several PAIs of *E. coli* (fig. 2). The 32-kb F plasmid-like transfer region of pAPEC-O1-ColBM is similar to that of pAPEC-O2-ColV, the F plasmid, and several F-like *E. coli* plasmids. pAPEC-O1-ColBM is a mosaic plasmid containing replicons and other genes typical to both IncI1 and IncFII groups [47].

The large virulence-related region of pAPEC-O1-ColBM comprises several genes that have been previously associated with APEC virulence. These genes include (i) the *colBM* operon, encoding the colicins B and M, (ii) the *iss* gene (increased serum survival) involved in complement resistance, (iii) the outer membrane protease-encoding *ompT* gene, (iv) *tsh*, a temperature-sensitive hemagglutinin-encoding

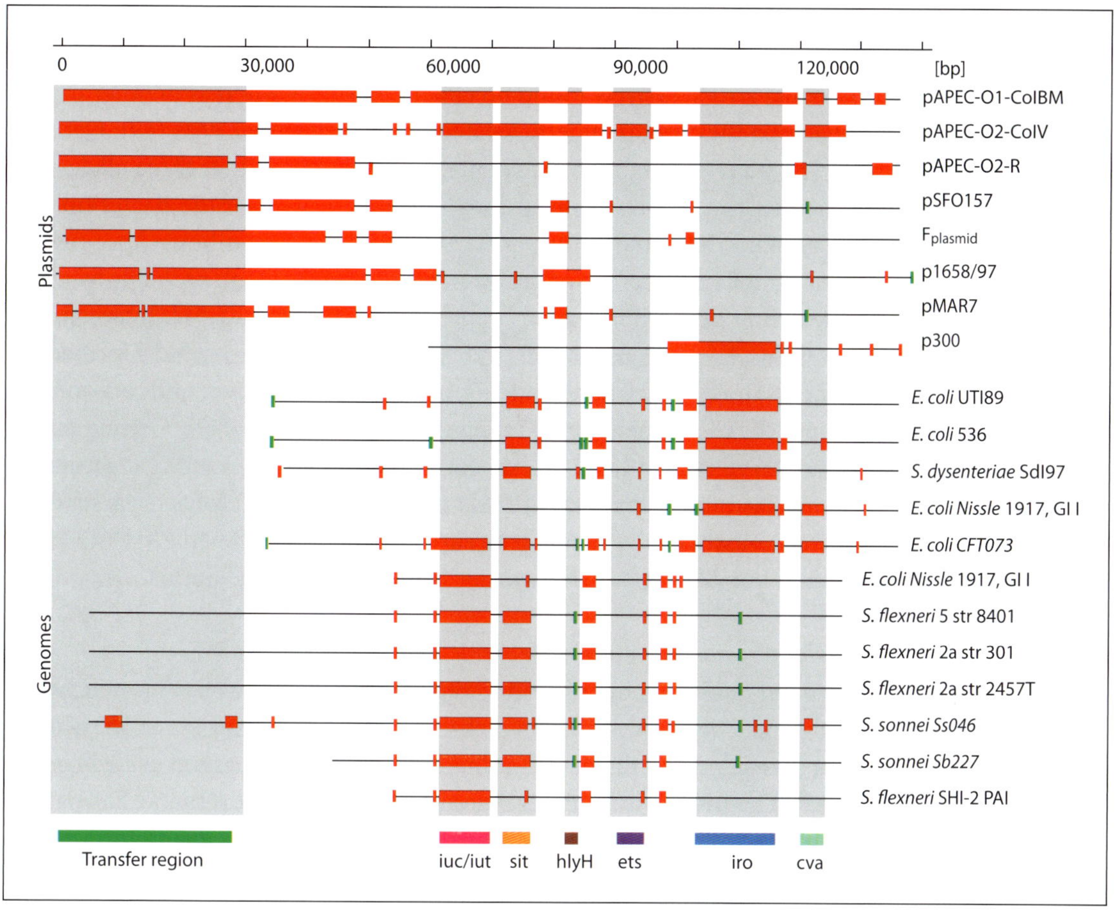

Fig. 2. Comparison of the genetic organization of colicin plasmids of extraintestinal pathogenic *E. coli* and other mobile genetic elements and genomic islands of *E. coli* and *Shigella* spp. Homologous regions of individual plasmids/islands are highlighted by red color. Functionally related DNA regions or gene clusters (plasmid transfer region; aerobactin siderophore determinant *(iuc)*; *Salmonella* iron transport siderophore determinant *(sit)*; putative hemolysin determinant *(hlyF)*; putative ABC transporter determinant *(ets)*; salmochelin siderophore determinant *(iro)*; microcin determinant *(cva)*) are indicated by different colors and their localization within the plasmids or genomic islands is also highlighted by grey areas.

gene, and (v) *hlyF* coding for a putative hemolysin. It also contains several operons associated with iron acquisition including the aerobactin system *(iuc/iut)*, the *iro* determinant coding for salmochelin, the *sit* operon, coding for an ABC transport system involved in iron and manganese transport and the *eitA-D* genes that code for a putative iron transport system [47]. Other genes identified as occurring in APEC were also found within this contiguous sequence, including the *etsA* and *etsB* genes of a putative ABC transport system or the *shiF* gene previously found on a PAI of *Shigella flexneri* [48].

Operons coding for the siderophore systems *sit*, *iut* and *iro* as well as the *iss* gene can be found on the bacterial chromosome as well as on the colicin plasmids. In APEC, these determinants are exclusively found on colicin plasmids whereas in other pathogenic enterobacteria they are frequently located on chromosomal PAIs [49]. Detailed analysis of the *iss* gene and its sequence context demonstrated that three alleles can be distinguished that may have evolved from the Bor protein of the bacteriophage lambda. Both proteins, Iss and Bor, are surface-exposed outer membrane lipoproteins and protect against the killing effect of the host complement system, probably by interfering with the action of the C5b-9 membrane attack complex [50, 51]. Interestingly, two *iss* types (alleles 2 and 3) are usually widespread and chromosomally located on prophage elements in ExPEC, whereas allele 1 has been exclusively found on conjugative plasmids of APEC and newborn meningitis *E. coli* isolates [49]. Consequently, the *iss* gene may serve as a suitable marker for diagnostics. The structural similarity between colicin plasmids and different PAIs of pathogenic enterobacteria suggests that these virulence-associated genes can be easily exchanged between PAIs and (colicin) plasmids and thus supports their transfer from one strain to another.

The mutS-rpoS Intergenic Region in Pathogenic and Non-Pathogenic E. coli

Although *mutS* and *rpoS* are generally conserved in Enterobacteriaceae, the *mutS-rpoS* intergenic region has been identified as a chromosomal region of extensive genetic variability that was subjected to genetic exchange during the evolution of pathogenic lineages [52, 53]. The intergenic region ranges in size from 40 kb in case of the pathogenicity island (SPI-1) [54] in *Salmonella enterica* and 12.6 kb in *S. typhimurium* LT2 [55] to 88 bp in *Yersinia pestis* (fig. 3).

Methyl-directed DNA mismatch repair (MMR) is important for maintenance of high DNA fidelity upon replication and recombination to ensure microbial fitness. However, genome plasticity due to increased mutation frequencies is also crucial for adaptation, pathogenicity and strain diversification [56]. The MMR system plays a key role in maintaining bacterial genomic stability. This system recognizes DNA mismatches and insertion/deletion nucleotide loops that result from DNA-polymerase errors during replication. In *E. coli* MMR, mismatch recognition involves the MutS protein [57]. MutS-dependent repair corrects not only mismatches in DNA, but also plays a role in maintaining fidelity of homologous recombination [58]. MutS mutants exhibit an increased mutation frequency and increased horizontal exchange of DNA [59]. The general stress response controlled by the sigma factor RpoS also protects bacteria under adverse growth conditions. RpoS is the sigma factor that regulates many stationary-phase and environmental stress response genes in *E. coli* [60].

A nearly identical 3-kb segment of DNA between the *mutS* and *rpoS* genes is found in *E. coli* serotype O157:H7 and other EHEC, *Shigella dysenteriae* type 1 and *S. flexneri* 2a strains, but it is absent in *E. coli* K-12 and many ExPEC in which a 6.9-kb DNA region exists (fig. 3). Further genetic polymorphisms in this region within different *E. coli* pathotypes could be of diagnostic interest: Many ExPEC lack at this

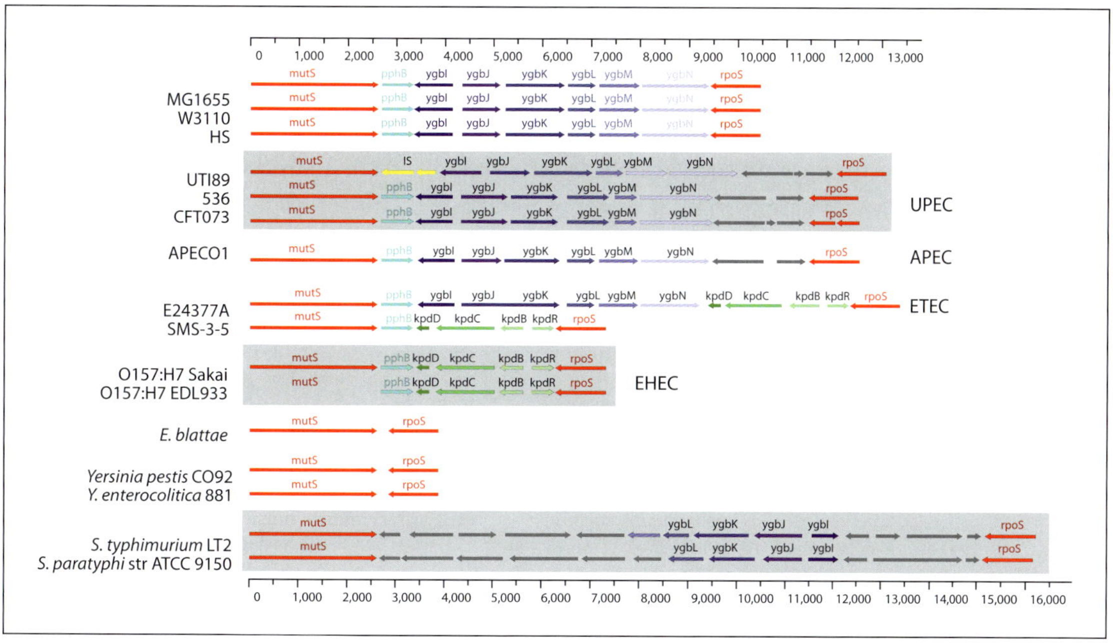

Fig. 3. Comparison of the genetic organization of the *mutS-rpoS* intergenic region in publicly available genome sequences of different Enterobacteriaceae. Identical regions are indicated by the same color. IS element-like DNA regions are highlighted in yellow. The phosphoprotein phosphatase gene *pphB* (turquoise), the 4-hydroxybenzoate decarboxylase determinant *kpd* (green) and additional putative ORFs (grey) as well as their orientations are indicated. (*E. blattae* genome sequence: Göttingen Genomics Laboratory, unpublished).

chromosomal position a 2.9-kb DNA stretch which is characteristic of EHEC strains. Instead, they harbor a 2.1-kb insertion of unknown origin. This insertion is shared by all members of the major *E. coli* phylogenetic lineage ECOR (*E. coli* collection of reference strains) group B2 [61], and larger intergenic regions exist in EPEC and EHEC strains [62]. Additionally, phylogenetic analysis supports the idea that the *mutS-rpoS* region is a recombination hot spot of the *E. coli* chromosome [63, 64] (fig. 3).

The polymorphisms in the *mutS-rpoS* intergenic region are considered to result from the close linkage of *mutS* and *rpoS*. These two genes are frequently mutated in *E. coli* evolution due to ecological specialization upon repeated shuttles between different environments, in which their inactivation as well as the re-acquisition of functional alleles has been of selective advantage (e.g. stress resistance, higher mutation rates and genome plasticity, stabilization of beneficial adaptive mutations) [65]. Horizontal gene transfer and multiple events of acquisition and loss of DNA segments from diverse sources played a crucial role in shaping the *mutS-rpoS* region. The genetic variability of this chromosomal region demonstrates the constantly changing demands of enterobacterial environments and the different selective pressures that operate for different genes.

Genome Plasticity and its Impact on Disease Severity

To adapt to the host immune defenses, pathogenic *E. coli* must possess mechanisms for rapid genome variation and diversification. In addition to genetic mechanisms involved in the genomic variability, DNA-repair mechanisms play an important role in genome dynamics. The severity of illness in *E. coli* serotype O157 outbreaks may vary considerably and this has been suggested to be associated with genome plasticity and differences in virulence gene expression [66]. Differences between O157 strains were so far considered to usually result from discrete insertions or deletions, rather than from single nucleotide polymorphisms (SNPs) [67]. Nevertheless, 500 EHEC O157 clinical isolates have been recently genotyped on the basis of 96 SNPs to analyze changes in the genome content in general and specific differences of individual O157 lineages with regard to clinical presentation and disease severity [68]. A particular O157:H7 clade (clade 8) was shown to be associated significantly more often with hemolytic uremic syndrome than other O157:H7 lineages. Furthermore, infection with such strains increased in frequency over the past five years. Comparative genome analysis of a clade 8 strain and the prototypic O157:H7 strains EDL933 and Sakai showed that the genomes of the latter two strains which belong to clade 3 and 1, respectively, are more similar to each other in gene content and nucleotide sequence identity than to the clade 8 strain. This suggests that an emergent subpopulation of the clade 8 lineage had time to change its genetic composition and to acquire traits that contribute to more severe disease relative to strains from other lineages.

Another study aimed at the identification of SNPs in *tir* and *eae*, coding for the translocated intimin receptor and intimin, respectively, in *E. coli* serotype O157 isolates which may correlate with human disease or carriage in cattle. Only *tir* polymorphisms could be correlated with the ability of O157 isolates to cause human disease. The distribution of different *tir* alleles in human patients or healthy cattle suggested that the *tir* allele harboring a T instead of an A at position 255 seems to be associated with disease in humans [69].

EHEC: Loss of the Shiga Toxin-Encoding Bacteriophage during Infection

The spread of (virulence-associated) genes by lysogenic phages is a general phenomenon in Gram-negative and -positive bacteria [70]. The different types of Shiga toxins (Stx), the major virulence factor of EHEC strains, are usually encoded on temperate bacteriophages [71, 72]. In addition to the *stx* determinants, several other putative virulence-associated genes are located on prophages [11, 73]. The Shiga toxin encoding genes *(stx)* are located on temperate lambdoid bacteriophages that are integrated in the host genome during lysogenic growth. The existence of *stx* genes in many different *E. coli* serotypes is attributable to transduction with *stx*-converting phages [71, 72]. Loss and transfer of the *stx* gene appear to occur during human infection and can lead to a change in the pathotype of the infecting strain [40, 74, 75]. Comparison of *stx* gene losses in sorbitol fermenting (SF) EHEC O157:NM and non-SF EHEC O157:H7 isolates showed a significantly higher proportion of *stx*-negative strains among SF *E. coli* serotype O157:NM [74]. The loss of *stx* genes has important diagnostic implications as *stx* detection is routinely used to screen for EHEC and thus *stx*-negative variants (which are still able to cause human diarrhoea and outbreaks) are not detected [75]. Furthermore, this may influence the outcome of the disease [74]. In SF *E. coli* serotype O157:NM, *yecE* is a hot spot for excision and integration of Shiga toxin 2-encoding bacteriophages. Consequently, SF EHEC O157:NM strains and their *stx*-negative derivatives can convert in both directions by the loss and gain of *stx2*-harboring phages [76, 77].

Asymptomatic Bacteriuria: Loss of Virulence Traits

Asymptomatic bacteriuria (ABU) is probably the most common form of urinary tract infection (UTI) and is frequently caused by *E. coli*. In ABU patients, *E. coli* establishes a carrier state, with more than 10^5 bacteria/ml of urine, but the patients do not develop symptoms [78]. Many ABU isolates belong to ECOR group B2, indicating a close relatedness to UPEC strains that cause symptomatic UTI. These ABU isolates do not express many classical UPEC virulence factors, but according to genotypic analysis they possess a large number of virulence-associated genes [79]. A recent genotypic and phenotypic analysis of selected pathogenicity factors of strain 83972 suggested that the loss of functional type 1-, F1C- and P fimbriae, as well as of α-hemolysin and long LPS O-side chain expression, was due to deletions or multiple point mutations, and it has been proposed that this might be essential for *E. coli* strain 83972 to cause ABU [78, 80].

The loss of virulence factors has been shown to reduce the host response to infection in animal models and specifically, the loss of fimbriae and long chain LPS expression decreases the innate host response and bacterial clearance from the urinary tract. P fimbriae enhance the establishment of bacteriuria and trigger the innate defense by stimulating the production of cytokines. Type 1 fimbriae have a similar function in mice and have also been shown to enhance intracellular persistence in the mouse bladder mucosa, but these effects have not been reproduced in the human urinary tract [79]. The weak host response to ABU is therefore consistent with the loss of adherence and functional fimbriae. These results thus suggest that the host response may drive co-evolution, and that virulence-associated genes with pro-inflammatory effects may be targeted for inactivation. In this way, ABU isolates may succeed in persisting without inducing a bactericidal inflammatory response.

Conclusions

The balance between sources of genetic variation, DNA repair and selective pressures defines the genetic diversity and fitness of an *E. coli* population. The *E. coli* genome is plastic and responsive to environmental changes. A variety of environmental stresses induce genomic alterations in bacteria, thus leading to the generation and selection of fitter mutants, and potentially accelerating adaptive evolution. Host-pathogen interactions are often driven by mechanisms, which involve genetic diversification, e.g. antigenic components of pathogenic *E. coli* are constantly under selective pressure. Thus, the high degree of inter- and intra-strain variability is not surprising. Many *E. coli* pathogens have evolved mechanisms to produce high mutation rates in specific regions of their genomes resulting in the rapid generation of variants, some of which will predominate during changing selective conditions. The analysis of genome plasticity can teach us a lot of pathogen evolution, adaptation and transmission dynamics of *E. coli*. Genomic research has already improved our understanding of microbial pathogenesis, but as this work also impacts on the development of accurate diagnostics, molecular epidemiological methods and the development of timely therapeutic interventions against *E. coli* infections, additional efforts are required in the future to further complete our picture of *E. coli* genome plasticity.

Acknowledgements

The work in Würzburg related to this topic was supported by the German Research Foundation (Sonderforschungsbereich 479). The work in Göttingen was supported by the Ministry of Science and Culture of the Lower Saxony (Niedersächsisches Ministerium für Wissenschaft und Kultur). This work was carried out in the frame of the European Virtual Institute for Functional Genomics of Bacterial Pathogens (CEE LSHB-CT-2005-512061) and the ERA-NET Pathogenomics project 'Deciphering the intersection of commensal and extraintestinal pathogenic *E. coli*' (Grant no. 0313937A).

References

1 Berg RD: The indigenous gastrointestinal microflora. Trends Microbiol 1996;4:430–435.

2 Kaper JB, Nataro JP, Mobley HL: Pathogenic *Escherichia coli*. Nat Rev Microbiol 2004;2:123–140.

3 Lawrence JG: Gene transfer, speciation, and the evolution of bacterial genomes. Curr Opin Microbiol 1999;2:519–523.

4 Ochman H, Lawrence JG, Groisman EA: Lateral gene transfer and the nature of bacterial innovation. Nature 2000;405:299–304.

5 Blattner FR, Plunkett G 3rd, Bloch CA, Perna NT, Burland V, et al: The complete genome sequence of *Escherichia coli* K-12. Science 1997;277:1453–1474.

6 Brzuszkiewicz E, Brüggemann H, Liesegang H, Emmerth M, Ölschläger T, et al: How to become a uropathogen: comparative genomic analysis of extraintestinal pathogenic *Escherichia coli* strains. Proc Natl Acad Sci USA 2006;103:12879–12884.

7 Chen SL, Hung CS, Xu J, Reigstad CS, Magrini V, et al: Identification of genes subject to positive selection in uropathogenic strains of *Escherichia coli*: a comparative genomics approach. Proc Natl Acad Sci USA 2006;103:5977–5982.

8 Durfee T, Nelson R, Baldwin S, Plunkett G 3rd, Burland V, et al: The complete genome sequence of *Escherichia coli* DH10B: insights into the biology of a laboratory workhorse. J Bacteriol 2008;190:2597–2606.
9 Hayashi K, Morooka N, Yamamoto Y, Fujita K, Isono K, et al: Highly accurate genome sequences of *Escherichia coli* K-12 strains MG1655 and W3110. Mol Syst Biol 2006;2:2006.0007.
10 Johnson TJ, Kariyawasam S, Wannemuehler Y, Mangiamele P, Johnson SJ, et al: The genome sequence of avian pathogenic *Escherichia coli* strain O1:K1:H7 shares strong similarities with human extraintestinal pathogenic *E. coli* genomes. J Bacteriol 2007;189:3228–3236.
11 Perna NT, Plunkett G 3rd, Burland V, Mau B, Glasner JD, et al: Genome sequence of enterohaemorrhagic *Escherichia coli* O157:H7. Nature 2001;409:529–533.
12 Welch RA, Burland V, Plunkett G 3rd, Redford P, Roesch P, et al: Extensive mosaic structure revealed by the complete genome sequence of uropathogenic *Escherichia coli*. Proc Natl Acad Sci USA 2002;99:17020–17024.
13 Dobrindt U: (Patho-)Genomics of *Escherichia coli*. Int J Med Microbiol 2005;295:357–371.
14 Bergthorsson U, Ochman H: Distribution of chromosome length variation in natural isolates of *Escherichia coli*. Mol Biol Evol 1998;15:6–16.
15 Dobrindt U, Hochhut B, Hentschel U, Hacker J: Genomic islands in pathogenic and environmental microorganisms. Nat Rev Microbiol 2004;2:414–424.
16 Gal-Mor O, Finlay BB: Pathogenicity islands: a molecular toolbox for bacterial virulence. Cell Microbiol 2006;8:1707–1719.
17 Ogura Y, Ooka T, Asadulghani, Terajima J, Nougayrede JP, et al: Extensive genomic diversity and selective conservation of virulence-determinants in enterohemorrhagic *Escherichia coli* strains of O157 and non-O157 serotypes. Genome Biol 2007;8:R138.
18 Ohnishi M, Terajima J, Kurokawa K, Nakayama K, Murata T, et al: Genomic diversity of enterohemorrhagic *Escherichia coli* O157 revealed by whole genome PCR scanning. Proc Natl Acad Sci USA 2002;99:17043–17048.
19 Tobe T, Beatson SA, Taniguchi H, Abe H, Bailey CM, et al: An extensive repertoire of type III secretion effectors in *Escherichia coli* O157 and the role of lambdoid phages in their dissemination. Proc Natl Acad Sci USA 2006;103:14941–14946.
20 Zhang Y, Laing C, Steele M, Ziebell K, Johnson R, et al: Genome evolution in major *Escherichia coli* O157:H7 lineages. BMC Genomics 2007;8:121.
21 Moulin-Schouleur M, Reperant M, Laurent S, Bree A, Mignon-Grasteau S, et al: Extraintestinal pathogenic *Escherichia coli* strains of avian and human origin: link between phylogenetic relationships and common virulence patterns. J Clin Microbiol 2007;45:3366–3376.
22 Grozdanov L, Raasch C, Schulze J, Sonnenborn U, Gottschalk G, et al: Analysis of the genome structure of the nonpathogenic probiotic *Escherichia coli* strain Nissle 1917. J Bacteriol 2004;186:5432–5441.
23 Hejnova J, Dobrindt U, Nemcova R, Rusniok C, Bomba A, et al: Characterization of the flexible genome complement of the commensal *Escherichia coli* strain A0 34/86 (O83:K24:H31). Microbiology 2005;151:385–398.
24 Janka A, Bielaszewska M, Dobrindt U, Greune L, Schmidt MA, Karch H: Cytolethal distending toxin gene cluster in enterohemorrhagic *Escherichia coli* O157:H- and O157:H7: characterization and evolutionary considerations. Infect Immun 2003;71:3634–3638.
25 Rendon MA, Saldana Z, Erdem AL, Monteiro-Neto V, Vazquez A, et al: Commensal and pathogenic *Escherichia coli* use a common pilus adherence factor for epithelial cell colonization. Proc Natl Acad Sci USA 2007;104:10637–10642.
26 Schubert S, Rakin A, Karch H, Carniel E, Heesemann J: Prevalence of the 'high-pathogenicity island' of *Yersinia* species among *Escherichia coli* strains that are pathogenic to humans. Infect Immun 1998;66:480–485.
27 Dobrindt U, Blum-Oehler G, Nagy G, Schneider G, Johann A, et al: Genetic structure and distribution of four pathogenicity islands (PAI I(536) to PAI IV(536)) of uropathogenic *Escherichia coli* strain 536. Infect Immun 2002;70:6365–6372.
28 Guyer DM, Kao JS, Mobley HL: Genomic analysis of a pathogenicity island in uropathogenic *Escherichia coli* CFT073:distribution of homologous sequences among isolates from patients with pyelonephritis, cystitis, and Catheter-associated bacteriuria and from fecal samples. Infect Immun 1998;66:4411–4417.
29 Castillo A, Eguiarte LE, Souza V: A genomic population genetics analysis of the pathogenic enterocyte effacement island in *Escherichia coli*: the search for the unit of selection. Proc Natl Acad Sci USA 2005;102:1542–1547.
30 Jores J, Rumer L, Kiessling S, Kaper JB, Wieler LH: A novel locus of enterocyte effacement (LEE) pathogenicity island inserted at *pheV* in bovine Shiga toxin-producing *Escherichia coli* strain O103:H2. FEMS Microbiol Lett 2001;204:75–79.

31 Jores J, Rumer L, Wieler LH: Impact of the locus of enterocyte effacement pathogenicity island on the evolution of pathogenic *Escherichia coli*. Int J Med Microbiol 2004;294:103–113.

32 Rumer L, Jores J, Kirsch P, Cavignac Y, Zehmke K, Wieler LH: Dissemination of *pheU*- and *pheV*-located genomic islands among enteropathogenic (EPEC) and enterohemorrhagic (EHEC) *E. coli* and their possible role in the horizontal transfer of the locus of enterocyte effacement (LEE). Int J Med Microbiol 2003;292:463–475.

33 Gophna U, Ron EZ, Graur D: Bacterial type III secretion systems are ancient and evolved by multiple horizontal-transfer events. Gene 2003;312:151–163.

34 Lacher DW, Steinsland H, Blank TE, Donnenberg MS, Whittam TS: Molecular evolution of typical enteropathogenic *Escherichia coli*: clonal analysis by multilocus sequence typing and virulence gene allelic profiling. J Bacteriol 2007;189:342–350.

35 Deng W, Li Y, Vallance BA, Finlay BB: Locus of enterocyte effacement from *Citrobacter rodentium*: sequence analysis and evidence for horizontal transfer among attaching and effacing pathogens. Infect Immun 2001;69:6323–6335.

36 Perna NT, Mayhew GF, Posfai G, Elliott S, Donnenberg MS, et al: Molecular evolution of a pathogenicity island from enterohemorrhagic *Escherichia coli* O157:H7. Infect Immun 1998;66:3810–3817.

37 Zhu C, Agin TS, Elliott SJ, Johnson LA, Thate TE, et al: Complete nucleotide sequence and analysis of the locus of enterocyte effacement from rabbit diarrheagenic *Escherichia coli* RDEC-1. Infect Immun 2001;69:2107–2115.

38 Donnenberg MS, Lai LC, Taylor KA: The locus of enterocyte effacement pathogenicity island of enteropathogenic *Escherichia coli* encodes secretion functions and remnants of transposons at its extreme right end. Gene 1997;184:107–114.

39 Elliott SJ, Wainwright LA, McDaniel TK, Jarvis KG, Deng YK, et al: The complete sequence of the locus of enterocyte effacement (LEE) from enteropathogenic *Escherichia coli* E2348/69. Mol Microbiol 1998;28:1–4.

40 Bielaszewska M, Sonntag AK, Schmidt MA, Karch H: Presence of virulence and fitness gene modules of enterohemorrhagic *Escherichia coli* in atypical enteropathogenic *Escherichia coli* O26. Microbes Infect 2007;9:891–897.

41 Gärtner JF, Schmidt MA: Comparative analysis of locus of enterocyte effacement pathogenicity islands of atypical enteropathogenic *Escherichia coli*. Infect Immun 2004;72:6722–6728.

42 Makino S, Tobe T, Asakura H, Watarai M, Ikeda T, et al: Distribution of the secondary type III secretion system locus found in enterohemorrhagic *Escherichia coli* O157:H7 isolates among Shiga toxin-producing *E. coli* strains. J Clin Microbiol 2003;41:2341–2347.

43 Ren CP, Chaudhuri RR, Fivian A, Bailey CM, Antonio M, et al: The ETT2 gene cluster, encoding a second type III secretion system from *Escherichia coli*, is present in the majority of strains but has undergone widespread mutational attrition. J Bacteriol 2004;186:3547–3560.

44 Ideses D, Gophna U, Paitan Y, Chaudhuri RR, Pallen MJ, Ron EZ: A degenerate type III secretion system from septicemic *Escherichia coli* contributes to pathogenesis. J Bacteriol 2005;187:8164–8171.

45 Moritz RL, Welch RA: The *Escherichia coli argW-dsdCXA* genetic island is highly variable, and *E. coli* K1 strains commonly possess two copies of *dsdCXA*. J Clin Microbiol 2006;44:4038–4048.

46 Cascales E, Buchanan SK, Duche D, Kleanthous C, Lloubes R, et al: Colicin biology. Microbiol Mol Biol Rev 2007;71:158–229.

47 Johnson TJ, Johnson SJ, Nolan LK: Complete DNA sequence of a ColBM plasmid from avian pathogenic *Escherichia coli* suggests that it evolved from closely related ColV virulence plasmids. J Bacteriol 2006;188:5975–5983.

48 Johnson TJ, Siek KE, Johnson SJ, Nolan LK: DNA sequence of a ColV plasmid and prevalence of selected plasmid-encoded virulence genes among avian *Escherichia coli* strains. J Bacteriol 2006;188: 745–758.

49 Johnson TJ, Wannemuehler YM, Nolan LK: Evolution of the *iss* gene in *Escherichia coli*. Appl Environ Microbiol 2008;74:2360–2369.

50 Barondess JJ, Beckwith J: *bor* gene of phage lambda, involved in serum resistance, encodes a widely conserved outer membrane lipoprotein. J Bacteriol 1995;177:1247–1253.

51 Binns MM, Mayden J, Levine RP: Further characterization of complement resistance conferred on *Escherichia coli* by the plasmid genes *traT* of R100 and *iss* of ColV,I-K94. Infect Immun 1982;35:654–659.

52 LeClerc JE, Li B, Payne WL, Cebula TA: High mutation frequencies among *Escherichia coli* and *Salmonella* pathogens. Science 1996;274:1208–1211.

53 LeClerc JE, Li B, Payne WL, Cebula TA: Promiscuous origin of a chimeric sequence in the *Escherichia coli* O157:H7 genome. J Bacteriol 1999;181:7614–7617.

54 Mills DM, Bajaj V, Lee CA: A 40 kb chromosomal fragment encoding *Salmonella typhimurium* invasion genes is absent from the corresponding region of the *Escherichia coli* K-12 chromosome. Mol Microbiol 1995;15:749–759.

55 Kotewicz ML, Li B, Levy DD, LeClerc JE, Shifflet AW, Cebula TA: Evolution of multi-gene segments in the *mutS-rpoS* intergenic region of *Salmonella enterica* serovar Typhimurium LT2. Microbiology 2002;148:2531–2540.

56 Tønjum T, Seeberg E: Microbial fitness and genome dynamics. Trends Microbiol 2001;9:356–358.
57 Horst JP, Wu TH, Marinus MG: *Escherichia coli* mutator genes. Trends Microbiol 1999;7:29–36.
58 Vulic M, Lenski RE, Radman M: Mutation, recombination, and incipient speciation of bacteria in the laboratory. Proc Natl Acad Sci USA 1999;96:7348–7351.
59 Radman M, Matic I, Taddei F: Evolution of evolvability. Ann N Y Acad Sci 1999;870:146–155.
60 Klauck E, Typas A, Hengge R: The sigmaS subunit of RNA polymerase as a signal integrator and network master regulator in the general stress response in *Escherichia coli*. Sci Prog 2007;90:103–127.
61 Culham DE, Wood JM: An *Escherichia coli* reference collection group B2- and uropathogen-associated polymorphism in the *rpoS-mutS* region of the *E. coli* chromosome. J Bacteriol 2000;182:6272–6276.
62 Herbelin CJ, Chirillo SC, Melnick KA, Whittam TS: Gene conservation and loss in the *mutS-rpoS* genomic region of pathogenic *Escherichia coli*. J Bacteriol 2000;182:5381–5390.
63 Brown J, Brown T, Fox KR: Affinity of mismatch-binding protein MutS for heteroduplexes containing different mismatches. Biochem J 2001;54:627–633.
64 Denamur E, Lecointre G, Darlu P, Tenaillon O, Acquaviva C, et al: Evolutionary implications of the frequent horizontal transfer of mismatch repair genes. Cell 2000;103:711–721.
65 Ferenci T: What is driving the acquisition of *mutS* and *rpoS* polymorphisms in *Escherichia coli*? Trends Microbiol 2003;11:457–461.
66 Jelacic JK, Damrow T, Chen GS, Jelacic S, Bielaszewska M, et al: Shiga toxin-producing *Escherichia coli* in Montana: bacterial genotypes and clinical profiles. J Infect Dis 2003;188:719–729.
67 Kudva IT, Evans PS, Perna NT, Barrett TJ, Ausubel FM, et al: Strains of *Escherichia coli* O157:H7 differ primarily by insertions or deletions, not single-nucleotide polymorphisms. J Bacteriol 2002;184:1873–1879.
68 Manning SD, Motiwala AS, Springman AC, Qi W, Lacher DW, et al: Variation in virulence among clades of *Escherichia coli* O157:H7 associated with disease outbreaks. Proc Natl Acad Sci USA 2008;105:4868–4873.
69 Bono JL, Keen JE, Clawson ML, Durso LM, Heaton MP, Laegreid WW: Association of *Escherichia coli* O157:H7 *tir* polymorphisms with human infection. BMC Infect Dis 2007;7:98.
70 Boyd EF, Brüssow H: Common themes among bacteriophage-encoded virulence factors and diversity among the bacteriophages involved. Trends Microbiol 2002;10:521–529.
71 Allison HE: Stx-phages: drivers and mediators of the evolution of STEC and STEC-like pathogens. Future Microbiol 2007;2:165–174.
72 Herold S, Karch H, Schmidt H: Shiga toxin-encoding bacteriophages – genomes in motion. Int J Med Microbiol 2004;294:115–121.
73 Hayashi T, Makino K, Ohnishi M, Kurokawa K, Ishii K, et al: Complete genome sequence of enterohemorrhagic *Escherichia coli* O157:H7 and genomic comparison with a laboratory strain K-12. DNA Res 2001;8:11–22.
74 Friedrich AW, Zhang W, Bielaszewska M, Mellmann A, Köck R, et al: Prevalence, virulence profiles, and clinical significance of Shiga toxin-negative variants of enterohemorrhagic *Escherichia coli* O157 infection in humans. Clin Infect Dis 2007;45:39–45.
75 Mellmann A, Bielaszewska M, Zimmerhackl LB, Prager R, Harmsen D, et al: Enterohemorrhagic *Escherichia coli* in human infection: in vivo evolution of a bacterial pathogen. Clin Infect Dis 2005;41:785–792.
76 Bielaszewska M, Middendorf B, Köck R, Friedrich AW, Fruth A, et al: Shiga toxin-negative attaching and effacing *Escherichia coli*: distinct clinical associations with bacterial phylogeny and virulence traits and inferred in-host pathogen evolution. Clin Infect Dis 2008;47:208–217.
77 Mellmann A, Lu S, Karch H, Xu JG, Harmsen D, et al: Recycling of Shiga toxin 2 genes in sorbitol-fermenting enterohemorrhagic *Escherichia coli* O157:NM. Appl Environ Microbiol 2008;74:67–72.
78 Zdziarski J, Svanborg C, Wullt B, Hacker J, Dobrindt U: Molecular basis of commensalism in the urinary tract: low virulence or virulence attenuation? Infect Immun 2008;76:695–703.
79 Dobrindt U, Agerer F, Michaelis K, Janka A, Buchrieser C, et al: Analysis of genome plasticity in pathogenic and commensal *Escherichia coli* isolates by use of DNA arrays. J Bacteriol 2003;185:1831–1840.
80 Klemm P, Roos V, Ulett GC, Svanborg C, Schembri MA: Molecular characterization of the *Escherichia coli* asymptomatic bacteriuria strain 83972: the taming of a pathogen. Infect Immun 2006;74:781–785.

Ulrich Dobrindt
Institut für Molekulare Infektionsbiologie
Röntgenring 11
DE–97070 Würzburg (Germany)
Tel. +49 931 312155, Fax +49 931 312578, E-Mail ulrich.dobrindt@mail.uni-wuerzburg.de

de Reuse H, Bereswill S (eds): Microbial Pathogenomics.
Genome Dyn. Basel, Karger, 2009, vol 6, pp 126–139

Role of Horizontal Gene Transfer in the Evolution of *Pseudomonas aeruginosa* Virulence

X. Qiu · B.R. Kulasekara · S. Lory

Department of Microbiology and Molecular Genetics, Harvard Medical School, Boston, Mass., USA

Abstract

The opportunistic pathogen *Pseudomonas aeruginosa* causes serious infections in immunocompromised patients and individuals with cystic fibrosis (CF). It is one of the most versatile organisms as illustrated by its ability to occupy a wide range of environmental niches. Comparative genomic analysis suggests that horizontal gene transfer (HGT) plays a significant role in determining the genetic repertoire of each strain. Genomic diversity is, in part, due to the acquisition of genetic material that has integrated into the chromosome at a relatively limited number of sites. The resulting genomic islands (GIs) contain genes specifying virulence traits as well as genes that may enhance fitness in a specific environmental niche. Several islands are integrative and conjugative elements (ICEs) that may have evolved from ancestral self-transmissible conjugative plasmids. For some genomic islands, the mechanism of acquisition is not apparent suggesting that the mechanisms utlized are either transformation or bacteriophage-mediated generalized transduction. It appears that HGT takes place primarily in the natural environment of *P. aeruginosa* and, conceivably, an uncharacterized host-pathogen interaction provides the selective pressures for acquisition and maintenance of the observed virulence phenotypes.

As a common inhabitant of diverse environments, *Pseudomonas aeruginosa* has become a major human opportunistic pathogen. The serious nature of *P. aeruginosa* infection is complicated by the poor efficacy of many common antibiotics. Given the absence of an effective vaccine and the rise of multiresistant strains, it is almost certain that this organism will continue to pose a serious threat to human health. Following the release of the first *P. aeruginosa* genome sequence in 2000, research efforts have been initiated to use genome-wide approaches to understand the fundamental basis of the virulence of this organism. This area of research is particularly interesting given the number of genes encoding an impressive 'armament' of virulence factors and the corresponding large number of regulatory elements, many of which are dedicated to controlling virulence gene expression. Although the likelihood that *P. aeruginosa* will encounter and successfully infect a compromised human host is relatively low, it is conceivable that many of these systems function in the context of a pathogenic interaction involving hosts encountered by *P. aeruginosa* in its natural environment. This review will provide an overview of genome dynamics of *P. aeruginosa* with a focus on the role of horizontal gene transfer (HGT) in shaping the pan-genome into a customized repertoire of genes that characterize individual *P. aeruginosa* strains.

P. aeruginosa as a Human Pathogen

P. aeruginosa, a Gram-negative bacterium, is a common inhabitant of soil and aquatic environments. It is also an important opportunistic pathogen for humans as it is responsible for causing severe infections in immunocompromised patients and is the major factor for morbidity and mortality in cystic fibrosis (CF) patients [1]. As a major nosocomial pathogen, it can cause a range of infections in hospital settings including a bacteremia with a mortality rate of nearly 40% [2]. Patients undergoing immunosuppression following organ transplantation and those with various malignancies are also at high risk for serious *P. aeruginosa* infections [3]. Clinical strains display high levels of antibiotic resistance, thus restricting therapeutic options. Moreover, the rise of multiresistant and pan-resistant strains represents a major challenge for effective management of *P. aeruginosa* infections in all clinical backgrounds [4].

Comparative Genomics of *P. aeruginosa*

Examination of the overall genomic architecture and evolutionary dynamics of *P. aeruginosa* is particularly interesting because of the broad environmental distribution of this organism and its highly variable and substantial genetic repertoire. Indeed, the genome sequences of *P. aeruginosa* strains available to date (PAO1, PA14, PACS2, C3719, PA7 and PA2192) clearly show that a large core genome of ca. 5,000 conserved genes is supplemented with genes from the accessory gene pool consisting of 2,000 additional genes, which are organized in a limited number of genomic islands (GIs) [5–7]. There appears to be little conservation in the composition of the accessory gene pool between sets of isolates that could explain a particular host tropism or type of infection caused by this organism. The genome of each strain carries a relatively modest number of unique sequences as no pair of strains shares more than 100 genes from the accessory genome [7].

Genome Evolution and Virulence

Analysis of the *P. aeruginosa* genomes shows that the core genome contains a large number of genes encoding determinants for survival in a variety of environments. Moreover, the majority of genes encoding virulence factors are highly conserved among strains. This observation was made previously in a DNA microarray-based comparative genomics study that included analyses of strains of both environmental and clinical origin [8]. Therefore, it appears that virulence traits are selected for and are maintained even in the absence of interaction with the human hosts. Given the ability of *P. aeruginosa* to infect a variety of eukaryotic organisms, it is likely that the interactions with hosts located in the environment such as amoeba or insects have been driving the evolution of virulence characteristics that are utilized during encounters with the human host.

The accessory genome shows a pattern of organization that is consistent with the co-evolution of blocks of genes. The majority (80%) of these genes are found in contiguous segments of four genes or more and are located at a limited number of loci. These strain-specific segments represent regions of genomic plasticity (RGP) and they include any region that is missing in at least one of the genomes analyzed [7]. The RGPs can consist of common or unique GIs and bacteriophage genomes or are the result of deletions of particular DNA segments in one or more strains. Comparison of the five sequenced *P. aeruginosa* genomes (as of June, 2008) characterized a total of 52 RGPs while an individual genome contains anywhere from 27 to 37.

Examination of the annotated RGPs of *P. aeruginosa* strains from different clinical backgrounds revealed that there was no obvious association between any particular RGP and clinical origin. Strikingly, none of the RGPs were particularly enriched in genes encoding virulence factors. However, in addition to containing a large number of genes encoding proteins of unknown function, various specialized metabolic enzymes and proteins involved in survival under oxidative stress conditions were present [7, 9]. This is consistent with the notion that the main function of the accessory genome is to enable *P. aeruginosa* to survive in the widest range of environmental niches. This evolutionary pattern is in contrast to the evolutionary adaptation of symbionts and obligate parasites to specialized niches through genome reduction. In a free-living environmental organism, evolution may favor genomic versatility by the progressive incorporation of accessory genes including GIs into the core genome. *P. aeruginosa* is a classical case of the 'mix and match' pattern of genome assembly dependent upon the need for a specialized function within a particular environment. Although rarely directly transmitted between individuals, its ubiquitous environmental distribution makes it more likely that this organism will encounter immunocompromised individuals than pathogens with a more restricted range of environmental habitats.

HGT Contributes Significantly to the Genomic Diversity in *P. aeruginosa*

The flexible genome of *P. aeruginosa* is composed of blocks of DNA that carry many signatures of horizontally acquired genes and are incorporated into the chromosome at a restricted number of sites. These elements often display significant sequence divergence and are, as a result, the major contributors to interstrain variation. A number of horizontally acquired DNA segments (either bacteriophage or plasmid origin) are located at identical sites on the chromosome, presumably due to the specificities of the enzymes such as integrases that catalyze the recombination between the *att* site on the chromosome and the corresponding sequence on the acquired element. Multiple tandem elements can be sequentially added to the same site provided the chromosomal *att* site is not destroyed during the previous integration event. In *P. aeruginosa*, individual genomic islands associated with a specific chromosomal integration site are postulated to be the result of evolutionary decays of an ancestral element where various insertions, deletions and rearrangements gave rise to strain-specific DNA segments [10]. A number of DNA elements, integrated into the *att* sites associated with different transfer RNA (tRNA) genes have been characterized in *P. aeruginosa* and will be described further. Moreover, those horizontally acquired GIs in which the mechanisms of transfer and retention are not yet understood will be also discussed.

The *P. aeruginosa* Genomic Island 1 (PAGI-1)

Comparative DNA hybridization was used to identify the first genomic island in *P. aeruginosa* [9]. An M13 library of DNA from strain X24509, isolated from a patient with a urinary tract infection, was screened using a DNA probe made from the reference strain PAO1 to facilitate identification of clones containing X24509-specific DNA. The inserts of these clones were used to identify cosmids encompassing a contiguous 48.9-kb region (51 open reading frames) of the X24509 chromosome termed PAGI-1 (*P. aeruginosa* genomic island 1). Examination of the incidence of PAGI-1 revealed that portions of the entire island are present in 85% of the strains from clinical sources. PAGI-1 is a composite island, consisting of two portions, with approximately one half of the island carrying sequences with a GC content significantly lower than the rest of the chromosome. Several of the genes on PAGI-1 encode insertion sequences, regulatory proteins, dehydrogenase gene homologs, proteins implicated in detoxification of reactive oxygen species (ROS). These genes may be responsible for enhancing fitness of the recipients under the conditions that generate ROSs and therefore provide the selective advantage for PAGI-1 acquisition and maintenance. PAGI-1 lacks any recognizable sequences associated with conjugation or transposition and it is not integrated near a tRNA gene. Although it is very likely that PAGI-1 was acquired by a large number of *P. aeruginosa* isolates through HGT, the genetic mechanism (conjugation, transformation or transduction) is not apparent.

The Genomic Islands Related to the Plasmid pKLC102

The large 103-kb *P. aeruginosa* plasmid pKLC102 capable of reversibly integrating into tRNA^{Lys} genes has been studied extensively because of its similarity to several evolutionarily-related genomic and pathogenicity islands. This plasmid was initially isolated from strains belonging to *P. aeruginosa* clone C, which is widely distributed in Europe. In the clone C strain SG17M, pKLC102 is found at up to thirty copies per cell in stationary phase bacterial cultures [11]. When integrating into the chromosome, pKLC102 favors the tRNA^{Lys} gene PA4541.1 (designated as PA4541.1 based on its location in the PAO1 genome) over the tRNA^{Lys} gene PA0976.1. Some of the genomic islands that are related to the integrated form of pKLC102 have been associated with the virulence of *P. aeruginosa*, while others, such as the PAGI-4 also found in clone C strains, contain genes of bacteriophage or plasmid origin and genes of unknown function [10–13]. Two such islands, termed *P. aeruginosa* pathogenicity island-1 and -2 (PAPI-1 and PAPI-2), have been studied in detail [10, 13].

The pathogenicity island PAPI-1 is a conserved genomic island found in the majority of *P. aeruginosa* strains [13]. In strain PA14, where it has been studied most extensively, the island encodes several virulence determinants and regulatory factors that play a role in biofilm formation and antibiotic resistance [13, 14]. In most but not all *P. aeruginosa* strains that carry this island, it has integrated into the chromosome at PA4541.1. The overall organization of genes within PAPI-1 and pKLC102 is highly conserved including a significant fraction of genes involved in conjugation, integration and maintenance. Therefore, PAPI-1 is a horizontally transmitted element that may share a common evolutionary ancestor with pKLC102.

A group of PAPI-1 carrying strains was probed with sets of PCR primers that can detect both the chromosomally integrated and the circular form of PAPI-1. Evidence of excision of the island was found in all strains [15]. Sequence analysis of the PCR products verified that the excision and circularization of PAPI-1 occurred via recombination between the *att* sites bordering the island. After excision, the sequence at the chromosomal site in strain PA14 was identical to that at the corresponding location in PAO1, a strain that does not naturally harbor PAPI-1. Furthermore, the circular PAPI-1 was observed to integrate into the chromosome at the second tRNA^{Lys} gene (PA0976.1), in which the *att* site is already occupied by PAPI-2. In the recently sequenced strain, PA7, the *att* site in PA4541.1 is unoccupied, however, a PAPI-1 like island is found immediately downstream of PA0976.1. Interestingly, a second island consisting of 24 genes is found adjacent to the PAPI-1 like island, suggesting that in this strain, PAPI-1 has been inserted into a previously occupied *att* site, analogous to its insertion into the tRNA^{Lys} PA0976.1 in PA14.

Given that circular forms of integrated GIs are often precursors for transfer, the mobility of PAPI-1 between PA14 (donor) and PAO1 (recipient) was characterized. Transfer of PAPI-1 was detected at frequencies ranging from 3.1×10^{-7} to 5.4×10^{-4}. The frequency was dependent upon mating conditions, with liquid media strongly promoting transfer, while minimal transfer was detected on surfaces of agar plates. In the recipient, PAPI-1 integrated into the chromosome at either of its *att* sites $tRNA^{Lys}$ PA0976.1 and PA4541.1. When the strain PAO1 carrying PAPI-1 was used as a donor in mating with a recipient PAO1, similar transfer efficiency was obtained. A significant decrease in transfer efficiency was observed when using a recipient that already carries this island [X. Qiu, unpublished] suggesting that PAPI-1 specifies a surface or mating exclusion system for preventing redundant acquisition.

Several genes in PAPI-1 encode functions typically associated with mobile genetic elements. Mutations in the PAPI-1 integrase gene *(int)* block its excision from the chromosome significantly. This is consistent with the role of integrases in recombination; to catalyze the excision and integration of mobile genetic elements. The gene *soj,* encoding a homologue of plasmid/chromosomal partition systems [16], is responsible for the maintenance of the circular PAPI-1. It is located at the end of the island opposite to *int*. Mutations in *soj* result in the elimination of PAPI-1 from cells. When an extra copy of *soj* was introduced into PA14 prior to its subsequent deletion from PAPI-1, this element behaved as it does in wild type PA14 existing as a circular form as well as an integrated form at both $tRNA^{Lys}$ sites (PA4541.1 and PA0976.1). Therefore, Soj is responsible for the maintenance of the circular PAPI-1, presumably by stabilizing it after excision. The *soj* gene is expressed only after PAPI-1 circularizes, transcribed from a promoter located on the opposite end of the island. In the absence of Soj, PAPI-1 excises from the chromosome but fails to be maintained as an episome, leading to its eventual loss from the entire population.

The ability of PAPI-1 to excise, transfer and integrate into the chromosome of the recipients strongly suggests that this island is an integrative and conjugative element (ICE), described in numerous bacterial species [17]. These ICEs, also known as conjugative transposons, represent a group of very well characterized GIs, which in many instances have retained mobility. A number of GIs appear to have originated from ancestral ICEs that became fixed in the bacterial chromosome. Most ICEs characterized to date contain specific features associated with conjugative plasmids and bacteriophages. In addition to carrying genes for antibiotic resistance, a number of ICEs may confer fitness traits such as promoting symbiosis or providing the ability to metabolize complex aromatic compounds. PAPI-1 represents the first *P. aeruginosa* ICE described to date which carries virulence factors.

Evolution of the PAPI-2 and the ExoU Islands

PAPI-2 and related islands are not as widely distributed among *P. aeruginosa* isolates as PAPI-1. In all strains examined to date, the location of the PAPI-2 like island, encoding a potent cytotoxin ExoU and its cognate chaperone for type III secretion, SpcU, is immediately downstream of the $tRNA^{Lys}$ gene PA0976.1 [7, 10, 13]. Unlike PAPI-1, PAPI-2 has undergone significant decay and deletions following its acquisition.

Yeast recombinational cloning was used to identify and sequence three additional islands evolutionarily related to PAPI-2 that are referred to as the ExoU Island family [10]. The largest of these ExoU Islands (ExoU Island A) was initially found in three strains from different clinical sources (ocular and urinary tract infections) as well as geographically distinct locations. ExoU Island A contains 77 ORFs and includes the coding sequence for an integrase, presumed to be responsible for incorporation of this island into the $tRNA^{Lys}$ gene. Several additional proteins encoded by ExoU Island A are clearly associated with transmissible genetic elements, including a putative plasmid stabilization factor, several helicases, and a TraG/TraD family protein. ExoU Island B (29.5-kb) was identified in the genome of another ocular isolate and the relatively short, 3.89-kb ExoU Island C, is carried by a *P. aeruginosa* blood isolate. Sequence comparisons of the various ExoU islands and the segments found at the same $tRNA^{Lys}$ gene in PAO1 suggest that these may have the same evolutionary origin and are likely the remnants of a common ancestral element. This element may be the same element as the ancestor of PAPI-1 and pKLC102 but at some period in its evolutionary history, it must have acquired additional genes and insertion sequences. Following integration into $tRNA^{Lys}$, several segments were deleted but retained variable sequences flanking the *exoU/spcU* genes. Based on conserved genes and their synteny, the possible evolutionary history of these islands can be deduced and is shown schematically in figure 1. Unlike PAPI-1, none of the ExoU islands examined appear to be excisable, presumably due to the absence of one out of two intact *att* sites that are needed for recombination.

Genomic Islands Integrated into the $tRNA^{Gly}$ Genes Adjacent to PA2819

The two $tRNA^{Gly}$ genes in the cluster of $tRNA^{Gly}$, $tRNA^{Gly}$, $tRNA^{Glu}$, designated as PA2819.1 and PA2819.2 in the genome of PAO1, serve as *att* sites for a variety of genomic islands. These include PAGI-2, PAGI-3 and RGP29 [7, 11, 18]. Although none of these islands encode virulence factors, they contribute to genomic diversity of *P. aeruginosa*.

Fig. 1. A model of the evolutionary history of the genomic islands located at the *P. aeruginosa* $tRNA^{Lys}$ PA0976.1. An ancestral transmissible integrative plasmid is postulated to have given rise to both the ExoU Island family as well as pKLC102-like elements and their genomic island derivatives. *exoU* was acquired through HGT where it then, with the invariantly associated IS407, inserted into the ancestral plasmid. This composite element subsequently integrated at the PA0976.1 $tRNA^{Lys}$. Alternatively, as indicated by the inset box, *exoU* and the linked IS407 were inserted into the chromosomally integrated ancestral plasmid giving rise to the various ExoU islands. The ancestral ExoU Island underwent insertions, inversions, and deletions to result in the presently observed ExoU encoding islands. The ancestral plasmid went through subsequent modifications to give rise to the pKLC102-like elements PAPI-1 and pKLK106. These elements, integrated at the same locus, underwent subsequent evolutionary events (insertions, inversions, and deletions) resulting in the elements PAGI-4 and the PAO1-associated island.

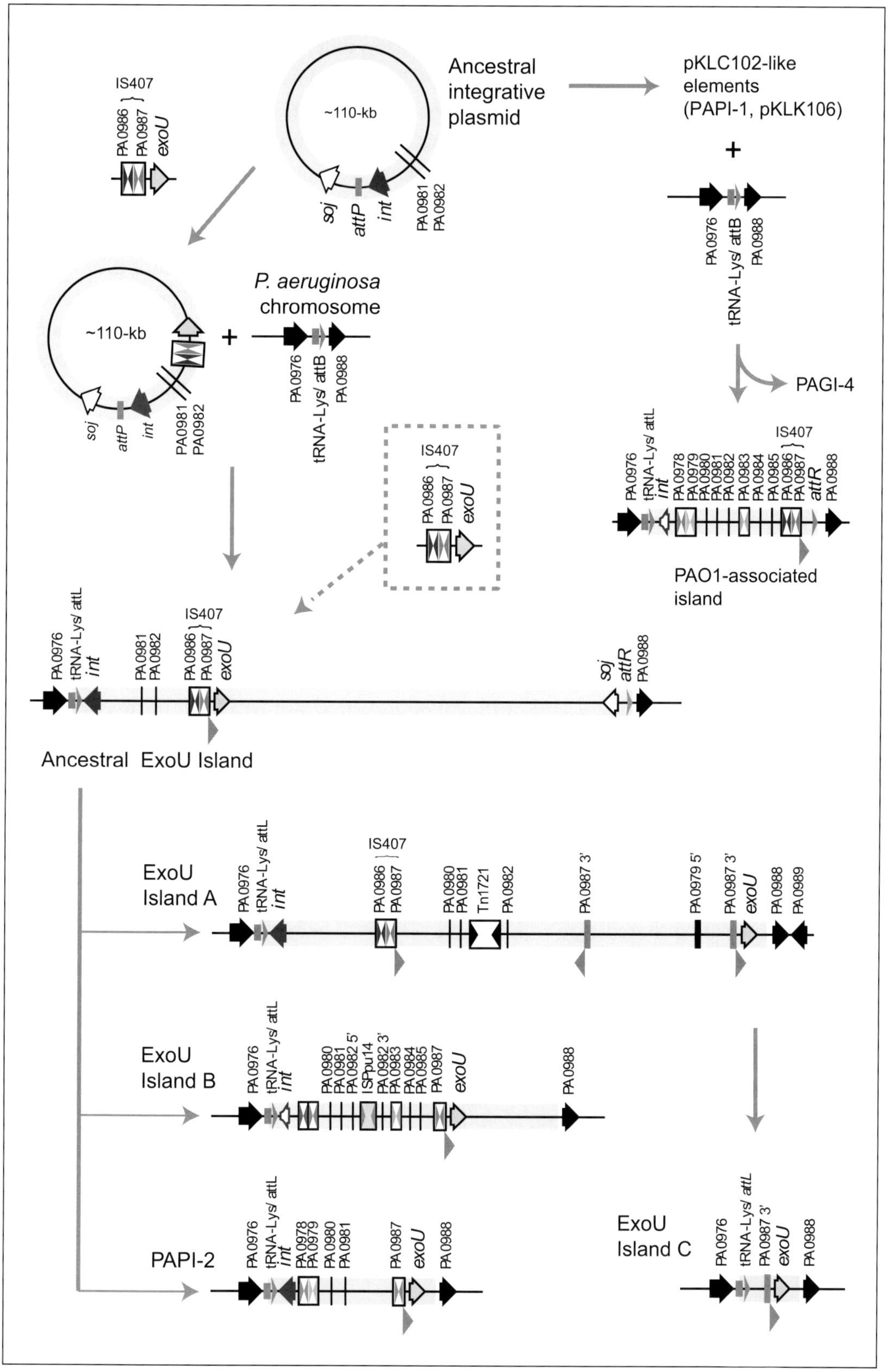

Ancestral integrative plasmid
~110-kb
soj
attP
int
PA0981
PA0982
IS407
PA0986
PA0987
exoU
pKLC102-like elements (PAPI-1, pKLK106)
+
PA0976
tRNA-Lys/ attB
PA0988
P. aeruginosa chromosome
PAGI-4
PA0978
PA0979
PA0980
PA0983
PA0984
PA0985
attR
PAO1-associated island
tRNA-Lys/ attL
Ancestral ExoU Island
ExoU Island A
Tn1721
PA0987 3'
PA0979 5'
PA0989
ExoU Island B
PA0982 5'
ISPpu14
PA0982 3'
PAPI-2
ExoU Island C

This tRNA trio serves as a target site for the integration of tandem elements giving rise to highly heterologous chromosomal segments in different strains.

RGP29, found in the genome of the CF isolate PA2192, is a 224-kb composite genomic island integrated into the chromosome at the $tRNA^{Gly}$ gene PA2819.1. Based on the comparison of direct repeats within RGP29, it was possible to deduce its evolutionary history. First, the so-called Dit Island was integrated into the 3′ end of the $tRNA^{Gly}$ gene PA2819.1, followed by the acquisition of the genomic island PAGI-2 [11, 18]. The Dit Island contains a cluster of 95 genes related to *dit* genes in other bacteria that encode abietane diterpenoid metabolism proteins. These compounds produced by wounded trees can be utilized as carbon source by several bacterial species, including *Pseudomonas abietaniphila* and *Burkholderia xenovorans.* Therefore, we can speculate that one of these organisms may have provided the ancestral source of the Dit Island. The presence of this element in a clinical isolate represents an example of environmentally driven expansion of a bacterial genome while retaining its full virulence potential.

PAGI-2 and PAGI-3 share several common features as well as a similar modular architecture, suggesting that they may have shared a distant ancestor. Both PAGI-2 and PAGI-3 contain, at their two opposite ends, the orthologues of *int* and *soj* genes [18]. Presumably the products of these genes specify an integrase/excisionase and a protein necessary for the maintenance of the circular forms of PAGI-2 or PAGI-3 that are utilized during the conjugal transfer event, however, excision has not been demonstrated for either PAGI-2 or PAGI-3.

In terms of genetic organization, PAGI-2 and PAGI-3 are more closely related to *clc*, the mobile genomic island of *Pseudomonas* sp. strain B13 [19]. In this organism, the transmissible *clc* element is also integrated into the $tRNA^{Gly}$ gene. The rest of the element is modular and includes genes specifying putative components of the type IV secretion/conjugation apparatus. The diversity between *clc*, PAGI-2, and PAGI-3 is the result of acquisition of unique blocks of genes which, in *clc*, encode the enzymes for the degradation of 3-chlorobenzoate. Based on nucleotide similarities, it has been suggested that *clc*, PAGI-2 and PAGI-3 belong to a larger superfamily of transmissible elements with a shared core architecture [11, 19]. The minimal arrangement includes the specific terminal locations of the *int* and *soj* genes and a block of genes involved in DNA processing and transfer likely via a conjugation mechanism. This arrangement is found in horizontally-acquired elements present not only in *clc*, PAGI-2, and PAGI-3, but also in the pKLC102 family and in islands described in other distally-related organisms, such as *Haemophilus* species (elements related to ice*hin*1056) and the SP1–7 island of *Salmonella typhi* [20].

The Flagellin Glycosylation Island

The flagellin protein of *P. aeruginosa,* the major subunit of the flagellar filament, can be classified as A-type or B-type. Each type is glycosylated and is dependent upon the presence of a distinct glycosylation island embedded within the chromosomal locus that contains a large number of structural and regulatory genes involved in flagellar assembly [21]. The A-type flagellins can be further divided into two sub-types, designated A1 and A2, based on sequence polymorphisms displayed by the flagellin proteins [22]. In a fraction of strains, the glycosylation island linked to the A1 flagellin consists of 14 open reading frames, *orfA–orfN*, while a shorter version of the island in which *orfD*, *-E* and *-H* are polymorphic and *orfI*, *-J*, *-K*, *-L*, and *-M* are absent is associated with strains expressing either A1 or A2 flagellin. In contrast, the glycosylation island linked to the B-type flagellin consists of only four genes.

The evolutionary history of the glycosylation island in *P. aeruginosa* cannot be deduced from sequence analysis. The glycosylation island is found in a region of the *P. aeruginosa* chromosome lacking tRNA genes and none of the glycosylation islands carry putative integrases or excisionases. Conceivably, the capture of this island was the result of acquisition of the corresponding DNA fragment by the recipient, followed by a homologous recombination between the conserved segments that flank this locus. Based on the sharp boundaries between the individual islands and the flanking chromosomal sequences in A- and B-type strains, one of the possible recombination points has been tentatively identified in the *fleP* gene located on the right side of the island. On the opposite side of the glycosylation islands, the *flgK* gene could provide the second homologous region for recombination. Transformation or generalized transduction would be the most logical mechanism of acquisition of these islands. *P. aeruginosa* has not been shown to be naturally competent for DNA uptake, however, a number of *P. aeruginosa* bacteriophages capable of generalized transduction have been identified [23, 24]. The GC content of this island is 63.3%, which is not significantly different from that of the PAO1 genome (66%). Therefore this island originated possibly from another *Pseudomonas* or a bacterium with comparable GC-rich DNA. A cluster of homologous genes corresponding to the shorter variant of the type A-associated glycosylation island is found in the genome of *Pseudomonas florescence* Pf-5. It is conceivable that the recent, and perhaps ongoing exchange of the flagellin genes and the linked glycosylation islands, occurs exclusively between *P. aeruginosa* strains and involves swapping of entire islands by double reciprocal recombination.

The LPS O-Antigen Genomic Islands

The minimal lipopolysaccharide (LPS) structure, consisting of lipid A and part of the core sugars, is an essential component of the outer membrane of Gram-negative bacteria. Although the O-side chain of LPS functions under certain circumstances in providing protection against serum killing, it is dispensable and its mutational loss is not lethal. Moreover, many strains express LPS with O-side chains varying markedly in sugar composition, sequence, and modifications. The genetic determinants that encode the various enzymes involved in building the O-side chain are highly divergent among bacterial pathogens and may be found on GIs [25].

LPS of *P. aeruginosa* is a recognized virulence factor. It stimulates a strong inflammatory response and is the target of humoral immunity [26]. Mutants lacking an LPS O-side chain display a significantly reduced infectivity in acute infection models. Furthermore antibodies directed to the O-side chain are protective against *P. aeruginosa* infections in almost all animal models. Interestingly, most *P. aeruginosa* CF isolates are serum sensitive because of a lack of O-side chains. This pathoadaptive mutation appears to be selected for during the adaptation of *P. aeruginosa* to chronic colonization of the respiratory tract.

The *P. aeruginosa* strains are grouped into twenty serotypes using the International Antigenic Typing System (IATS). The unique serotype of an individual strain is based on the presence of a distinct gene cassette located at the same chromosomal site [27]. Each cassette encodes one or several enzymes involved in LPS synthesis or modification. In total, eleven cassettes account for the twenty serotypes, with certain cassettes providing novel serotypes because of mutations. For example, serotype O17 contains the same cassette as serotype O11 with two insertions and one deletion relative to O11. Similarly, the gene cassettes in serotypes O13 and O14 are identical with the exception of a frameshift mutation in a hypothetical gene located in the gene cassette conferring serotype O14.

Although it is clear that the unique O-side chain gene cassettes are a result of HGT, the precise mechanism for acquisition of this gene cluster and the selective pressure for their stable maintenance are not understood yet. In addition to atypical GC content, ranging from 48–54%, the LPS O-side chain locus is found near a tRNA gene, a common location of various GIs. Only one serotype (O15 serotype) lacks a gene cassette at this location, however it possesses remnants of the core O-side chain gene cassette (a partial insertion sequence element and a portion of the *wbpM* gene suggests that the original cassette was present in this location but then was deleted at an unspecified time in the evolutionary history of this strain). Although the various LPS gene cassettes have many signatures of HGT, their origin and the mechanism of acquisition and insertion into the identical chromosomal site are unclear.

Where Does Horizontal Gene Transfer Take Place?

Presence of horizontally acquired blocks of DNA in the genome of an organism requires the presence of another organism to serve as a source (donor), a functional genetic mechanism for DNA transfer and selective conditions that assure maintenance of the genes in the recipient by contributing to its fitness in a particular environment. There are limited studies on HGT in natural environments of microorganisms. When considering the genetic requirements for *P. aeruginosa* to function as a pathogen, evidence from comparative genomics and limited studies of virulence in animal models suggest that environmental organisms are as virulent as clinical isolates. In the case of *P. aeruginosa* isolates from chronically infected CF patients, pathoadaptive mutations occur in those genes that have been implicated in infectivity [28]. Therefore bacteria adapted to long-term survival in the lung environment may in fact be less virulent compared to free-living bacteria. Clearly, compensatory mutations can occur in certain circumstances, as highly virulent, epidemic CF isolates of *P. aeruginosa* have been described [29]. Another important finding from the comparisons of genome sequences of clonal isolates from a chronically infected CF patient is the complete absence of new gene acquisition in this particular lineage [28], although infections with different strains, or transient infections, are not uncommon. Therefore, it appears that evolution of the *P. aeruginosa* genome including the acquisition of virulence traits takes place in the natural environment of these organisms. Genes required for survival in a particular niche very likely specify the same determinants that benefit the pathogen during a successful infection in a human host. Although host-pathogens interactions in the environment have received little attention, in laboratory conditions, *P. aeruginosa* can infect a wide range of organisms that it may routinely encounter in its environment, including plants, insects, fungi and nematodes. It is these interactions that may provide the selective environment for acquisition and maintenance of virulence traits [30]. Moreover, analysis of the composition of the flexible gene pool strongly argues for ongoing evolution and customization of the genetic repertoire that favor niche expansion. Preferential survival of *P. aeruginosa* in a wide range of environments also enhances the opportunities for this organism to infect compromised human hosts. Future works should therefore focus more on studies of *P. aeruginosa* in its natural environment which would undoubtedly provide new insights into an important aspect of bacterial evolution that shapes the pathogenic potential of not only *P. aeruginosa* but also other pathogens.

Acknowledgements

The work in S.L.'s laboratory was supported by the grant GM068516 from the NIH. X.Q. was supported by a postdoctoral fellowship from the Cystic Fibrosis Foundation.

References

1 Gómez MI, Prince A: Opportunistic infections in lung disease: *Pseudomonas* infections in cystic fibrosis. Curr Opin Pharmacol 2007;7:244–251.

2 Wisplinghoff H, Bischoff T, Tallent SM, Seifert H, Wenzel RP, Edmond MB: Nosocomial bloodstream infections in US hospitals: analysis of 24,179 cases from a prospective nationwide surveillance study. Clin Infect Dis 2004;39:309–317.

3 Chatzinikolaou I, Abi-Said D, Bodey GP, Rolston KV, Tarrand JJ, Samonis G: Recent experience with *Pseudomonas aeruginosa* bacteremia in patients with cancer: Retrospective analysis of 245 episodes. Arch Intern Med 2000;160:501–509.
4 Mutlu GM, Wunderink RG: Severe pseudomonal infections. Curr Opin Crit Care 2006;12:458–463.
5 Stover CK, Pham XQ, Erwin AL, Mizoguchi SD, Warrener P, et al: Complete genome sequence of *Pseudomonas aeruginosa* PAO1, an opportunistic pathogen. Nature 2000;406:959–964.
6 Lee DG, Urbach JM, Wu G, Liberati NT, Feinbaum RL, et al: Genomic analysis reveals that *Pseudomonas aeruginosa* virulence is combinatorial. Genome Biol 2006;7:R90.
7 Mathee K, Narasimhan G, Valdes C, Qiu X, Matewish JM, et al: Dynamics of *Pseudomonas aeruginosa* genome evolution. Proc Natl Acad Sci USA 2008;105:3100–3105.
8 Wolfgang MC, Kulasekara BR, Liang X, Boyd D, Wu K, et al: Conservation of genome content and virulence determinants among clinical and environmental isolates of *Pseudomonas aeruginosa*. Proc Natl Acad Sci USA 2003;100:8484–8489.
9 Liang X, Pham XQ, Olson MV, Lory S: Identification of a genomic island present in the majority of pathogenic isolates of *Pseudomonas aeruginosa*. J Bacteriol 2001;183:843–853.
10 Kulasekara BR, Kulasekara HD, Wolfgang MC, Stevens L, Frank DW, Lory S: Acquisition and evolution of the exoU locus in *Pseudomonas aeruginosa*. J Bacteriol 2006;188:4037–4050.
11 Klockgether J, Würdemann D, Reva O, Wiehlmann L, Tümmler B: Diversity of the abundant pKLC102/PAGI-2 family of genomic islands in *Pseudomonas aeruginosa*. J Bacteriol 2007;189:2443–2459.
12 Klockgether J, Reva O, Larbig K, Tümmler B: Sequence analysis of the mobile genome island pKLC102 of *Pseudomonas aeruginosa* C. J Bacteriol 2004;186:518–534.
13 He J, Baldini RL, Déziel E, Saucier M, Zhang Q, et al: The broad host range pathogen *Pseudomonas aeruginosa* strain PA14 carries two pathogenicity islands harboring plant and animal virulence genes. Proc Natl Acad Sci USA 2004;101:2530–2535.
14 Drenkard E, Ausubel FM: *Pseudomonas* biofilm formation and antibiotic resistance are linked to phenotypic variation. Nature 2002;416:740–743.
15 Qiu X, Gurkar AU, Lory S: Interstrain transfer of the large pathogenicity island (PAPI-1) of *Pseudomonas aeruginosa*. Proc Natl Acad Sci USA 2006;103:19830–19835.
16 Ebersbach G, Gerdes K: Plasmid segregation mechanisms. Annu Rev Genet 2005;39:453–479.
17 Burrus V, Marrero J, Waldor MK: The current ICE age: biology and evolution of SXT-related integrating conjugative elements. Plasmid 2006;55:173–183.
18 Larbig KD, Christmann A, Johann A, Klockgether J, Hartsch T, et al: Gene islands integrated into tRNA(Gly) genes confer genome diversity on a *Pseudomonas aeruginosa* clone. J Bacteriol 2002;184:6665–6680.
19 Gaillard M, Vallaeys T, Vorhölter FJ, Minoia M, Werlen C, et al: The *clc* element of *Pseudomonas* sp. strain B13, a genomic island with various catabolic properties. J Bacteriol 2006;188:1999–2013.
20 Mohd-Zain Z, Turner SL, Cerdeño-Tárraga AM, Lilley AK, Inzana TJ, et al: Transferable antibiotic resistance elements in *Haemophilus influenzae* share a common evolutionary origin with a diverse family of syntenic genomic islands. J Bacteriol 2004;186:8114–8122.
21 Arora SK, Bangera M, Lory S, Ramphal R: A genomic island in *Pseudomonas aeruginosa* carries the determinants of flagellin glycosylation. Proc Natl Acad Sci USA 2001;98:9342–9347.
22 Arora SK, Wolfgang MC, Lory S, Ramphal R: Sequence polymorphism in the glycosylation island and flagellins of *Pseudomonas aeruginosa*. J Bacteriol 2004;186:2115–2122.
23 Budzik JM, Rosche WA, Rietsch A, O'Toole GA: Isolation and characterization of a generalized transducing phage for *Pseudomonas aeruginosa* strains PAO1 and PA14. J Bacteriol 2004;186:3270–3273.
24 Beumer A, Robinson JB: A broad-host-range, generalized transducing phage (SN-T) acquires 16S rRNA genes from different genera of bacteria. Appl Environ Microbiol 2005;71:8301–8304.
25 Reeves PP, Wang L: Genomic organization of LPS-specific loci. Curr Top Microbiol Immunol 2002;264:109–135.
26 Pier GB: *Pseudomonas aeruginosa* lipopolysaccharide: a major virulence factor, initiator of inflammation and target for effective immunity. Int J Med Microbiol 2007;297:277–295.
27 Raymond CK, Sims EH, Kas A, Spencer DH, Kutyavin TV, et al: Genetic variation at the O-antigen biosynthetic locus in *Pseudomonas aeruginosa*. J Bacteriol 2002;184:3614–3622.
28 Smith EE, Buckley DG, Wu Z, Saenphimmachak C, Hoffman LR, et al: Genetic adaptation by *Pseudomonas aeruginosa* to the airways of cystic fibrosis patients. Proc Natl Acad Sci USA 2006; 103:8487–8492.

29 Salunkhe P, Smart CH, Morgan JA, Panagea S, Walshaw MJ, et al: A cystic fibrosis epidemic strain of *Pseudomonas aeruginosa* displays enhanced virulence and antimicrobial resistance. J Bacteriol 2005; 187:4908–4920.

30 Rahme LG, Ausubel FM, Cao H, Drenkard E, Goumnerov BC, et al: Plants and animals share functionally common bacterial virulence factors. Proc Natl Acad Sci USA 2000;97:8815–8821.

Stephen Lory
Department of Microbiology and Molecular Genetics, Harvard Medical School
200 Longwood Avenue, 363 Warren Alpert Building
Boston, MA 02115 (USA)
Tel. +1 617 432 5099, Fax +1 617 738 7664, E-Mail stephen_lory@hms.harvard.edu

de Reuse H, Bereswill S (eds): Microbial Pathogenomics.
Genome Dyn. Basel, Karger, 2009, vol 6, pp 140–157

The Genus *Burkholderia*: Analysis of 56 Genomic Sequences

D.W. Ussery[a] · K. Kiil[a] · K. Lagesen[b] · T. Sicheritz-Pontén[a] · J. Bohlin[c] · T.M. Wassenaar[a,d]

[a]Center for Biological Sequence Analysis, Technical University of Denmark, Lyngby, Denmark; [b]Department of Informatics, University of Oslo, Blindern, Oslo, and the Centre for Molecular Biology and Neuroscience and Institute of Medical Microbiology, University of Oslo, Oslo, [c]Norwegian School of Veterinary Science, Oslo, Norway; [d]Molecular Microbiology and Genomics Consultants, Zotzenheim, Germany

Abstract

The genus *Burkholderia* consists of a number of very diverse species, both in terms of lifestyle (which varies from category B pathogens to apathogenic soil bacteria and plant colonizers) and their genetic contents. We have used 56 publicly available genomes to explore the genomic diversity within this genus, including genome sequences that are not completely finished, but are available from the NCBI database. Defining the pan- and core genomes of species results in insights in the conserved and variable fraction of genomes, and can verify (or question) historic, taxonomic groupings. We find only several hundred genes that are conserved across all *Burkholderia* genomes, whilst there are more than 40,000 gene families in the *Burkholderia* pan-genome. A BLAST matrix visualizes the fraction of conserved genes in pairwise comparisons. A BLAST atlas shows which genes are actually conserved in a number of genomes, located and visualized with reference to a chosen genome. Genomic islands are common in many *Burkholderia* genomes, and most of these can be readily visualized by DNA structural properties of the chromosome. Trees that are based on relatedness of gene family content yield different results depending on what genes are analyzed. Some of the differences can be explained by errors in incomplete genome sequences, but, as our data illustrate, the outcome of phylogenetic trees depends on the type of genes that are analyzed.

The genus *Burkholderia* belongs to the beta sub-division of Proteobacteria and contains a wide variety of Gram-negative species that occupy very different niches. Some are zoonotic pathogens, others are opportunistic human pathogens whilst yet others live harmless in the environment. Some species are able to degrade industrial waste compounds. Plant pathogens are also represented, and in contrast others protect plants against pathogens or promote plant growth. *Burkholderia* genomes consist of two or three chromosomes and frequently contain plasmids as well. Their genomes are large, variable, and extremely interesting as they can provide important insights to the evolutionary processes that shape bacterial genomes. The two species that attract attention because of their potential in bio-terrorism are *B. mallei* and *B. pseudomallei*. With multiple genome sequences available for these species and for a number of related species, comparative genomics of the genus *Burkholderia* is now en vogue. Here we will compare 56 sequenced *Burkholderia* genomes and present observations to illustrate that presumed evolutionary relatedness depends on which fraction of the genome is analyzed. First, *B. mallei, B. pseudomallei* and the diseases they cause are introduced.

***Burkholderia mallei* Causes Glanders and *B. pseudomallei* Causes Melioidosis**

B. mallei is a nonmotile, nonsporulating, obligate aerobe organism previously known as *Pseudomonas mallei*. It causes glanders in horses and several other animal species. Animals contract the disease by ingestion of contaminated food or water. Traditionally, the disease is divided into nasal, pulmonary or cutaneous cases. The disease frequently progresses to septicaemia that will be fatal within days. A chronic form can occur in horses where nasal and subcutaneous nodules develop; such animals can be carriers for months or years before death occurs. The disease was once widespread, but by the mid-1900s it was eradicated in many countries by isolating and eradicating infected animals. It is still endemic in regions in Africa, Asia, the Middle East and Central and South America. A vaccine does not exist.

Human infections caused by *B. mallei* are rare although exceptionally few organisms are needed for human infection. Transmission from animal to man is inefficient and human-to-human spread is extremely rare. Cases result from direct and prolonged contact with infected domestic animals or from direct contamination with the infectious agent in the laboratory, presumably resulting from aerosols forming during routine handling. The low infectious dose, and the usual fatal outcome in humans, makes *B. mallei* a potential agent for biological warfare and bio-terrorism. Symptoms in humans depend on whether it is a localized cutaneous, pulmonary or bloodstream infection. Bloodstream infections have a fatality rate of 95% within a few days.

B. pseudomallei causes melioidosis, also known as Whitmore disease. The disease is similar to glanders but is restricted to the tropics and is endemic in tropical parts of Southeast Asia (notably Thailand), Australia and China. It is also found in tropical Africa and India. Occasionally, travelers import the disease into Europe or the US. In contrast to *B. mallei*, which is not frequently detected outside a host, *B. pseudomallei* survives in soil and water and it has a broader host range. As a consequence, human melioidosis is far more common than glanders and in some regions it accounts for 20 to 40% of community-acquired septicaemia. Melioidosis can be transmitted through contaminated water, notably during the rainy season, or by inhalation of contaminated dust. Human infections have a high mortality. The latent phase between infection and disease can be extremely long, up to months or even years and relapse is quite common.

B. mallei has most probably evolved from *B. pseudomallei*. This was concluded from multilocus sequence typing (MLST), a technique that assesses allelic variation in a number of household genes [1]. In recognition of this close relationship, *B. pseudomallei* and *B. mallei* are both taxonomically included in what is called the Pseudomallei group.

Other *Burkholderia* Species Have a Variety of Lifestyles

In addition to *B. mallei* and *B. pseudomallei*, the genus *Burkholderia* contains more than 40 other species. Only those for which a genome sequence is available are listed here. Two of these belong to the Pseudomallei group: *B. thailandensis* also lives in tropical environments but is not pathogenic to mammals. *B. oklahomensis* has been described as '*B. pseudomallei*-like', but MLST and DNA-DNA hybridization have identified it as a novel species [2]. *B. oklahomensis* has been isolated from wounds associated with soil contamination.

Another important group of closely related species is the *B. cepacia* complex (BCC), wherein each species is also known as a genomovar, with *B. cepacia* as genomovar I. (There are more than nine species within BCC, with recent novel additions [3], but their genomes have not yet been sequenced). They are all opportunistic pathogens, frequently causing infections in cystic fibrosis patients where the infection can be fatal. Besides this relevance to human medicine, a number of species of the BCC also have other interesting properties. *B. cenocepacia* (genomovar III) is ubiquitous in the environment as a phytopathogen. *B. dolosa* was formerly known as *B. cepacia* genomovar IV. *B. multivorans* cannot transmit from patient to patient, in contrast to the other BCC species. *B. ambifaria* (genomovar VII) has attracted interest since it lives in the rhizosphere of pea plants where it can protect the plants against pathogens. *B. vietnamiensis* is also beneficial to plants and has been studied as a growth-promoting bacterium. It has also bioremediation properties as it can degrade aromatic hydrocarbons such as benzene and toluene. *B. ubonensis* (also known as *B. uboniae*) is a common soil bacterium that is proposed as a new member of the BCC [4]. The latest addition of the BCC for which a genome sequence is available is *B. lata*, first described in 2009 [5].

The remainder of species for which a genome species is available are not pathogenic to humans and do not belong to a particular subgroup. *B. xenovorans* is an environmental organism of economic importance as it can degrade polychlorinated biphenyl (PCB) compounds. In contrast, *B. phymatum* lives in symbiotic relationship with tropical legumes. *B. phytofirmans* is also beneficial to its plant host, and lives outside the tropics. *B. graminis* is found in the rhizosphere of Gramineae plants, such as wheat and corn.

The First *Burkholderia* Genome Sequences

The potential use in biological warfare raised a scientific interest that resulted in a relatively large number of published genome sequences. The genome of *B. mallei* contains two chromosomes and the first complete sequence was published in 2004 (*B. mallei* strain ATCC 23344) [6]. At the same time the sequence for both chromosomes of *B. pseudomallei* strain K96243 was published [7]. A large number of insertion sequences were found in the *B. mallei* genome that have mediated multiple deletions and rearrangements compared to the genome of *B. pseudomallei*. The genome of the latter contained 16 genomic islands that appeared absent in the smaller genome of *B. mallei*. The authors speculated that these genomic islands had been absent from the genetic repertoire of the *B. pseudomallei* ancestral clone that produced *B. mallei* [7]. Gene loss would be consistent with the reduced adaptive potential and restricted host specificity of *B. mallei* compared to *B. pseudomallei*. Other differences between the two species observed related to the fact that *B. pseudomallei* is motile but *B. mallei* is not (a few of its motility genes have undergone mutations as a result of release of selective pressure), and that *B. pseudomallei* can secrete a number of toxins that *B. mallei* produces but cannot secrete, due to a mismatch in a secretory system component. Finally, the *B. mallei* genome contains two type III secretion systems on chromosome 2, which contributes to its virulence potential.

The two species share an exceptionally high number of local direct repeat sequences, covering more than 20% of the total length of the chromosomes. We classify repeats as 'local' when they are found by searching with a 15 nucleotide (nt) window within a 100 nt region, and as 'global' when determining the frequency of 100 nt-long sequences repeated anywhere on the genome [8]. The two chromosomes of each species also showed significant functional partitioning, with the large chromosome 1 (4.1 Mb in *B. pseudomallei*, 3.5 Mb in *B. mallei*) encoding many genes involved in metabolism and growth, the smaller chromosome 2 (3.2 Mb and 2.3 Mb, respectively) containing genes related to adaptation and survival in different niches.

The genome of *B. thailandensis* was sequenced in 2006 but already in 2004 it was recognized that its genome had also undergone gene reduction compared to *B. pseudomallei* [9]. This work was based on microarray analysis using partial genome sequences of *B. pseudomallei* K96243. The authors concluded that genome reduction of *B. thailandensis* occurred independent of that of *B. mallei*, possibly by different mechanisms, as the deleted genes were not found present in clusters in *B. pseudomallei*, but rather dispersed over its genome. When the *B. thailandensis* genome sequence became available, it was obviously compared to *B. pseudomallei* [10]. The authors concentrated on *B. mallei* genes that are up- or downregulated during colonization in a mouse model, and found that down-regulated genes were more strongly conserved in *B. thailandensis* than in *B. pseudomallei*.

Over time more *Burkholderia* genome sequences have been finished, such as that of *B. xenovorans* LB400 [11]. Its genome contains three chromosomes, totaling 9.73 Mb, though other strains can have smaller genomes with 7.4 Mb being the currently known minimum. As in the other *Burkholderia* species, the chromosomes have undergone functional specialization and the two smaller chromosomes have undergone less selective pressure, allowing for more variation. As the number of genome sequences grew, including multiple genomes for a number of species, the comparison within and between species became truly interesting. A database especially dedicated to *Burkholderia* genomes has recently been established at www.burkholderia.com [12].

Genome sequences do not have to be complete (with each chromosome in a single, contiguous piece) to be used for comparative analysis. Incomplete genome sequences are frequently released into the public domain as multiple contigs, and sometimes it is left to that. Here we perform comparative genomic analysis of partial and complete genome sequences within the *Burkholderia* genus that are publicly available.

Practicalities of Large-Scale Comparative Genomics: Introducing the BLAST Matrix

The 56 *Burkholderia* genome sequences available at the time of writing are summarized in table 1. The number of contigs is given for all genomes. Working with such large number of genomes one can soon be overwhelmed with data: the interpretation and graphical representation of findings becomes a real issue. We largely concentrate on coding regions, and here we zoom in on the degree of gene conservation between genomes, ignoring gene location, chromosome separation or gene synteny. We did not perform a detailed analysis of gene function, nor did we relate individual genes to the characteristics of that particular strain or species (thus respecting the objectives of any sequencing project). This simplified approach allowed us to do large-scale analysis of gene conservation and chromosome evolutionary processes.

The approach is quite straightforward: Starting with one chromosome as a query, every gene is compared by BLAST to a second genome and conserved genes are scored. After all genes of the query genome are checked, the next genome is chosen to compare with the query genome until all genomes have been screened. Then the next genome is used as a query source, again checking all its individual genes against all other genomes. This way every genome in the analysis set will serve as a query against all others, and will also be queried by all other genomes [8].

Comparison of amino acid sequences of coding regions requires a standardized gene finding process, in order to rule out differences introduced by various (automated) gene identification programs. Genomes are frequently over- or under-annotated and occasionally the wrong strand of a gene is annotated [13]. Over-annotation is frequently seen in very short open reading frames, which can be erroneously recognized as genes if the cut-off for gene finding is taken too low (although some very short open reading frames can indeed be true genes). Under-annotation is sometimes observed for non-translated genes, such as tRNA or even rRNA genes that can be missing in a genome annotation. In our analysis only amino acid sequences were used, and non-translated RNA genes were excluded. In order to avoid artificial variation in our analysis, all used *Burkholderia* genomes were annotated by a standard gene finding and annotation program, so that arbitrarily chosen cut-offs would be consistent and not influence comparative analyses [14, 15].

Table 1. Genome sequences included in this study. All genomes used are publicly available for analysis

Group	Species	Strain[a]	No. of contigs[b]	PID	Sequence Source[c]
Pseudomallei group	*B. pseudomallei*	1106a	2	16182	TIGR
	B. pseudomallei	1710b	2	13954	TIGR
	B. pseudomallei	668	2	13953	TIGR
	B. pseudomallei	K96243	2	178	Sanger Institute
	B. pseudomallei	576	21	31091	LANL
	B. pseudomallei	305	36	18775	TIGR
	B. pseudomallei	S13	169	13951	TIGR
	B. pseudomallei	1655	194	13949	TIGR
	B. pseudomallei	1106b	202	16181	TIGR
	B. pseudomallei	1710a	209	13950	TIGR
	B. pseudomallei	Pasteur 52237	217	13952	TIGR
	B. pseudomallei	406e	271	16231	TIGR
	B. pseudomallei	BCC215	1030	19491	NMRC
	B. pseudomallei	NCTC 13177 (WKo97)	1077	19493	NMRC
	B. pseudomallei	112	1274	19495	NMRC
	B. pseudomallei	B7210	1424	19499	NMRC
	B. pseudomallei	7894	1568	19497	NMRC
	B. pseudomallei	91	1690	19505	NMRC
	B. pseudomallei	9	1762	19503	NMRC
	B. pseudomallei	14	1888	19507	NMRC
	B. pseudomallei	DM98 (BCC11)	2371	19509	NMRC
	B. mallei	ATCC 23344	2	171	TIGR
	B. mallei	NCTC 10229	2	13943	TIGR
	B. mallei	NCTC 10247	2	13946	TIGR
	B. mallei	SAVP1	2	13947	TIGR
	B. mallei	ATCC 10399	106	13944	TIGR
	B. mallei	GB8 horse 4	181	13945	TIGR
	B. mallei	JHU	184	13988	TIGR
	B. mallei	FMH	205	13987	TIGR
	B. mallei	2002721280	208	16352	TIGR
	B. mallei	PRL-20	272	19147	TIGR
	B. thailandensis	E264[d] (ATCC 700388)	2	10774	TIGR
	B. thailandensis	Bt4	803	19533	NMRC
	B. thailandensis	TXDOH	810	19541	NMRC
	B. thailandensis	MSMB43	1230	19501	NMRC
	B. oklahomensis	C6786[d]	633	19535	NMRC
	B. oklahomensis	EO147	886	19537	NMRC
Complex (BCC)	*B. cenocepacia*	J2315	4	339	Sanger Institute
	B. cenocepacia	AU 1054	3	13919	DOE
	B. cenocepacia	H12424	4	13918	DOE
	B. cenocepacia	MC0-3	3	17929	DOE
	B. cenocepacia	PC184	174	16169	Broad Institute

Table 1. Continued

Group	Species	Strain[a]	No. of contigs[b]	PID	Sequence Source[c]
	B. multivorans	ATCC 17616	4	17407	DOE
	B. ambifaria	AMMD[4]	4	13490	DOE
	B. ambifaria	MC40-6	4	17411	DOE
	B. ambifaria	IOP40-10	629	20669	DOE
	B. ambifaria	MEX-5	706	20667	DOE
	B. dolosa	AU0158	233	16168	Broad Institute
	B. vietnamiensis	G4	8	10696	DOE
	B. ubonensis	Bu	1143	19539	NMRC
	B. lata	383	3	10695	DOE
None	*B. phymatum*	STM815	4	17409	DOE
None	*B. phytofirmans*	PsJN[d]	3	17463	DOE
None	*B. xenovorans*	LB400	3	254	DOE
None	*B. graminis*	C4D1M[d]	70	20537	DOE
None	*Burkholderia spp.*	H160	310	29197	DOE

[a] Alternative names appear between parentheses.
[b] Number of contigs below 10 indicate that all chromosomes and plasmids are in one piece.
[c] DOE = US Department of Energy Joint Genome Institute; TIGR = The Institute of Genome Research; NMRC = Naval Medical Research Center/Defense Research Directorate, Genomics, USA. LANL = Los Alamos National Laboratory. Inst = Institute.
[d] Type strain of the species.

Another difficulty of comparisons of coding sequences is to decide when to call a pair of genes 'conserved'. This balancing act has two opposing risks. One can set very strict rules of identity, so that genes have to be highly similar in order to be screened as 'conserved' (in gene sequence and thus presumably in biological function). Consequentially, this may result in a very high number of genes without homologs, which decreases the significance of the findings. Alternatively, one can set relatively loose requirements for conservation, but then genes may be grouped together that have different biological functions as a result of divergent evolutionary processes, which also results in questionable results. As a rule-of-thumb, we have found that two genes need to have at least 50% identity over at least 50% of their lengths in order to be scored as conserved. This 50–50 rule has been found satisfactory for a number of species and genera that we analyzed. By varying these parameters (for instance 40% identity over at least 70% of sequence length) we observed that the analysis was quite robust.

The next challenge faced is how to represent the findings. BLAST produces long lists summarizing the findings that are obviously not conceivable or interpretable in their raw form. The data were instead condensed to two numbers per genome, indicating how many genes were tested as query and what fraction of these found

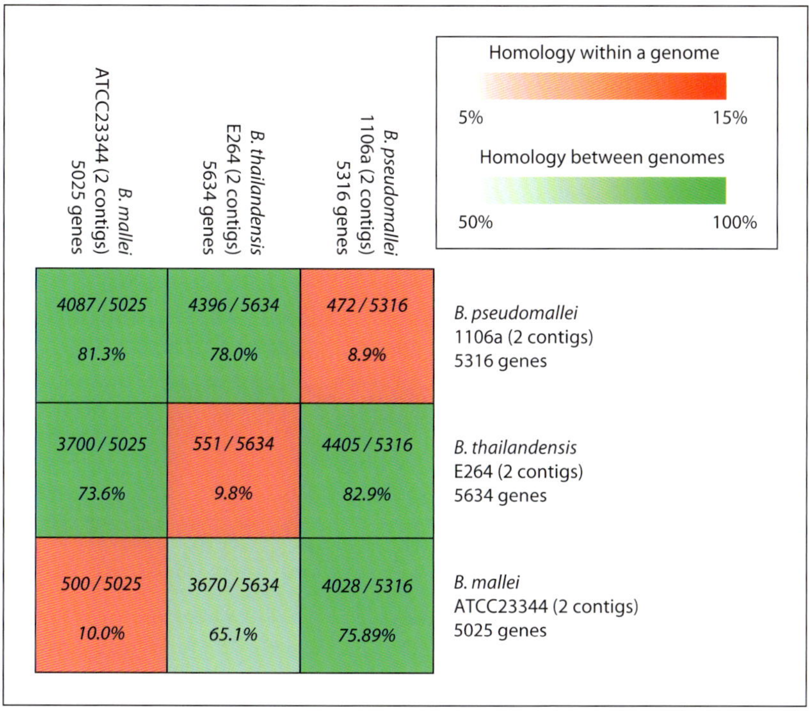

Fig. 1. BLAST matrix of *Burkholderia* genomes of three species. The scores in each field give the number of homologous genes per number of total genes in the tested genome, followed by percentage. The coloring of the cells depends on this fraction. The red cells represent homologous genes detected within one genome. The color scales can be adjusted according to the spread of the percentages in the analyzed genomes.

homologs in the blasted genome. These numbers can be shown in a matrix [16] of which figure 1 shows a simplified example.

The cells of the matrix are colored according to the fractions of homology: the higher this percentage, the more intense a color is used. In this way even very large BLAST comparisons can still be captured in a figure that immediately reveals its information by visual inspection. An example is given in figure 2, where 28 genomes are compared of 4 *B. mallei*, 4 *B. thailandensis* and 20 *B. pseudomallei* strains. For this matrix the color scale has been adjusted to cover a wider range. From this matrix it is obvious (even without being able to read the actual numbers) that 9 *B. pseudomallei* genomes form a group within this species, and these are less homologous to the others, indicated by the lighter color of the matrix cells. The four *B. mallei* genomes are quite similar, as they report similar homology percentages (similar color intensities) for all comparisons. In contrast, the four *B. thailandensis* genomes differ considerably. It should be noted, however, that the *B. thailandensis* genome indicated by the arrow still consists of >1200 contigs; this indicates its sequence is still incomplete, and that may explain why fewer homologous genes are detected in this genome.

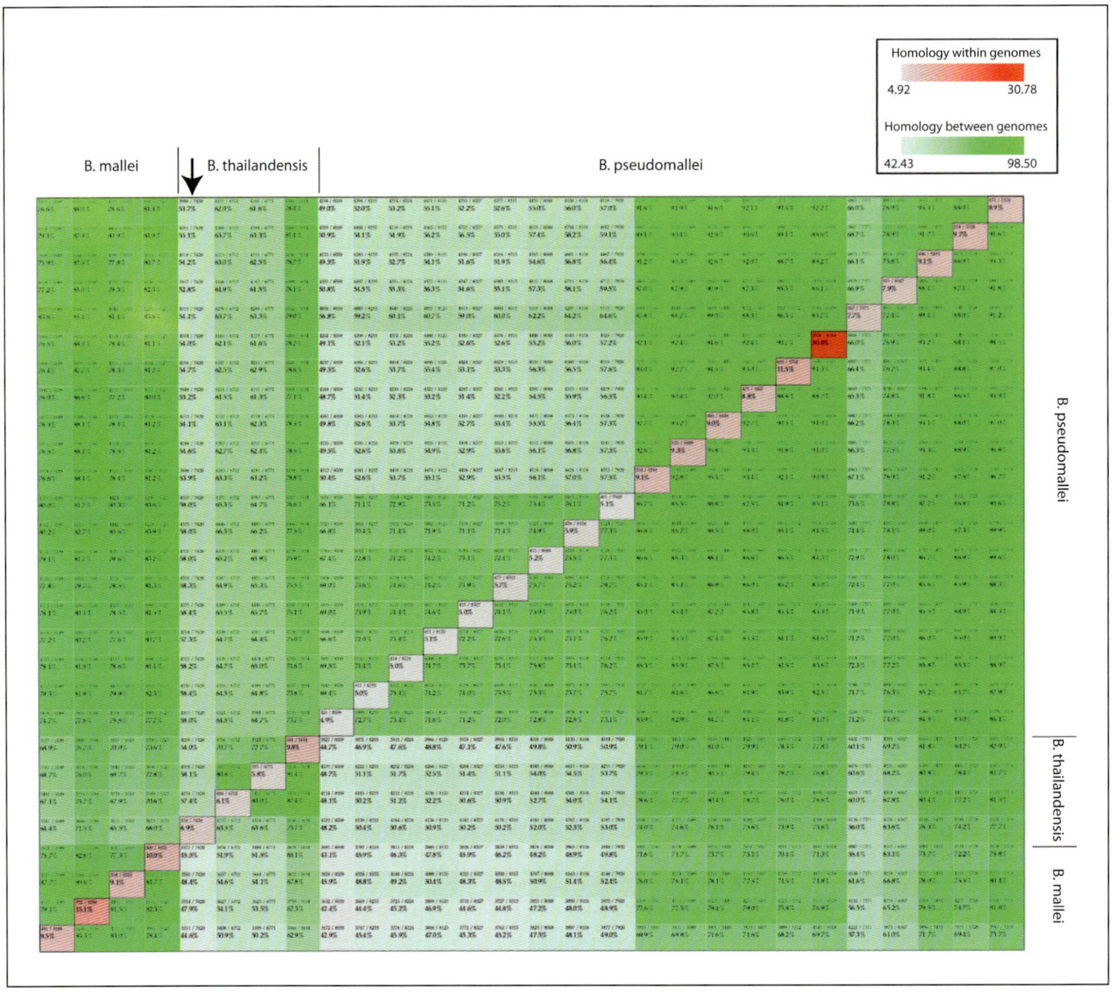

Fig. 2. BLAST matrix of 28 *Burkholderia* genomes, belonging to 4 *B. mallei*, 4 *B. thailandensis* and 20 *B. pseudomallei* strains. The arrow identifies the *B. thailandensis* MSMB43 genome whose sequence is still relatively incomplete.

Zooming in at Genes: Comparing Genomes in a BLAST Atlas

Although a BLAST matrix as shown in figure 2 gives valuable insights into which genomes are more and which are less closely related, it only reports information on the number of homologous genes. The matrix does not contain information about the identity of these genes, or whether the same set of genes is conserved in the next pairwise alignment. To capture such data, an atlas is more suitable [17].

Figure 3 shows a Genome Atlas of *B. cenocepacia* strain J2315, for all three chromosomes and the plasmid. Although the sequence had been finished a few years ago, it has only recently been published [18]. Three lanes have been added to a classical Genome

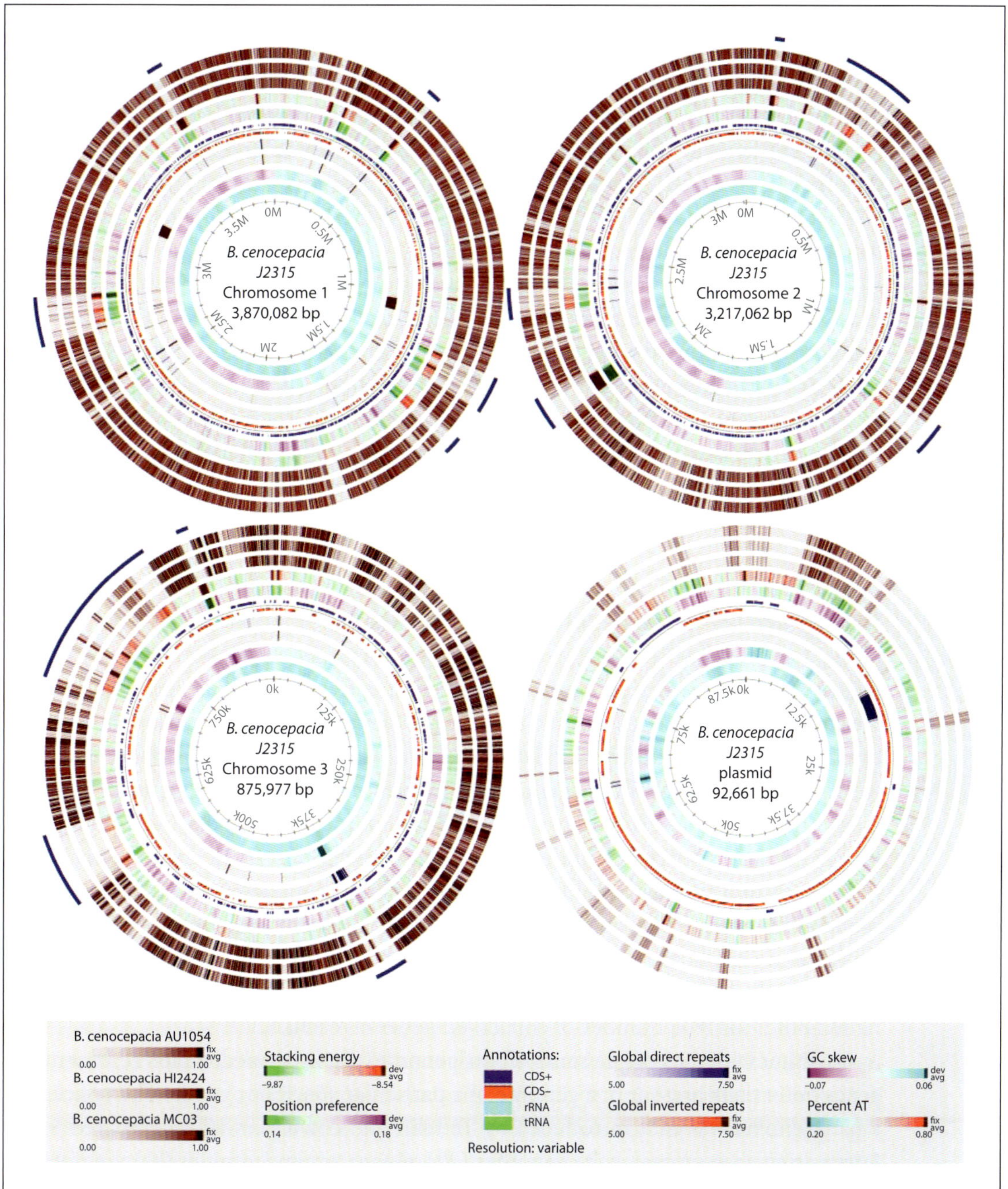

Fig. 3. Genome Atlases for the genome of *B. cenocepacia* strain *J2315*, with three BLAST lanes added for other *B. cenocepacia* genomes. The scale of the three chromosomes and the plasmid obviously differ. The location of genome islands present in J2315, recognizable by DNA structural properties and by their absence in the other genomes, is indicated by blocks around each chromosomal atlas.

Atlas (as already introduced in the first chapter of this book [19]): the outer three lanes show which genes of the J2315 genome are conserved (as identified by BLAST) in other sequenced *B. cenocepacia* strains. The figure illustrates that the largest chromosome is the most conserved of the four DNA entities, and that the plasmid is the least conserved. The BLAST lanes identify regions in the J2315 chromosomes that are not conserved in the other *B. cenocepacia* genomes. Some of these regions (marked in fig. 3) also report DNA structural properties that are unique from the rest of the chromosomes, and these happen to be the genomic islands for strain J2315. Genes present in the plasmid of strain J2315 are not found in the other three strains, except for a locus around 4–10 kb, which contains a few genes including a DNA polymerase III subunit. This kind of analysis does not reveal whether the BLAST matches are also plasmid-encoded in the other strains; in fact, neither *B. cenocepacia* AU1054 nor MC03 do carry plasmids.

Given that genomic islands are frequent in *Burkholderia* genomes [20], and most of these are species or even isolate-specific, we asked the question whether the species or even the genus can still be considered as a more-or-less uniform group, to which the concept of an evolutionary tree would still hold.

The Pan- and Core Genomes of *Burkholderia* Species

Figure 3 identifies which genes that are present in one particular *Burkholderia* genome are conserved in other genomes of the species. Such analysis can be extended to identify the fraction of genes that is always present in every *Burkholderia* genome, which we call the core genome of the genus. (A core genome was previously introduced with a less strict definition to comprise genes that are present in most individuals [21], but we use here a stricter definition). The conserved core genome can be determined for a genus or a species, provided sufficient genome sequences are available, and the sequenced strains truly represent the diversity that is out there. A core genome will decrease in size as more genomes are added, as genes that were found conserved in one lot of genomes may be lacking in a next added genome. Eventually, the curve will flatten out if the true number of conserved genes is reached.

Together with the core genome, a pan-genome can be defined, which represents all genes potentially present in a genome of a particular species or genus. The concept of a pan-genome was first introduced by Tettelin and coworkers who compared 8 different *Streptococcus agalactiae* genomes [22]. Genes or gene families that are not part of the core genome are called 'accessory' or 'auxiliary'. The pan-genome will increase with each added genome, as novel genes are discovered for each added genome. Again, this curve is expected to flatten out when the true pan-genome of a species (genus) is covered. More about pan- and core genomes is described in [8].

When the pan- and core genomes of one species (say, *B. pseudomallei*) have thus been established, a genome of a different species could be added, say a *B. mallei*, to see what effect this new species has to the pan- and core genome curves. This is illustrated

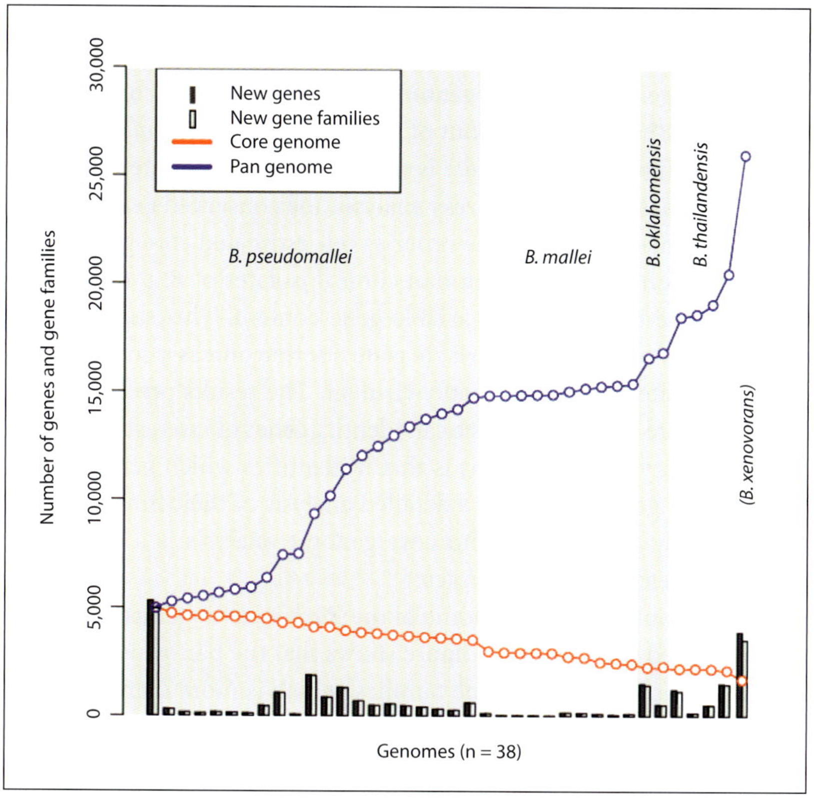

Fig. 4. Pan- and core genome plot of the Pseudomallei group currently consisting of 21 *B. pseudomallei*, 10 *B. mallei*, 4 *B. thailandensis* and 2 *B. oklahomensis* genomes. A *B. xenovorans* genome is added at the end for comparison. Within the species, the genomes are ordered for increasing numbers of genes.

in figure 4, where the Pseudomallei group is analyzed. As can be seen, the pan-genome curve for *B. pseudomallei* does not yet reach a plateau after 21 genomes; apparently, the true diversity of this species has not yet been covered. Compared to this, the curves of *B. mallei* are much more flattened, indicating less genetic diversity within this species. Note the drop in the core genome curve when leaving *B. pseudomallei* and entering *B. mallei*. This drop is caused by genes conserved in *B. pseudomallei* but not in *B. mallei*. Addition of the two *B. oklahomensis* genomes and after that the four *B. thailandensis* genomes adds quite a few genes to the pan-genome but hardly influences the core genome. In contrast, addition of *B. xenovorans* (which does not belong to the Pseudomallei group) causes a significant increase in the pan-genome and drop in the core-genome curve. This illustrates how far removed *B. xenovorans* is from the Pseudomallei group, in terms of the fraction of shared genes. Plots like these can thus assess the relatedness of isolates within and between taxonomic divisions.

From figure 4 we can see that the core genome of *B. pseudomallei* covers only approximately 4,000 of the 5,000 genes or gene families (80%) in a single genome whereas the pan-genome easily comprises 15,000 genes (remember that the pan-genome is an artificial sum of all genes encountered in the analyzed genomes and by far exceeds the number of genes in a single genome). For *B. mallei*, the core genome comprises approximately 58% (2,800 genes out of 4,800) of a small *B. mallei* genome (this cannot be read from figure 4 as *B. mallei* is not the first species listed here). In an experimental approach based on micro-array analysis, the conserved gene fraction of *B. pseudomallei* was estimated in the same order as our estimated core genome, as 85% [23]. Their findings pointed out that human clinical isolates of *B. pseudomallei* clustered together on a tree based on the variable gene content. This suggests that virulence potential is largely coded in the variable gene fraction and as a consequence not all *B. pseudomallei* isolates would be equally virulent. The results presented here illustrate how a pan- and core genome analysis can identify genes of interest for pathogenicity research. The beauty of this analysis is that it identifies which genes belong to the variable fraction of a genome, so that a detailed analysis of their functions and interrelationships can easily follow. Pan- and core genome analysis is a promising strategy to include in the field of pathogenomics.

Figure 5 represents the pan- and core genome of the *Burkholderia* genus, extracted from all currently sequenced genomes. The figure shows that the pan-genome of the genus *Burkholderia* contains over 40,000 gene families, which is more than the number of genes present in a human genome. The large number of gene families of this genus is most likely due to the enormous diversity within this genus. The core genome of the genus, however, has decreased to only a few hundred genes that are conserved across all *Burkholderia* genomes.

Phylogenetic Trees

One simple analysis to perform for any complete or incomplete genome is to extract the 16S rRNA *(rrn)* gene(s) and to produce a tree including related isolates or species, as this can be used as confirmation that the correct DNA was sequenced. Examples of the 'wrong' organism being sequenced exist, and can arise from contamination during cultivation, DNA extraction, cloning and sequencing or even due to contamination (over-writing) of sequencing files. Incomplete genome sequences do not always include the *rrn* genes, as these are often repeated on a chromosome, and such repeats complicate the assembly process, so that they are temporarily removed from the raw sequences.

Figure 6 shows a phylogenetic tree based on 16S rRNA extracted from 56 genomes. As expected, there is little resolution within a species, due to the high degree of similarity of the 16S rRNA sequences from the same species. In light of the assumed ancestry of *B. mallei*, it is not surprising that the *B. pseudomallei* and *B. mallei* genes are somewhat mixed up, as nearly all of these are very similar (the long branch of *B. pseudomallei* 305 is probably an artefact due to a sequencing error, as this genome is not finished yet), and they are clearly separated from the BCC group (which are all

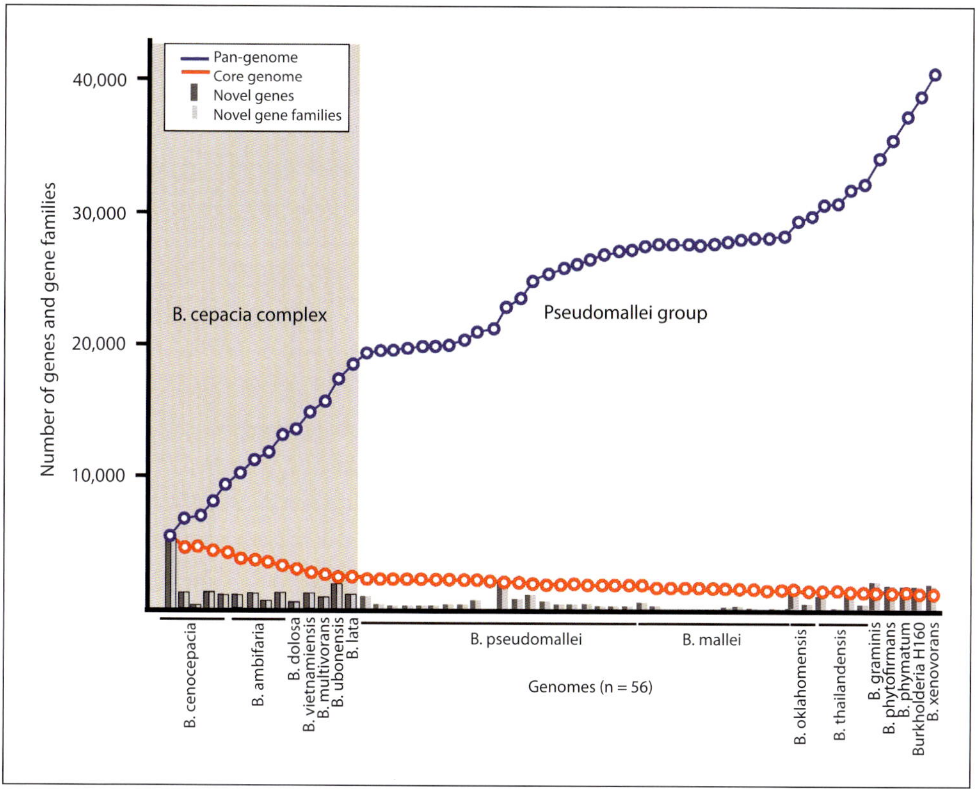

Fig. 5. Pan- and core genome plot of all 56 genome sequences from table 1, sorted for group and species. The BCC complex is plotted first, followed by the Pseudomallei group and last the species that do not belong to any group.

depicted in shades of blue). However, the *B. thailandensis* 16S rRNA genes are positioned as outliers of the Pseudomallei group, and one of them is somewhat in between that and the BCC group (indicated by an arrow). Moreover, the two *B. oklahomensis* 16S rRNA genes do not cluster within the Pseudomallei group, where they would be if their 'Pseudomallei-like' nature was reflected by their 16S rRNA. Finally, *B. ubonensis* is an outlier, and not positioned within the BCC group where it was reported previously [24]. Note, however, that the *rrn* sequence was extracted from a rather premature genome sequence (it was still in 1143 contigs) so it may still contain sequencing errors. Matching our expectations are *B. xenovorans*, *B. phytofirmans* and *B. phymatum* that are only distantly related to the other species. The unspecified genome, of isolate H160, has a ribosomal gene quite different to all other *Burkholderia* genes analyzed.

The method of MLST is used to analyze population genetics within a species, or between members of closely related species. For *Burkholderia*, partial sequences of 7 genes are usually analyzed but different schemes exist [25, 26]. We extracted the DNA fragments described in reference 24 from the genomes and analyzed these as one

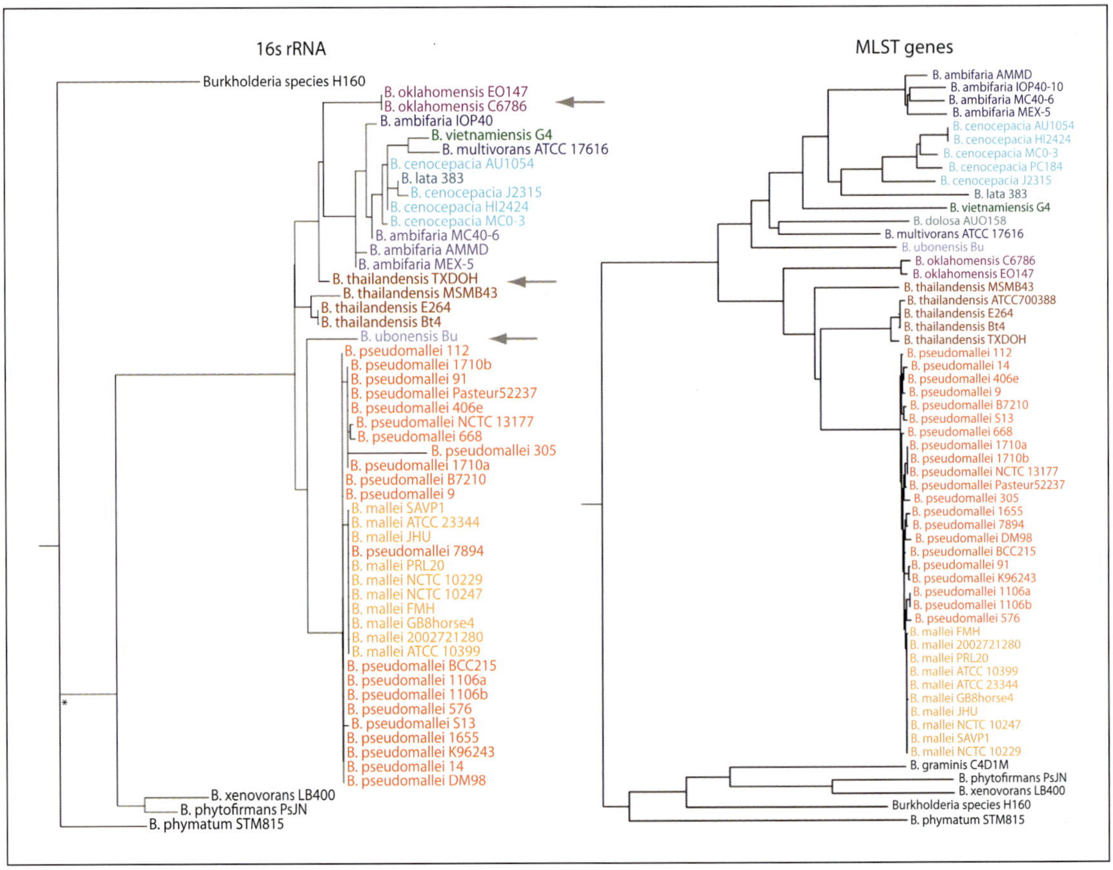

Fig. 6. To the left: a phylogenetic tree of the 16S rRNA gene (*rrn*) extracted from 53 genome sequences. One gene per genome was analyzed. *B. cenocepacia* PC184, *B. graminis* and *B. dolosa* were excluded, due to the lack of a full length 16S rRNA gene in these partially sequenced genomes. Genomes are color-coded according to species. Grey arrows indicate genes positioned different from expectations. The node for *B. phymatum* produced low bootstrap values (<500/1,000), indicated by an asterisk. To the right: phylogenetic tree of 7 concatenated MLST genes [24] extracted from 56 genomes.

artificially concatenated piece. This produced a tree (by neighbor joining) as shown to the right of figure 6. In this tree all proposed members of the Pseudomallei group cluster together with *B. thailandensis* and *B. oklahomensis* as closely related, and all members of the BCC group cluster as well. So this tree, based on all MLST genes combined, matches the currently used grouping better than the tree based on the *rrn* gene. *Burkholderia* species H160 could not be analyzed as its MLST genes were not yet completely sequenced.

Would the addition of more genes produce a similar tree? After all, MLST genes are supposed to be marker genes for the genetic relationship of most of the genome. The problem is that genes can be exchanged between (and within) species by horizontal gene transfer, so that they no longer produce consistent trees. To get around

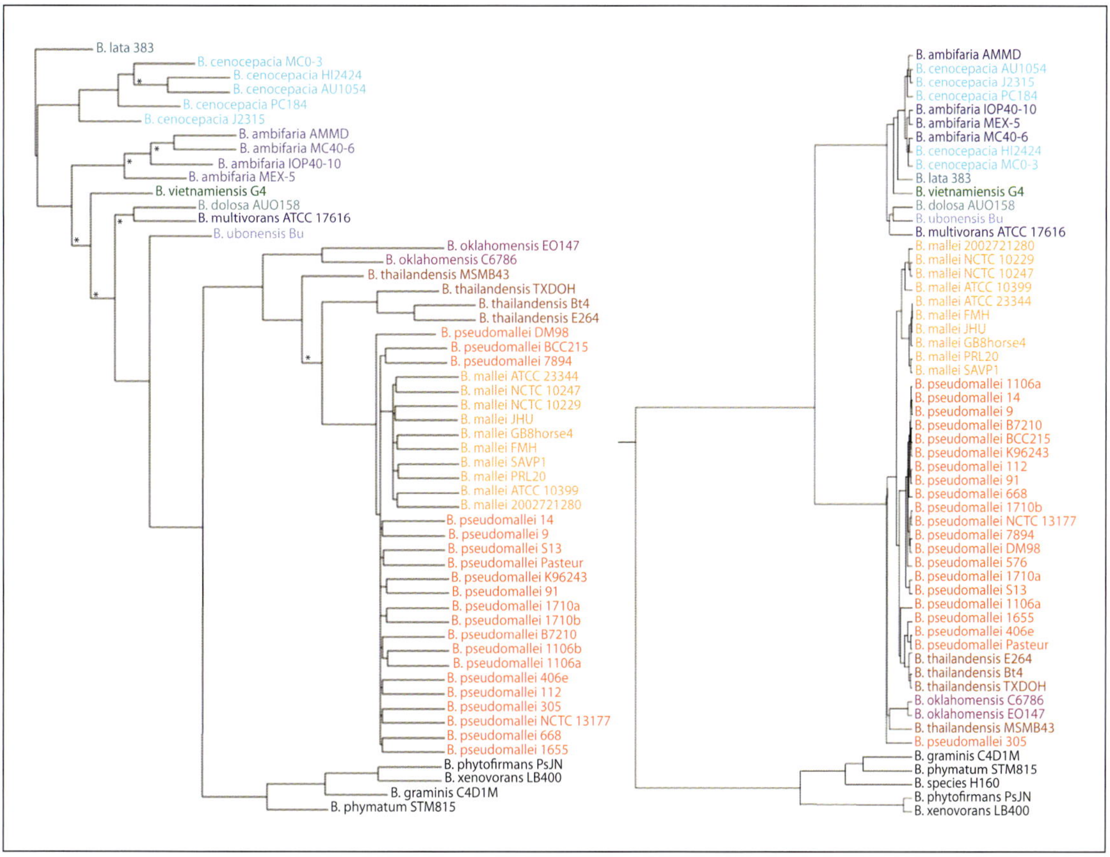

Fig. 7. The tree on the left is based on 612 protein genes that gave consistent trees when individually analyzed. Bootstrap values below 50/100 are indicated with an asterisk. The clustering on the right is based on the observed frequency of tetranucleotides compared to expected values, using a first-order Markov chain model. Such a clustering is independent of genes.

this, we identified those genes that produce consistent trees, so as to concentrate on genes to be least influenced by horizontal gene transfer. The tree to the left of figure 7 is based on 612 genes that are part of the *Burkholderia* core genome and produced consistent trees. Note that this is only about 12–15% of all genes in a given genome. The tree clearly separates the BCC group, the Pseudomallei group and those species not dedicated to any group. The biggest difference between the tree in figure 7 and the MLST tree in figure 6 is that the genomes now produce branches within a species, as there is more intra-species variation between 612 genes than between 7 (MLST) genes. We believe that figure 7 is a more complete representation of the true similarity and differences of these investigated organisms than the MLST tree provides.

All analyses presented so far concentrated on RNA or protein-coding genes, but it is also possible to compare the complete DNA sequence of the genome, irrespective of what the nucleotides code for. One way to do so is to compare the frequency of oligomers, such as tetranucleotides, and compare this distribution to statistically expected values. The latter can be calculated in various ways, for example based on a first-order Markov chain model. The result is a genomic signature that is likely to be reflective of an organism's environment, as well as reflective of relatedness [27]. This 'genomic signature' is not affected by the number of contigs of a genome sequence and is independent of where on the genome it is searched for. The panel to the right of figure 7 shows such a clustering, and in general the observed arrangement is in agreement with the groupings of the other trees. It is reassuring that two completely independent methods result in similar clusters, and this suggests that these groupings are a true reflection of biological relationship.

In summary, we find that determining the taxonomic grouping of several of the *Burkholderia* species, based on their genomic sequences, is possible, but we suggest not to base this on a single (as in *rrn* analysis) or a few (as in MLST) genes, but rather to analyze a large number of genes or the complete DNA sequence, in order to optimally reflect the true genetic relationship between organisms. With the number of bacterial genome sequences steadily increasing, this approach will become more and more applicable to other species as well.

Acknowledgement

We thank the several sequencing centers that have deposited unfinished genomic data into the RefSeq database at NCBI. In particular, we would like to thank Tim Reed for kindly providing us with permission to use the as yet unpublished sequences of 15 *Burkholderia* genomes.

References

1 Godoy D, Randle G, Simpson AJ, Aanensen DM, Pitt TL, et al: Multilocus sequence typing and evolutionary relationships among the causative agent of melioidosis and glanders, *Burkholderia pseudomallei* and *Burkholderia mallei*. J Clin Microbiol 2003; 41:2068–2079.

2 Glass MB, Steigerwalt AG, Jordan JG, Wilkins PP, Gee JE: *Burkholderia oklahomensis* sp. nov., a *Burkholderia pseudomallei*-like species formerly known as the Oklahoma strain of *Pseudomonas pseudomallei*. Int J Syst Evol Microbiol 2006;56:2171–2176.

3 Vanlaere E, Lipuma JJ, Baldwin A, Henry D, De Brandt E, et al: *Burkholderia latens* sp. nov., *Burkholderia diffusa* sp. nov., *Burkholderia arboris* sp. nov., *Burkholderia seminalis* sp. nov. and *Burkholderia metallica* sp. nov., novel species within the *Burkholderia cepacia* complex. Int J Syst Evol Microbiol 2008;58:1580–1590.

4 Yabuuchi E, Kawamura Y, Ezaki T, Ikedo M, Dejsirilert S, et al: *Burkholderia uboniae* sp. nov., L-arabinose-assimilating but different from *Burkholderia thailandensis* and *Burkholderia vietnamiensis*. Microbiol Immunol 2000;44:307–317.

5 Vanlaere E, Baldwin A, Gevers D, Henry D, De Brandt E, et al: Taxon K, a complex within the *Burkholderia cepacia* complex, comprises at least two novel species, *Burkholderia contaminans* sp. nov. and *Burkholderia lata* sp. nov. Int J Syst Evol Microbiol 2009;59:102–111.

6 Nierman WC, DeShazer D, Kim HS, Tettelin H, Nelson KE, et al: Structural flexibility in the *Burkholderia mallei* genome. Proc Natl Acad Sci USA 2004;101:14246–14251.

7 Holden MT, Titball RW, Peacock SJ, Cerdeño-Tárraga AM, Atkins T, et al: Genomic plasticity of the causative agent of melioidosis, *Burkholderia pseudomallei*. Proc Natl Acad Sci USA 2004;101: 14240–14245.
8 Ussery DW, Borini S, Wassenaar TM: Computing for Comparative Microbial Genomics: Bioinformatics for Microbiologists (Computational Series). Springer Verlag London, 2008.
9 Ong C, Ooi CH, Wang D, Chong H, Ng KC, et al: Patterns of large-scale genomic variation in virulent and avirulent *Burkholderia* species. Genome Res 2004;14:2295–2307.
10 Kim HS, Schell MA, Yu Y, Ulrich RL, Sarria SH, et al: Bacterial genome adaptation to niches: divergence of the potential virulence genes in three *Burkholderia* species of different survival strategies. BMC Genomics 2006;6:174.
11 Chain PS, Denef VJ, Konstantinidis KT, Vergez LM, Agulló L, et al: *Burkholderia xenovorans* LB400 harbors a multi-replicon, 9.73-Mbp genome shaped for versatility. Proc Natl Acad Sci USA 2006;103:15280–15287.
12 Winsor GL, Khaira B, Rossum TV, Lo R, Whiteside MD, Brinkman FS: The *Burkholderia* Genome Database: facilitating flexible queries and comparative analysis. Bioinformatics 2008;24:2803–2804.
13 Fukuchi S, Nishikawa K: Estimation of the number of authentic orphan genes in bacterial genomes. DNA Res 2004;11:219–231.
14 Nielsen P, Krogh A: Large-scale prokaryotic gene prediction and comparison to genome annotation. Bioinformatics 2005;21:4322–4329.
15 Larsen TS, Krogh A: EasyGene – a prokaryotic gene finder that ranks ORFs by statistical significance. BMC Bioinformatics 2003;4:21.
16 Binnewies TT, Hallin PF, Staerfeldt HH, Ussery DW: Genome Update: proteome comparisons. Microbiology 2005;151:1–4.
17 Hallin PF, Binnewies TT, Ussery DW: The genome BLAST atlas – a GeneWiz extension for visualization of whole-genome homology. Mol Biosyst 2008; 4:363–371.
18 Holden MT, Seth-Smith HM, Crossman LC, Sebaihia M, Bentley SD, et al: The genome of *Burkholderia cenocepacia* J2315, an epidemic pathogen of cystic fibrosis patients. J Bacteriol 2009;191:261–277.
19 Wassenaar TM, Bohlin J, Binnewies TT, Ussery DW: Genome comparison of bacterial pathogens. Genome Dyn 2009;6:1–20.
20 Tuanyok A, Leadem BR, Auerbach RK, Beckstrom-Sternberg SM, Beckstrom-Sternberg JS, et al: Genomic islands from five strains of *Burkholderia pseudomallei*. BMC Genomics 2008;9:566.
21 Lan R, Reeves PR: Intraspecies variation in bacterial genomes: the need for a species genome concept. Trends Microbiol 2000;8:395–401.
22 Tettelin H, Masignani V, Cieslewicz MJ, Donati C, Medini D, et al: Genome analysis of multiple pathogenic isolates of *Streptococcus agalactiae*: implications for the microbial 'pan-genome'. Proc Natl Acad Sci USA 2005;102:13950–13955.
23 Sim SH, Yu Y, Lin CH, Karuturi RK, Wuthiekanun V, et al: The core and accessory genomes of *Burkholderia pseudomallei*: implications for human melioidosis. PLoS Pathogens 2008;4:e1000178.
24 Tayeb LA, Lefevre M, Passet V, Diancourt L, Brisse S, Grimont PA: Comparative phylogenies of *Burkholderia, Ralstonia, Comamonas, Brevundimonas* and related organisms derived from *rpoB*, *gyrB* and *rrs* gene sequences. Res Microbiol 2008; 159:169–177.
25 Godoy D, Randle G, Simpson AJ, Aanensen DM, Pitt TL, et al: Multilocus sequence typing and evolutionary relationships among the causative agents of melioidosis and glanders, *Burkholderia pseudomallei* and *Burkholderia mallei*. J Clin Microbiol 2003;41:2068–2079. Erratum in: J Clin Microbiol 2003;41:4913.
26 Baldwin A, Mahenthiralingam E, Thickett KM, Honeybourne D, Maiden MC, et al: Multilocus sequence typing scheme that provides both species and strain differentiation for the *Burkholderia cepacia* complex. J Clin Microbiol 2005;43:4665–4673.
27 Bohlin J, Skjerve E, Ussery DW: Reliability and applications of statistical methods based on oligonucleotide frequencies in bacterial and archaeal genomes. BMC Genomics 2008;9:104.

David W. Ussery
Center for Biological Sequence Analysis, Department of Systems Biology
Building 208, Technical University of Denmark
DK–2800 Lyngby (Denmark)
Tel. +45 45 25 24 88, Fax +45 45 93 15 85, E-Mail dave@cbs.dtu.dk

de Reuse H, Bereswill S (eds): Microbial Pathogenomics.
Genome Dyn. Basel, Karger, 2009, vol 6, pp 158–169

Genomics of Host-Restricted Pathogens of the Genus *Bartonella*

P. Engel · C. Dehio

Biozentrum, University of Basel, Basel, Switzerland

Abstract

The α-proteobacterial genus *Bartonella* comprises numerous arthropod-borne pathogens that share a common host-restricted life-style, which is characterized by long-lasting intraerythrocytic infections in their specific mammalian reservoirs and transmission by blood-sucking arthropods. Infection of an incidental host (e.g. humans by a zoonotic species) may cause disease in the absence of intraerythrocytic infection. The genome sequences of four *Bartonella* species are known, i.e. those of the human-specific pathogens *Bartonella bacilliformis* and *Bartonella quintana*, the feline-specific *Bartonella henselae* also causing incidental human infections, and the rat-specific species *Bartonella tribocorum*. The circular chromosomes of these bartonellae range in size from 1.44 Mb (encoding 1,283 genes) to 2.62 Mb (encoding 2,136 genes). They share a mostly synthenic core genome of 959 genes that features characteristics of a host-integrated metabolism. The diverse accessory genomes highlight dynamic genome evolution at the species level, ranging from significant genome expansion in *B. tribocorum* due to gene duplication and lateral acquisition of prophages and genomic islands (such as type IV secretion systems that adopted prominent roles in host adaptation and specificity) to massive secondary genome reduction in *B. quintana*. Moreover, analysis of natural populations of *B. henselae* revealed genomic rearrangements, deletions and amplifications, evidencing marked genome dynamics at the strain level.

Until the early 1990s, the genus *Bartonella* comprised a single species, *B. bacilliformis*. Since then, the reclassification of previously described bacteria based on 16S rRNA sequences (i.e., *Grahamella* and *Rochalimea*) and the description of novel *Bartonella* species isolated from various animal reservoirs resulted in a major expansion of the genus to currently 19 approved species, one of which *(Bartonella vinsonii)* is split into 3 subspecies. Among those, nine have been associated with human diseases (fig. 1) [1, 2]. The arthropod-borne bartonellae are widespread pathogens that colonize mammalian endothelial cells and erythrocytes as major target cells [3]. While endothelial cells and potentially other nucleated cells may get infected in both reservoir and incidental hosts, erythrocyte invasion takes place exclusively in the reservoir host,

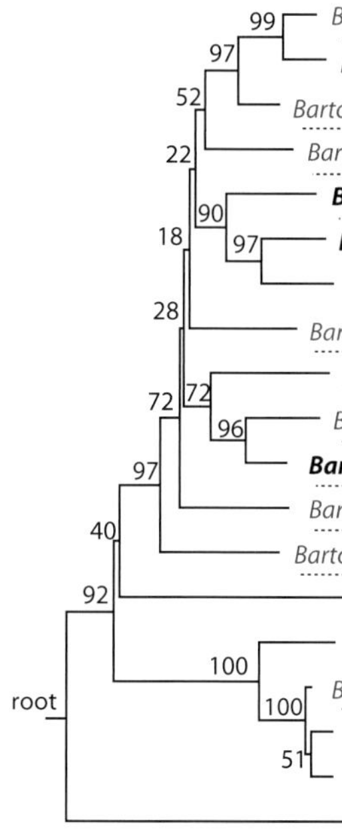

		Important GI			
	Hosts	*vbh*	*virB*	*trw*	Flagellae
Bartonella vinsonii ssp. berkhoffi	Dog (Human)	–	+	+	–
Bartonella vinsonii ssp. vinsonii	Vole	–	+	+	–
Bartonella vinsonii ssp. arupensis	Mouse (Human)	–	+	+	–
Bartonella taylorii	Mouse, Vole	–	+	+	–
Bartonella quintana	**Human**	–	+	+	–
Bartonella henselae	**Cat (Human)**	–	+	+	–
Bartonella koehlerae	Cat (Human)	–	+	+	–
Bartonella alsatica	Rabbit (Human)	–	+	+	–
Bartonella grahamii	Mouse, Vole (Human)	+	+	+	–
Bartonella elizabethae	Rat (Human)	+	+	+	–
Bartonella tribocorum	**Rat**	+	+	+	–
Bartonella birtlesii	Mouse	+	+	+	–
Bartonella doshiae	Vole	+	+	+	–
Bartonella clarridgeiae	Cat (Human)	–	+	–	+
Bartonella bovis	Cattle	+	–	–	–
Bartonella capreoli	Roe Deer	+	–	–	+
Bartonella chomelii	Cattle	+	–	–	+
Bartonella schoenbuchensis	Roe Deer	+	–	–	+
Bartonella bacilliformis	**Human**	–	–	–	+

Fig. 1. Phylogeny and epidemiology of the genus *Bartonella*, distribution of important genomic islands (GI) encoding virulence factors, and presence/absence of flagella. For zoonotic species, man as an incidental host is indicated in brackets. Species with known genome sequences are highlighted in bold. The phylogenetic tree was calculated on the basis of protein sequences of *rpoB*, *groEL*, *ribC*, and *gltA* as described by [9]. Numbers at the nodes of the tree indicate bootstrap values for 1,000 replicates. Except for *Bartonella talpae* and *Bartonella peromysci*, for which no type strains exist, all approved species are included in the tree.

resulting in the establishment of a long-lasting intraerythrocytic bacteremia. Despite the fact that most *Bartonella* species are restricted to one reservoir host, there is an increasing body of evidence that some species can infect several different mammalian hosts [4–8]. The bartonellae represent an interesting model to study the evolution of host adaptation/host restriction as most mammals infested by blood-sucking arthropods serve as a reservoir host for at least one *Bartonella* species [9].

The highly virulent human-specific pathogen *B. bacilliformis* (causing life-threatening Oroya fever and verruga peruana) holds an isolated position in the *Bartonella* phylogeny as sole representative of an ancestral lineage. All other species evolved in a separate 'modern' lineage by radial speciation. These modern species represent host-adapted pathogens of rather limited virulence potential within their diverse mammalian reservoirs. Examples are the human-specific species *B. quintana* causing trench

Table 1. General features of *Bartonella* genome sequences. PCG, protein-coding genes; n.d., not determined. The coding content of *B. bacilliformis* and *B. tribocorum* were (re-)calculated by dividing the total length of all protein-coding genes and tRNA/rRNA coding regions by the chromosome length. In addition, the average length of PCG was calculated for *B. bacilliformis* by dividing the total length of all PCG by the number of PCG.

	B. bacilliformis	*B. tribocorum*	*B. henselae*	*B. quintana*
Chromosome size	1,445,021 bp	2,619,061 bp	1,931,047 bp	1,581,384 bp
G+C content	38.2%	38.8% (35.0%)[a]	38.2%	38.8%
Total number of PCG	1,283	2,136 (18)[a]	1,488	1,142
Average length of PCG	909 bp	906 bp	942 bp	999 bp
Integrase remnants	n.d.	47 (0)[a]	43	4
Number of rRNA operons	2	2 (0)[a]	2	2
Number of tRNA genes	44	42 (0)[a]	44	44
Percentage coding	81.6%	74.6% (69.8%)[a]	72.3%	72.7%
Plasmid	0	1 (23,343 bp)[a]	0	0

[a] Numbers in brackets refer to the plasmid

fever, the cat-adapted zoonotic pathogen *B. henselae* causing cat-scratch-disease and various other disease manifestations in the incidental human host, and the rat-specific pathogen *B. tribocorum* not yet associated with human infection (fig. 1). Over the last decade, the availability of animal and cell culture infection models in combination with powerful bacterial genetics has facilitated research aiming at understanding the cellular and molecular interactions that contribute to the complex relationship between *Bartonella* and its mammalian hosts [1–3]. More recently, *Bartonella* has entered the post-genomic era by the release of several complete genome sequences. Here, we summarize the comparative and functional genomic studies on *Bartonella* that have been reported to date.

General Features of *Bartonella* Genomes

Complete genome sequences are presently available for four *Bartonella* species, i.e., *B. henselae* and *B. quintana* [10], *B. tribocorum* [9], and *B. bacilliformis* (GenBank accession no. CP000525). Additionally, the genome composition of *Bartonella koehlerae* has been analyzed by comparative genomic hybridization profiling (CGH) based on the genome sequence of the closely related species *B. henselae* [11]. The four available *Bartonella* genomes are composed of single circular chromosomes (plus one plasmid in *B. tribocorum*), which display a uniformly low G+C content of 38.2% to 38.8%, and a noteworthy low coding density of 72.3% to 81.6% (table 1). The chromosome sizes range from 1,445 kb (encoding 1,283 genes) for *B. bacilliformis* to 2,619 kb (encoding

2,136 genes) for *B. tribocorum* (table 1, fig. 2). Orthologous gene assignments resulted in the identification of a core genome of 959 genes [9], which is encoded by a rather well conserved chromosomal backbone in a largely synthenic manner (fig. 2, see dotplots).

The relatively small core genome of the bartonellae reflects specific adaptations to the genus-specific lifestyle. For instance, a striking example of host-integrated metabolism is represented by hemin. This important source for iron and porphyrin is particularly abundant in the host niches colonized by bartonellae, i.e. the intracellular space of erythrocytes and the midgut lumen of blood-sucking arthropods. The strict hemin requirement for growth of *B. quintana* (and probably other bartonellae) in vitro correlates with the presence of multiple genes encoding hemin binding and hemin uptake proteins, while no hemin biosynthesis enzyme is encoded by this organism [10]. A large-scale mutagenesis screen in the *B. tribocorum*-rat model identified several of the hemin-uptake genes as essential for establishing intraerythrocytic infection. Moreover, this screen revealed that the majority of pathogenicity factors required for establishing intraerythrocytic bacteremia is encoded by the core genome inferred from the four available *Bartonella* genome sequences (66 of 97 pathogenicity genes) [9], indicating that this genus-specific infection strategy is to a large extent dependent on a conserved set of core genome-encoded pathogenicity factors.

Genome Dynamics by Lineage-Specific Expansion and Reduction

Despite of a largely synthenic core genome, the known *Bartonella* genomes are diversified by the variable size and composition of their accessory genomes. These were shaped in evolution by massive expansions (due to lateral gene transfer and gene duplication) and reductions (due to gene decay and deletion), which mostly occurred in a lineage-specific manner.

A marked example for genome reduction is *B. quintana*, which shares 1,106 orthologous genes with *B. henselae* as its closest relative (fig. 1). *B. henselae* codes for 382 genes without orthologs in *B. quintana*, while only 36 genes are unique to *B. quintana* [9, 10]. Interestingly, *Rickettsia prowazekii* representing another pathogen transmitted by the human body louse has also undergone recent genome reduction, suggesting that the extensive genome decay in the *B. quintana* lineage may be related to the biology of this arthropod vector [10]. However, *B. bacilliformis*, a pathogen vectored by the sandfly *Lutzomyia verrucarum*, displays also a remarkably small genome sequence, indicating that adaptation to humans could be accompanied by reductive genome evolution. Consistently, several of the more recently evolved human-specific pathogens display marked genome decay, e.g. *Salmonella typhi* and *Mycobacterium leprae* [12].

With an accessory genome exceeding the size of the core genome (1,195 vs. 959 genes), *B. tribocorum* represents a remarkable example of lineage-specific genome

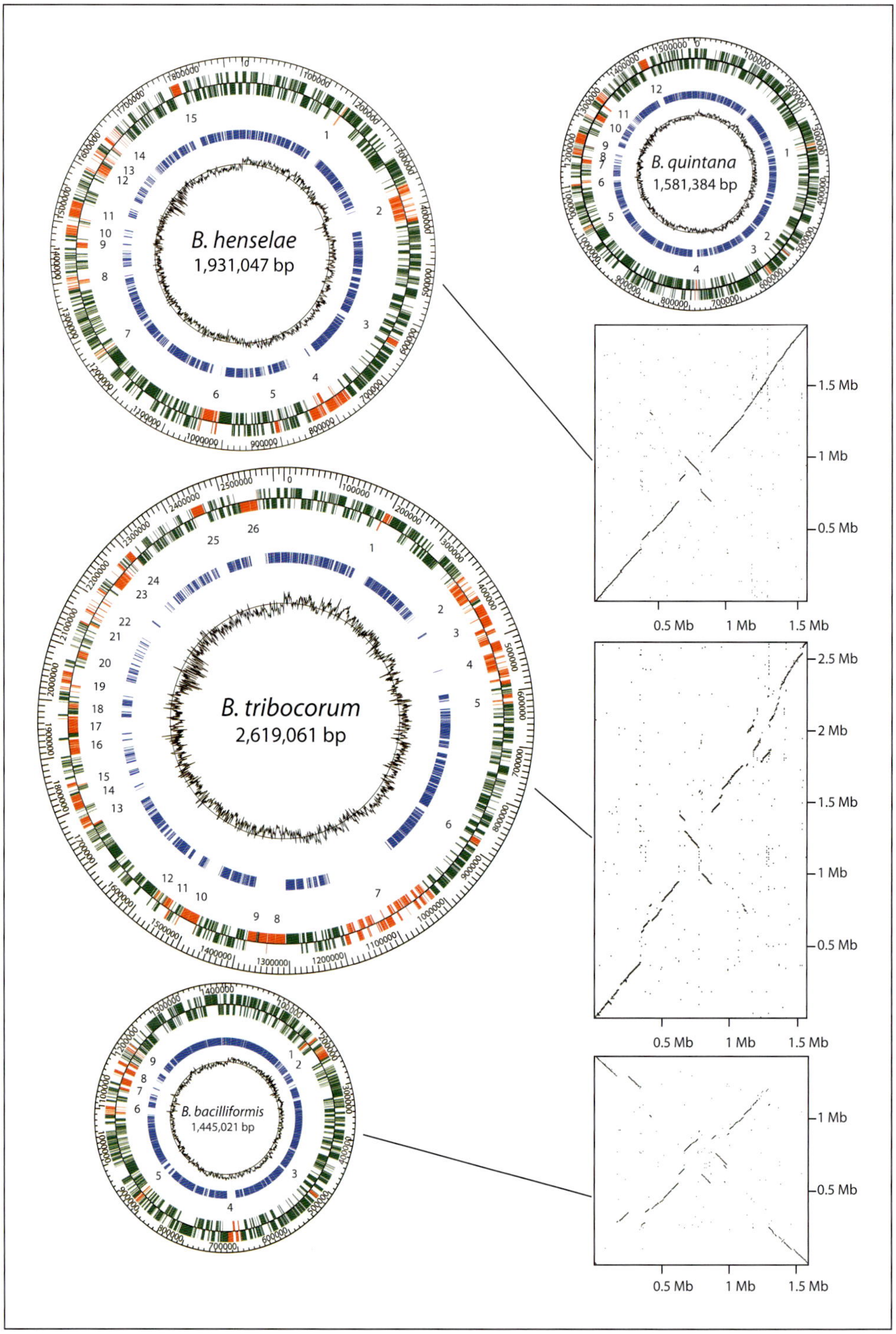
B. henselae
1,931,047 bp
B. quintana
1,581,384 bp
B. tribocorum
2,619,061 bp
B. bacilliformis
1,445,021 bp
0.5 Mb
1 Mb
1.5 Mb
2 Mb
2.5 Mb

expansion, and to a lesser extent genome expansion is also evident in *B. henselae* (accessory genome of 529 genes). The primary source for these genome expansions are prophages and other laterally acquired genomic islands (GIs, table 2 and fig. 2). One phage-related GI is conserved in all known *Bartonella* genomes (table 2; BB-GI2 and homologs). *B. tribocorum* and *B. henselae* encode in addition large (>50 kb) prophage regions (table 2; BH-GI2, BT-GI2/4) that are homologous but highly plastic in their genetic organization [10]. These mosaic prophage regions and the related GIs encoding homologous phage genes were probably shaped during evolution by a consecutive acquisition of different prophages, followed by duplication, excision, reintegration, and reduction of prophage segments of different size and origin. Exclusively *B. tribocorum* encodes another large prophage (>30 kb) that, moreover, is present in multiple copies (table 2; BT-GI8/10/17/26). The different copies of this prophage display a strictly conserved gene order (fig. 3a) and a marked similarity to the genetic organization and sequence of P2- and Mu-like prophages described in other bacterial taxa. GIs encoding two-partner secretion systems, which often also carry phage genes, have also contributed to the large accessory genomes of *B. tribocorum* and *B. henselae* (table 2; BH-GI4/6, and BT-GI3/7/9/11). Remnants of these GIs are found in the reduced genome of *B. quintana*, while they are absent from the ancestral *B. bacilliformis* lineage and closely related α-proteobacterial taxa. A prototype of these GIs was thus likely acquired by the common ancestor of the modern *Bartonella* lineage, followed by lineage-specific expansions and reductions. At present it is unknown whether the prophages, phage-related GIs and GIs encoding two-partner secretion systems, that contributed to the remarkable genome expansion exemplified by *B. tribocorum* and *B. henselae*, have any beneficial role in host interaction, or whether these two species are just not under the selective pressure that resulted in massive genome reduction in *B. quintana*.

Some other GIs constituting the accessory genomes of the bartonellae are well established pathogenicity factors with important roles in the process of host colonization. Unlike *B. bacilliformis*, all species of the modern lineage encode at least one of the closely related type IV secretion systems (T4SSs) VirB/VirD4 or Vbh (VirB homolog) (fig. 1), which likely emanated from an ancestral duplication event and which are redundant in function. These VirB-like T4SSs are considered to represent major host adaptability factors that contributed to the remarkable evolutionary success of the modern lineage [9]. T4SSs are transporters ancestrally related to bacterial conjugation systems that mediate the vectorial translocation of virulence factors across the

Fig. 2. Circular genome maps of the four *Bartonella* genome sequences and Dot-plot representation of genome colinearity (micro-syntheny). The genome maps indicate (outside circles to inside circles) the genes on the + and – strands (genes located on genomic islands which are >5 kb or encoding more than five CDS are colored in red, all other genes in green), the genes belonging to the core genome (in blue), and the GC skew (black). Dot-plots were plotted for the *B. quintana* genome against any other genome for a sliding window of 20 nucleotides. Numbers in the genome circles refer to the different genomic islands (see also table 2).

Table 2. List of genomic islands (GIs) >10 kb of the four known *Bartonella* genomes. The first and last gene of each island is indicated by its locus tag (only the number of each locus tag is shown). The length refers to the start and end of the first and last gene of the island, respectively.

GI#	Similar genomic Islands	Description	tRNA	Begin	End	Length
B. bacilliformis						
BB-GI2	BT-GI20/23, BH-GI6/12, BQ-GI10	*Bartonella*-specific island encoding phage genes	yes	0217	0240	22115
BB-GI4		duplicated genomic region encoding housekeeping genes	no	0679	0710	26295
BB-GI5		conserved exported protein and transporter encoding genes	yes	0883	0894	10151
BB-GI6		conserved exported protein and phage genes	yes	1055	1080	17466
BB-GI8	BT-GI13, BH-GI10, BQ-GI8	flagella genes and inducible *Bartonella* autotransporter (*iba*) genes	yes	1116	1160	46499
BB-GI9		conserved exported protein and phage-related genes	no	1180	1190	12068
B. tribocorum						
BT-GI1		BT-specific helicase and phage-related genes	yes	0156	0167	15612
BT-GI2	BH-GI2, BQ-GI1	phage island	yes	0303	0377	51254
BT-GI3	BH-GI4/6	type II secretion system island	yes	0387	0422	44997
BT-GI4	BH-GI2/6	phage island	yes	0423	0564	110682
BT-GI5		BT-specific island encoding predicted membrane proteins	yes	0577	0596	17292
BT-GI6	BH-GI3, BQ-GI2, BB-GI3	putative membrane proteins not present in other alphaproteobacteria	no	0832	0834	11826
BT-GI7	BH-GI2/4/5/6	phage genes, type II secretion systems and helicase genes	yes	0941	1122	181527
BT-GI8		BT-specific phage island I	yes	1218	1283	53256
BT-GI9		BT-specific type II secretion systems and hypothetical genes	yes	1292	1301	18348
BT-GI10		BT-specific phage island II	yes	1382	1429	37682
BT-GI11	BH-GI4	type II secretion system island	no	1446	1464a	18888
BT-GI13	BH-GI10, BQ-GI8, BB-GI8	inducible *Bartonella* autotransporter (*iba*) genes	no	1650	1663	21879

Table 2. Continued

GI#	Similar genomic Islands	Description	tRNA	Begin	End	Length
BT-GI14	BH-GI11, BQ-GI9	VirB T4SS and *Bartonella* effector protein (Bep) genes	no	1689	1710	25598
BT-GI16	BH-GI9, BQ-GI7, BB-GI7	conserved *Bartonella*-specific autotransporter encoding genes	no	1785	1796	28492
BT-GI17		BT-specific phage island III	yes	1810	1849	32182
BT-GI19	BH-GI8, BQ-GI6	transporter-associated genes, and restriction system specific to BT	yes	1897	1930	35415
BT-GI20	BH-GI6/12, BQ-GI10, BB-GI2	*Bartonella*-specific island encoding phage genes	yes	1965	1983	12384
BT-GI22	BH-GI14, BQ-GI11, BB-GI1	*Bartonella*-specific island encoding *yopP* gene(s) in BQ and BT	yes	2113	2225	53002
BT-GI23	BH-GI6/12, BQ-GI10, BB-GI2	*Bartonella*-specific island encoding phage genes	yes	2263	2306	37989
BT-GI24		VirB-homologous (Vbh) T4SS	no	2331	2351	13874
BT-GI25	BH-GI15, BQ-GI12	Trw T4SS	no	2507	2533	22519
BT-GI26		BT-specific phage island IV	yes	2603	2646	35567
B. henselae						
BH-GI2	BT-GI2/4/7, BQ-GI1	phage island	yes	02730	03760	65723
BH-GI4	BT-GI3/7/11	type II secretion system island	yes	06500	07260	75441
BH-GI6	BT-GI3/4/7/20/23, BQ-GI10, BB-GI2	phage genes and type II secretion	yes	08980	09500	33315
BH-GI8	BT-GI19, BQ-GI6	transporter-associated genes	yes	12470	12600	20850
BH-GI10	BT-GI13, BQ-GI8, BB-GI8	inducible *Bartonella* autotransporter (*iba*) genes	no	13120	13190	19100
BH-GI11	BT-GI14, BQ-GI9	VirB T4SS and *Bartonella* effector protein (Bep) genes	no	13250	13440	28575
BH-GI12	BT-GI20/23, BQ-GI10, BB-GI2	*Bartonella*-specific island encoding phage genes	yes	13900	14090	21639
BH-GI14	BT-GI22, BQ-11, BB-GI1	*Bartonella*-specific island	yes	14450	14630	29125
BH-GI15	BT-GI25, BQ-GI12	Trw T4SS	no	15530	15760	16156
B. quintana						
BQ-GI1	BT-GI2/4, BH-GI2	Remnants of phage island present in BH and BT	yes	02600	02760	12764
BQ-GI6	BT-GI19, BH-GI8	Transporter-associated genes	yes	09850	09930	10161
BQ-GI8	BT-GI13, BH-GI10, BB-GI8	inducible *Bartonella* autotransporter (*iba*) genes	no	10360	10410	12121

Table 2. Continued

GI#	Similar genomic Islands	Description	tRNA	Begin	End	Length
BQ-GI9	BT-GI14, BH-GI11	VirB T4SS and *Bartonella* effector protein (Bep) genes	no	10510	10680	22110
BQ-GI10	BT-GI20/23, BH-GI6/12, BB-GI2	*Bartonella*-specific island encoding phage genes	yes	11020	11160	17399
BQ-GI11	BT-GI22, BH-GI14, BB-GI1	*Bartonella*-specific island encoding *yopP* gene(s) in BQ and BT	yes	11400	11630	20809
BQ-GI12	BT-GI25, BH-GI15	Trw T4SS	no	12450	12680	16587

two Gram-negative bacterial membranes and the host cell plasma membrane directly into the host cell cytoplasm [1]. The VirB/VirD4 T4SS of *B. henselae* was shown to translocate several effector proteins, termed Beps, into endothelial cells that subvert cellular functions, such as apoptosis and the inflammatory response, that are considered critical for establishing chronic infection [13–15]. The molecular mechanism by which VirB-like T4SSs mediate host adaptability is probably also dependent on the translocated Beps. Comparison of the *virB/virD4/bep* T4SS loci of *B. henselae, B. quintana* and *B. tribocorum* revealed that the *virB/virD4* genes encoding the 11 essential T4SS components are highly conserved, while the *bep* genes encoding the translocated Beps displayed a higher degree of sequence variation (fig. 3b), suggesting an increased rate of evolution as the result of positive selection for adaptive functions in the infected host [9].

A third T4SS, Trw, is present in a sub-branch of the modern lineage (fig. 1) and essential for the process of erythrocyte invasion [16]. Interestingly, the presence of Trw by the modern lineage correlates with the loss of flagella (fig. 1), which are required for the invasion of erythrocytes by *B. bacilliformis* and probably also the flagellated bacteria of the modern lineage [1]. Trw does not translocate any known effectors, but produces multiple variant pilus subunits due to tandem gene duplication and diversification (by combinatorial sequence shuffling and point mutations) of *trwL* (encoding the major pilus subunit TrwL) and *trwJ* (encoding the minor pilus-associated subunit TrwJ) (fig. 3c) [17]. The variant pilus subunits exposed on the bacterial surface are thought to facilitate the interaction with different erythrocyte receptors or blood group antigens, and may thus represent major determinants of host specificity [1].

Genome Dynamics on the Strain Level

Evidence for genome dynamics on the intra-species level is accumulating for different *Bartonella* species. To access the natural variation in gene content and genome

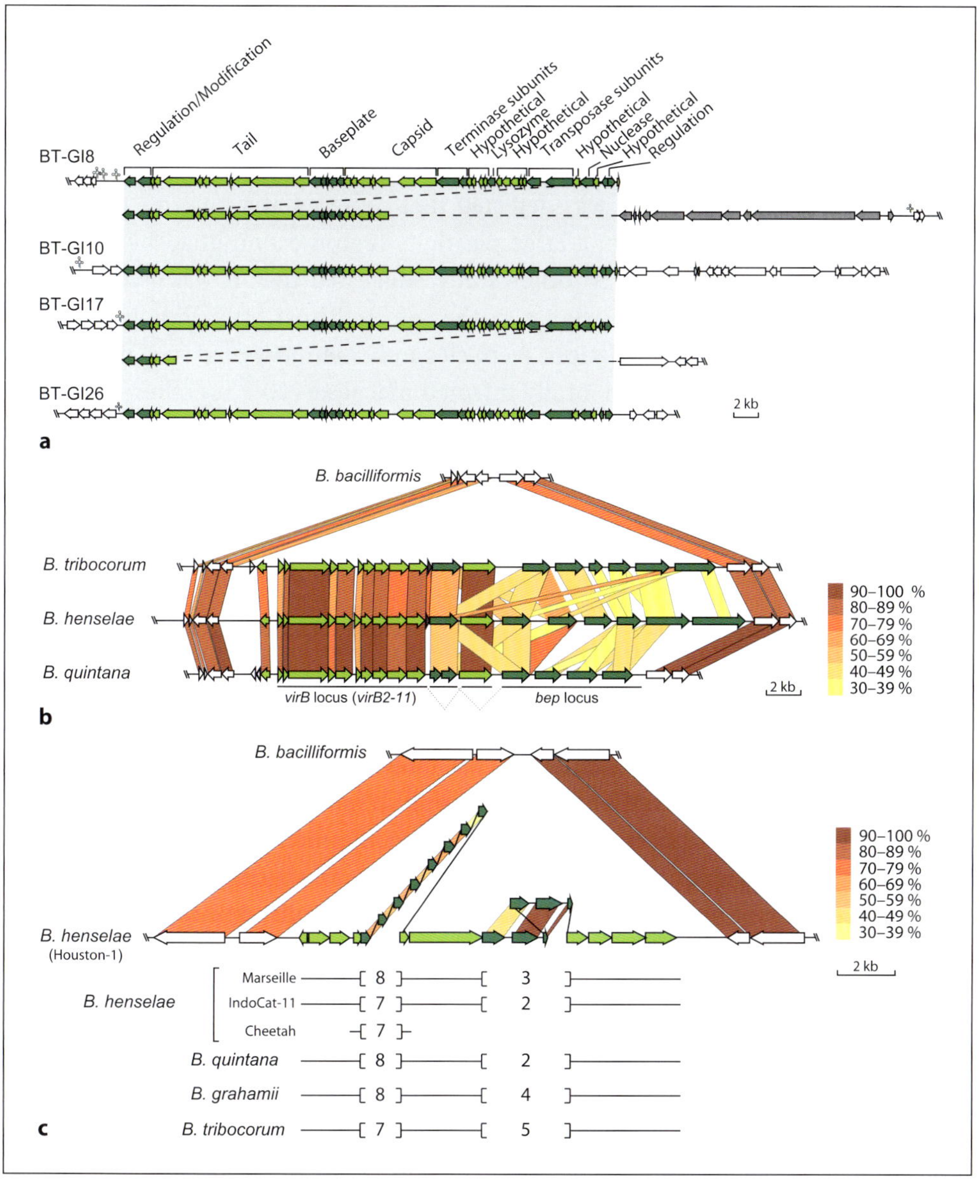

Fig. 3. Representation of selected GIs encoded in *Bartonella* genomes. Genes belonging to the GIs are shown in green, flanking genes are shown in white. (**a**) Alignment of the GIs encoding a *B. tribocorum*-specific prophage. Genes belonging to the prophage are located within the gray area. Noteworthy, BT-GI8 is flanked on one side by another island (gray gene symbols); (**b**) Alignment of the GI encoding the conserved T4SS VirB/VirD4 (*virB2–11* and *virD4* genes, colored in light green) and the highly variable translocated effectors (*bep* genes, colored in dark green); (**c**) Alignment of the GI (and flanking genes) encoding the T4SS-locus *trw*. The number of tandem repeats of *trwL* and *trwIJH* is indicated by gene symbols (colored in dark green) for the sequenced Houston-1 strain of *B. henselae* and by numbers in brackets for further *B. henselae* strains and the other species with known gene sequences. For (**b**) and (**c**), sequence similarity is shown with the percent identity indicated according to the color scales.

structure of *B. henselae*, a set of 38 strains isolated from cats and humans was analyzed by comparative genome hybridization [18]. The variation in gene content was modest and confined to the mosaic prophage region and other GIs, whereas extensive rearrangements were detected across the terminus of replication with breakpoints frequently locating to GIs. Moreover, in some strains a growth-phase dependent DNA-amplification was detected that centered at a putative phage replication initiation site located in a large plasticity region exemplified by a particularly low coding density [18]. Another study suggested that *B. henselae* exists as a mosaic of different genetic variants in the infected host [19]. Finally, genomic rearrangements due to gene deletions were elegantly demonstrated in serial isolates of *B. quintana* from an experimentally infected macaque [20]. Together, these data strongly suggest that various mechanisms contribute to a dynamic genome variation on the strain level.

Conclusions

Comparative and functional analysis of the four available complete genome sequences of species belonging to the genus *Bartonella* yielded first insights into the evolution, ecology and host interaction of this largely understudied group of bacterial pathogens. The small core genome reflects a host-integrated metabolism and codes for the majority of genes involved in the genus-specific infection strategy characterized by long-lasting intraerythrocytic infections in specific mammalian reservoir hosts. However, it is also evident that the accessory genomes contribute significantly to this infection strategy, e.g. flagella serving in the process of erythrocyte invasion by more ancestral species are considered to be functionally replaced by a laterally-acquired T4SS in more recently evolved species. Other laterally-acquired T4SSs were associated with the remarkable host adaptability exemplified by the radiating modern lineage. Genome expansion by lateral gene transfer in combination with secondary genome reduction has shaped the variable accessory genomes of the known *Bartonella* genomes. Additional *Bartonella* genome sequences expected to get available in the near future should result in a better understanding of the evolutionary processes that facilitated the emergence of a radiating group of host-restricted pathogens adapted to colonize a large variety of mammalian species that is infested by blood-sucking arthropods.

Acknowledgements

We are grateful to Arto Pulliainen for critically reading of the manuscript. The work was supported by grant 3100A0–109925/1 from the Swiss National Science Foundation (SNF), and grant 55005501 from the Howard Hughes Medical Institute (HHMI).

References

1 Dehio C: Infection-associated type IV secretion systems of *Bartonella* and their diverse roles in host cell interaction. Cell Microbiol 2008;10:1591–1598.

2 Dehio C: Molecular and cellular basis of *Bartonella* pathogenesis. Annu Rev Microbiol 2004;58:365–390.

3 Dehio C: *Bartonella*-host-cell interactions and vascular tumour formation. Nat Rev Microbiol 2005; 3:621–631.

4 Harms C, Maggi RG, Breitschwerdt EB, Clemons-Chevis CL, Solangi M, et al: *Bartonella* species detection in captive, stranded and free-ranging cetaceans. Vet Res 2008;39:59.

5 Jones SL, Maggi R, Shuler J, Alward A, Breitschwerdt EB: Detection of *Bartonella henselae* in the blood of 2 adult horses. J Vet Intern Med 2008;22:495–498.

6 Maggi RG, Harms CA, Hohn AA, Pabst DA, McLellan WA, et al: *Bartonella henselae* in porpoise blood. Emerg Infect Dis 2005;11:1894–1898.

7 Bown KJ, Bennet M, Begon M: Flea-borne *Bartonella grahamii* and *Bartonella taylorii* in bank voles. Emerg Infect Dis 2004;10:684–687.

8 Engbaek K, Lawson PA: Identification of *Bartonella* species in rodents, shrews and cats in Denmark: detection of two *B. henselae* variants, one in cats and the other in the long-tailed field mouse. Apmis 2004;112:336–341.

9 Saenz HL, Engel P, Stoeckli MC, Lanz C, Raddatz G, et al: Genomic analysis of *Bartonella* identifies type IV secretion systems as host adaptability factors. Nat Genet 2007;39:1469–1476.

10 Alsmark CM, Frank AC, Karlberg EO, Legault BA, Ardell DH, et al: The louse-borne human pathogen *Bartonella quintana* is a genomic derivative of the zoonotic agent *Bartonella henselae*. Proc Natl Acad Sci USA 2004;101:9716–9721.

11 Lindroos HL, Mira A, Repsilber D, Vinnere O, Naslund K, et al: Characterization of the genome composition of *Bartonella koehlerae* by microarray comparative genomic hybridization profiling. J Bacteriol 2005;187:6155–6165.

12 Pallen MJ, Wren BW: Bacterial pathogenomics. Nature 2007;449:835–842.

13 Schmid MC, Scheidegger F, Dehio M, Balmelle-Devaux N, Schulein R, et al: A translocated bacterial protein protects vascular endothelial cells from apoptosis. PLoS Pathog 2006;2:e115.

14 Schulein R, Guye P, Rhomberg TA, Schmid MC, Schroder G, et al: A bipartite signal mediates the transfer of type IV secretion substrates of *Bartonella henselae* into human cells. Proc Natl Acad Sci USA 2005;102:856–861.

15 Schmid MC, Schulein R, Dehio M, Denecker G, Carena I, Dehio C: The VirB type IV secretion system of *Bartonella henselae* mediates invasion, proinflammatory activation and antiapoptotic protection of endothelial cells. Mol Microbiol 2004;52:81–92.

16 Seubert A, Hiestand R, de la Cruz F, Dehio C: A bacterial conjugation machinery recruited for pathogenesis. Mol Microbiol 2003;49:1253–1266.

17 Nystedt B, Frank AC, Thollesson M, Andersson SG: Diversifying selection and concerted evolution of a type IV secretion system in *Bartonella*. Mol Biol Evol 2008;25:287–300.

18 Lindroos H, Vinnere O, Mira A, Repsilber D, Naslund K, Andersson SG: Genome rearrangements, deletions, and amplifications in the natural population of *Bartonella henselae*. J Bacteriol 2006;188: 7426–7439.

19 Berghoff J, Viezens J, Guptill L, Fabbi M, Arvand M: *Bartonella henselae* exists as a mosaic of different genetic variants in the infected host. Microbiology 2007;153:2045–2051.

20 Zhang P, Chomel BB, Schau MK, Goo JS, Droz S, et al: A family of variably expressed outer-membrane proteins (Vomp) mediates adhesion and autoaggregation in *Bartonella quintana*. Proc Natl Acad Sci USA 2004;101:13630–13635.

Christoph Dehio
Biozentrum, University of Basel
Klingelbergstrasse 70
CH–4056 Basel (Switzerland)
Tel. +41 61 267 2140, Fax +41 61 267 2118, E-Mail christoph.dehio@unibas.ch

de Reuse H, Bereswill S (eds): Microbial Pathogenomics.
Genome Dyn. Basel, Karger, 2009, vol 6, pp 170–186

Legionella pneumophila – Host Interactions: Insights Gained from Comparative Genomics and Cell Biology

M. Lomma · L. Gomez Valero · C. Rusniok · C. Buchrieser

Institut Pasteur, Unité Biologie des Bactéries Intracellulaires and CNRS URA 2171, Paris, France

Abstract

Legionella pneumophila is the etiological agent of Legionnaires' disease and of the less acute disease Pontiac fever. It is a Gram-negative bacterium present in fresh and artificial water environments that replicates in protozoan hosts and is also found in biofilms. Replication within protozoa is essential for the survival of the bacterium. The last years have seen a giant step forward in the genomics of *L. pneumophila*. The establishment and publication of the complete genome sequences of three clinical *L. pneumophila* isolates in 2004 and a fourth in 2007 has paved the way for major breakthroughs in understanding the biology of *L. pneumophila* in particular and *Legionella* in general. Sequence analysis identified several specific features of *Legionella*: (i) an extraordinary genetic diversity among the different isolates and (ii) the presence of an unexpected high number and variety of eukaryotic-like proteins, predicted to be involved in the exploitation of the host cellular processes by mimicking specific eukaryotic functions. In this chapter, we will first discuss the insights gained from genomics by highlighting the characteristic features and common traits of the four *L. pneumophila* genomes obtained through genome analysis and comparison and then we will focus on the newest results obtained by functional analysis of different eukaryotic-like proteins and describe their involvement in the pathogenicity of *L. pneumophila*.

Pathogens that are able to enter and multiply within human cells are responsible for multiple diseases and millions of deaths worldwide. Thus, the challenge is to elucidate these pathogen-specific and cell biological mechanisms involved in intracellular growth and spread. Many different techniques, such as molecular genetics, tissue culture systems, high-resolution microscopy, in vivo infection models, and recently also in vivo imaging techniques have been applied to the study of the mechanisms of intracellular pathogenesis. Since the publication of the first bacterial genome sequence in 1995 [1] a tremendous increase in genomic information has substantially altered our view on bacterial pathogenesis and has led to the application of many different genomics and post genomics approaches in microbial research. Here, we will discuss the insights gained from genomics and post genomics studies of the intracellular pathogen *Legionella pneumophila*.

Legionella pneumophila belongs to the genus *Legionella,* a group of Gram-negative bacteria of the class of γ-proteobacteria. The bacterium's natural environment is water where its survival and spread depend on the ability to replicate inside eukaryotic phagocytic cells like the aquatic protozoa *Acanthamoeba castellani, Hartmanella* sp. or *Naeglaria* sp. [2, 3]. *Legionella* are environmental bacteria but they are also serious human pathogens. The two main clinical forms of infection are Legionnaires' disease and Pontiac fever. Legionnaires' disease is a severe atypical pneumonia that can be fatal if not promptly treated. Pontiac fever is a mild, non-pneumonia influenza-like illness [4].

A particular feature of *Legionella* is its dual host system allowing the intracellular growth in protozoa and, during infection, in human alveolar macrophages. The capacity of pathogens like *Legionella* to infect eukaryotic cells is intimately linked to the ability to manipulate host cell functions to establish an intracellular niche for their replication. It is tempting to assume that the interaction of *L. pneumophila* with aquatic protozoa has generated a pool of virulence traits during evolution, which allow *Legionella* to infect also human cells. Upon internalization into the eukaryotic cell, *L. pneumophila* guarantees its survival by manipulating host cell functions such as disturbing vesicle trafficking, therewith reprogramming the endosomal-lysosomal degradation pathway of the phagocytic cell. One of the virulence factors indispensable for *L. pneumophila*'s intracellular survival is a type IV secretion system (T4SS) called Dot/Icm [5, 6], which translocates a large repertoire of bacterial effectors into the host cell. These effectors modulate multiple host cell processes and in particular, redirect trafficking of the *L. pneumophila* phagosome and mediate its conversion into an ER-derived organelle competent for intracellular bacterial replication [7].

Despite the elucidation of important players necessary for entry and intracellular replication of *L. pneumophila* already during the pregenomic era, many questions remained to be answered. An important step forward in *Legionella* research was the establishment and publication of the first three complete *L. pneumophila* genome sequences in 2004 [8, 9], (http://genolist.pasteur.fr/LegioList/). Three years later an additional *L. pneumophila* sequence was published [10]. The availability of these complete sequences paved the way for major breakthroughs in understanding the biodiversity and biology of *L. pneumophila* in particular and *Legionella* in general.

The *L. pneumophila* Genomes Show a Conserved Organization but Each Has Many Unique Interspersed Regions and Single Genes

At present the complete genome sequences of four strains of *L. pneumophila* serogroup 1 (Sg 1) are completed and published: strains Paris, Lens, Philadelphia and Corby [8–10]. Phylogenetic analysis using the Neighbour-Joining method based

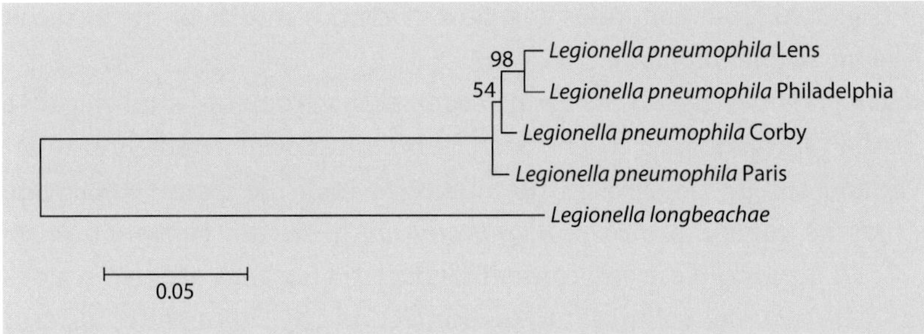

Fig. 1. Phylogenetic tree of the sequenced *L. pneumophila* strains based on the *proA* sequence. The *proA* gene is a fast evolving gene that encodes a zinc metalloprotease. The tree was constructed by using the Neighbor-Joining method. The *proA* gene sequence of *Legionella longbeachae* was used as out-group. Bootstrap values are indicated next to the corresponding node (1,000 replicates).

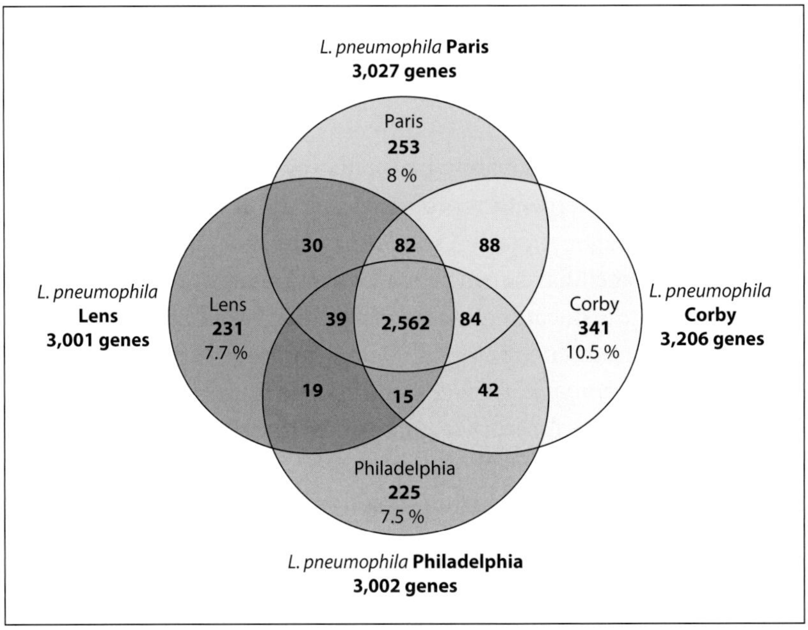

Fig. 2. Diagram showing the core genome and the unique gene complement of strains *L. pneumophila* Paris, Lens, Philadelphia and Corby. Orthologous genes were defined by reciprocal best-match FASTA comparisons. The threshold was set to a minimum of 80% sequence identities and a ratio of the length of 0.75 to 1.33.

on the *proA* gene sequence shows that the four strains are phylogenetically closely related, with the strains Philadelphia and Lens showing the closest phylogenetic relationship (fig. 1).

The genome of these strains is composed of a single circular chromosome, with a size of 3.35 Mb (strain Lens) to 3.58 Mb (strain Corby). One circular plasmid has been

Table 1. General features of the sequenced *L. pneumophila* genomes. Data for plasmids are in parentheses

	Paris	Lens	Philadelphia	Corby
Chromosome size (kb)	3,504 (0.131)	3,345 (0.060)	3,397	3,576
G+C content (%)	38.3 (37.4)	38.4 (38)	38.27	38
G+C content of CDS (%)	39.1	39.4	38.6	38.6
No. of total CDS[a]	3,136 (142)	3,001(57)	3,002	3,259
No. of protein coding genes[a]	3,027 (139)	2,878 (56)	2,942	3,206
Percentage of CDS (%)	87.9	88,00	90.2	86.8
Average length of CDS (bp)	994.6	935.9	960.7	959.4
No. of 16S/23S/5S	3/3/3	3/3/3	3/3/3	3/3/3
No. transfer RNA	44	43	43	43
Plasmids	1	1	0	0

[a] Updated annotation; CDS = coding sequence.

detected in strains Lens and Paris (table 1). The genomes show a high homogeneity regarding GC content (approximately 39%), coding percentage and average length of the coding sequences (table 1). The particular features of the *Legionella* genomes as deduced from the sequence analyses are: (i) high genome plasticity as many pathogenicity islands and mobility genes were discovered, (ii) high genetic diversity, as 7.5 to 10.5% of the genes of each strain are specific. This is a considerable number given the fact that these four strains belong to the same species and to the same Sg 1 (fig. 2). The high genome diversity is further underlined by a recent study comparing the gene content of over 200 *L. pneumophila* strains. Except for known and putative virulence factors, which are highly conserved among the investigated strains, *L. pneumophila* is a genetically diverse species [11]. The most intriguing feature of the *L. pneumophila* genomes, discovered through genome sequencing and genome analysis, is the presence of a (iii) high number and a wide variety of eukaryotic-like proteins (ELP) or eukaryotic protein domains (EPD). These proteins are good candidates for being involved in manipulating host cell functions to the bacterium's advantage [8, 12, 13].

Presence and Distribution of Eukaryotic-Like Proteins and Eukaryotic Motifs among the Four *L. pneumophila* Genomes

According to our definition, eukaryotic-like proteins are defined as proteins that have their best BLASTp hit with at least 20% amino acid identity over more than a third of the length of a eukaryotic protein or contain motifs mostly or uniquely present in eukaryotes [8]. De Felipe and collaborators (2005) do not distinguish between these two categories but define their EPD analysis as protein motifs that are widespread in eukaryotic species and significantly underrepresented in archaeal and prokaryotic species and having cellular functions associated with eukaryotes. However, the results may change with the progressive changes in the database and the analysis should thus be done in parallel with a phylogenetic analysis to confirm the closer evolutionary relationship to eukaryotic than to prokaryotic sequences. Our analysis had identified 30 ELP and 33 EPD in the *L. pneumophila* strain Paris genome [8, 14].

Based on our original definition of ELP and EPD we undertook a comparative analysis of the four sequenced genomes. This reveals a high conservation of the ELP proteins with two exceptions: one plasmid encoded protein similar to a hypersensitive induced response protein and one genome encoded protein similar to a nuclear membrane binding protein, which are specific for strain Paris. Additionally, except for one protein similar to an RNA binding protein precursor that is missing in strain Lens, all ELPs are conserved (table 2). The situation is very similar for the EPDs as there is only heterogeneity among the ankyrin protein family and the F- and U-box containing proteins, whereas all other EPDs are conserved among the genomes (table 3). This result is also seen when investigating the presence of ELP and EPD coding genes by DNA/DNA hybridization. Nearly all of them are conserved among over 200 *L. pneumophila* genomes, but they are absent or highly divergent in other non-*pneumophila Legionella* species [11].

Possible Functions of Eukaryotic-Like Proteins and Proteins Containing Eukaryotic Domains

The abundance and high conservation of ELPs and EPDs in the *L. pneumophila* genomes suggest that they are important for the *L. pneumophila* life cycle. Function prediction based on similarity searches makes many to promising candidates for modulating host cell functions to the pathogen's advantage. An example is *lpp2128* coding for a protein similar to sphingosine-1-phosphate lyase (Spl). Except in the bacterium *Porphyromonas gingivalis* (a pathogenic bacterium that causes periodontal disease), the metabolic pathway for sphingomyelin metabolism is not present in prokaryotes [15]. In contrast in *L. pneumophila* we identified the genes coding for proteins highly similar to sphingomyelinase, sphingosinekinase and sphingosine-1-phosphate lyase (Spl), all of which are part of the sphingomyelin degradation pathway. Sphingosine kinase phosphorylates the catabolite of ceramide, sphingosine into sphingosine-1-phosphate, which is cleaved irreversibly by sphingosine-1-phosphate lyase. Spl is a bioactive metabolite of the sphingolipid metabolism, that is known for its influence on a wide range of physiological functions, including cell survival and apoptosis, proliferation, migration, differentiation, platelet aggregation, angiogenesis, vascular permeability, cardioprotection, inflammation, lymphocyte trafficking and development [16]. In the parasitic protozoa *Leishmania,* Spl has been shown to be

Table 2. Proteins with the highest similarity to eukaryotic proteins and their distribution in the four sequenced strains

	L. pneumophila strains and G+C content of the respective genes							
	Paris	G+C	Lens	G+C	Philadelphia	G+C	Corby	G+C
PurC	*lpp1647*	38%	*lpl1640*	39%	*lpg1675*	40%	*lpc1106*	40%
ExoA exoDNase III	*lpp0702*	39%	*lpl0684*	39%	*lpg0648*	39%	*lpc2646*	40%
RNA binding protein precursor	*lpp0321*	34%	–	–	*lpg0251*	37%	*lpc0328*	35%
Pyruvate decarboxylase	*lpp1157*	39%	*lpl1162*	39%	*lpg1155*	40%	*lpc0618*	40%
Thiamine biosynthesis protein NMT-1	*lpp1522*	38%	*lpl1461*	39%	*lpg1565*	40%	*lpc0988*	39%
NuoE NADH dehydrogenase I chain E	*lpp2832*	38%	*lpl2701*	38%	*lpg2785*	37%	*lpc3071*	37%
Hypersensitive induced response protein	*plpp0050*	36%	–	–	–	–	–	–
Hypothetical protein	*lpp0634*	39%	*lpl0618*	39%	*lpg0584*	39%	*lpc2719*	39%
DegP protease	*lpp0965*	39%	*lpl0935*	40%	*lpg0903*	40%	*lpc2388*	39%
Phytanoyl-coA dioxygenase	*lpp2748*	36%	*lpl2621*	36%	*lpg2694*	36%	*lpc0442*	36%
Sphingosine-1-phosphate lyase	*lpp2128*	41%	*lpl2102*	41%	*lpg2176*	40%	*lpc1635*	41%
Glucoamylase	*lpp0489*	39%	*lpl0465*	39%	*lpg0422*	39%	*lpc2921*	39%
Cytokinin oxidase	*lpp0955*	39%	*lpl0925*	39%	*lpg0894*	40%	*lpc2399*	39%
Phytanoyl coA dioxygenase	*lpp0578*	36%	*lpl0554*	37%	*lpg0515*	37%	*lpc2829*	37%
Hypothetical protein	*lpp0379*	39%	*lpl0354*	40%	*lpg0301*	40%	*lpc0380*	40%
Ectonucleoside triphosphate diphosphohydrolase (apyrase)	*lpp1033*	40%	*lpl1000*	39%	*lpg0971*	40%	*lpc2316*	40%
6-pyruvoyl-tetrahydropterin synthase	*lpp2923*	34%	*lpl2777*	35%	*lpg2865*	35%	*lpc3150*	36%
Zinc metalloproteinase	*lpp3071*	38%	*lpl2927*	38%	*lpg2999*	38%	*lpc3315*	38%
SAM dependent methyltransferase	*lpp2134*	35%	*lpl2109*	36%	*lpg2182*	36%	*lpc1642*	36%
Ectonucleoside triphosphate diphosphohydrolase (apyrase)	*lpp1880*	39%	*lpl1869*	39%	*lpg1905*	40%	*lpc1359*	40%
SAM dependent methyltransferase	*lpp2747*	35%	*lpl2620*	35%	*lpg2693*	36%	*lpc0443*	36%
Cytochrome P450	*lpp2468*	39%	*lpl2326*	39%	*lpg2403*	38%	*lpc2075*	40%
Nuclear membrane binding protein	*lpp1824*	34%	–	–	–	–	–	–

Table 2. Continued

	L. pneumophila strains and G+C content of the respective genes							
	Paris	G+C	Lens	G+C	Philadelphia	G+C	Corby	G+C
Uracyl DNA glycosylase	*lpp1665*	36%	*lpl1659*	36%	*lpg1700*	37%	*lpc1129*	37%
Chromosome condensation 1-like	*lpp1959*	41%	*lpl1953*	38%	*lpg1976*	43%	*lpc1462*	42%
Hypothetical protein	*lpp0358*	38%	*lpl0334*	38%	*lpg0282*	39%	*lpc0359*	39%
Ca^{2+}-transporting ATPase	*lpp1127*	37%	*lpl1131*	37%	*lpg1126*	38%	*lpc0584*	38%
Uridine kinase	*lpp1167*	33%	*lpl1173*	34%	*lpg1165*	34%	*lpc0630*	34%
Serine/threonine protein kinase domain	*lpp2626*	32%	*lpl2481*	32%	*lpg2556*	32%	*lpc1906*	32%
Serine/threonine protein kinase	*lpp1439*	36%	*lpl1545*	35%	*lpg1483*	36%	*lpc0898*	36%

necessary for virulence and development [17], and in the amoeba *Dictyostelium discoideum,* the disruption of this gene results in aberrant actin distribution, an abnormal morphogenetic phenotype and increased viability during stationary phase [18]. It is thus tempting to assume that Spl of *L. pneumophila* may modulate the sphingomyelin degradation pathway of the host cell, perhaps by influencing cell survival and apoptosis of its host.

Another example is the presence of a predicted protein similar to the zinc metalloproteinase ZmpC. In pneumococci, it was shown to specifically cleave human MMP-9 (matrix metalloproteinase 9) [19]. Furthermore, the presence of this gene correlates with strains isolated from pneumonia cases and with virulence in a murine pneumonia model. Thus it has been suggested that ZmpC plays a role in pneumococcal virulence and pathogenicity in the lung [19]. As *L. pneumophila* also causes pneumonia, it is possible that the *L. pneumophila* zinc metalloprotease plays a role in infection of the lung.

Typical eukaryotic motifs that are present in the *Legionella* genomes are ankyrin repeats, Sel-1 motifs, SET, Sec7, U- and F-box domains and serine threonine kinase domains (STPK) (table 3). Ankyrin repeats are also present in a few other bacterial genomes such as *Coxiella burnetii* [20], *Wolbachia pipitentis* [21] or *Rickettsia felis* [22]. Proteins carrying serine threonine kinase domains, SET, and F-box domains have not been investigated yet in *L. pneumophila.* However, in other pulmonary pathogens such as *Mycobacterium tuberculosis,* which like *L. pneumophila* blocks phagosome lysosome fusion, the STPK PknG is implicated in the inhibition of the phagosome-lysosome fusion and promotes intracellular survival [23]. The STPK PknB is essential for sustaining mycobacterial growth [24] and STPK PknD alters

Table 3. *L. pneumophila* proteins encoding domains preferentially found within eukaryotic proteins and their distribution

L. pneumophila strains and G+C content of the respective genes								
Paris	G+C	Lens	G+C	Philadelphia	G+C	Corby	G+C	
EnhC (Lpp2692)	39%	EnhC (Lpl2564)	39%	EnhC (Lpg2639)	39%	EnhC (Lpc0501)	39%	21 sel-1 domains
LidL (Lpp1174)	38%	LidL (Lpl1180)	39%	LidL (Lpg1172)	39%	LidL (Lpc0638)	38%	6 sel-1 domains
Lpp1310	41%	Lpl1307	41%	Lpg1356	41%	Lpc0770	42%	4 sel-1 domains
Lpp2174	40%	Lpl1303	39%	Lpg2222	41%	Lpc1689	40%	3 sel-1 domains
–	45%	Lpl1059	45%	Lpl1062	44%	Lpc2212	44%	7 sel-1 domains
RalF (Lpp1932)	34%	RalF (Lpl1919)	34%	RalF (Lpp1950)	35%	RALF (Lpc1423)	35%	Sec7 domain
Lpp0267	38%	Lpl0262	39%	Lpg0208	38%	Lpc0283	39%	Ser/thr protein kinase domain
Lpp2626	32%	Lpl2481	32%	Lpg2556	32%	Lpc1906	32%	Ser/thr protein kinase domain
Lpp1439	36%	Lpl1545	35%	Lpg1483	36%	Lpc0898	36%	Ser/thr protein kinase domain
Lpp2065	37%	Lp2055	37%		42%	Lpc1573	38%	Ankyrin repeat
Lpp0037	38%	Lpl0038	39%	Lpg0038	38%	Lpc0039	39%	Ankyrin repeat
Plpp0098	37%	–	–	–	–	–	–	Ankyrin repeat
Lpp2058	38%	Lpl2048	38%	–	–	Lpc1566	39%	Ankyrin repeat
Lpp0750	35%	Lpl0732	35%	Lpg0695	36%	Lpc2599	36%	Ankyrin repeat
Lpp2061	39%	Lpl2051	39%	–	–	Lpc1569	39%	Ankyrin repeat
Lpp2270	34%	Lpl2242	34%	Lpg2322	35%	Lpc1789	35%	Ankyrin repeat
Lpp0503	38%	Lpl0479	36%	Lpg0436	37%	Lpc2906	37%	Ankyrin repeat
Lpp1905	35%	–	–	–	–	–	–	Ankyrin repeat
Lpp1683	33%	Lpl1682	34%	Lpg1718	34%	Lpc1152	34%	Ankyrin repeat + SET domain
Lpp2248	39%	Lpl2219	39%	Lpg2300	39%	Lpc1765	39%	Ankyrin repeat
Lpp0202	38%	–	–	–	–	–	–	Ankyrin repeat
Lpp0469	38%	Lpl0445	38%	Lpg0403	39%	Lpc2941	39%	Ankyrin repeat
Lpp2517	36%	Lpl2370	37%	Lpg2452	37%	Lpc2026	37%	Ankyrin repeat
Lpp1100	48%	–	–	–	–	–	–	Ankyrin repeat
Lpp0126	39%	Lpl0111	39%	Lpg0112	39%	Lpc0131	38%	Ankyrin repeat

Table 3. Continued

L. pneumophila strains and G+C content of the respective genes								
Paris	G+C	Lens	G+C	Philadelphia	G+C	Corby	G+C	
Lpp0356	38%	–	–	–	–	–	–	Ankyrin repeat
Lpp2522	39%	Lpl2375	39%	Lpg2456	40%	Lpc2020	39%	Ankyrin repeat
Lpp0547	40%	Lpl0523	41%	Lpg0483	42%	Lpc2861	41%	Ankyrin repeat
–	34%	Lpl1681	34%	–	–	Lpc1151	34%	Ankyrin repeat
–	35%	Lpl2344	35%	–	–	–	–	Ankyrin repeat
–	40%	Lpl2058	40%	Lpg2128	37%	–	–	Ankyrin repeat
–	38%	–	–	Lpg0402	38%	–	–	Ankyrin repeat
–	39%	–	–	Lpg2131	39%	–	–	Ankyrin repeat
Lpp2082	36%	Lpl2072	36%	Lpg2144	37%	Lpc1593	38%	F-Box domain + ankyrin repeat
Lpp2486	34%	–	–	–	–	–	–	F-Box domain + coiled-coil
–	–	–	–	Lpg2224	43%	–	–	F-Box domain
Lpp0233	39%	Lpl0234	39%	Lpg0171	40%	–	–	F-Box domain
Lpp2887	35%	–	–	Lpg2830	35%	–	–	Two U-Box domains

sel = Suppressor and/or enhancer of lin-12; Sec7 = domain similar to yeast sec7; Ser/thr = Serine/Threonine; SET = Su(var)3-9, Enhancer-of-zeste and Trithorax; F-box = occurrence in cyclin F; U-box = Ubiquitin ligase domain.

the transcriptional program of *M. tuberculosis* in response to an unknown signal by stimulating phosphorylation of a sigma factor regulator [25]. Thus the presence of three Ser/Thr protein kinases (STPKs) in *L. pneumophila* suggests that these proteins are also implicated in influencing trafficking in the host cell.

Interestingly coiled-coil domains are also frequently found in the *L. pneumophila* genomes. Coiled coil domains consist of two to five amphipathic alpha-helices that twist around one another to form a supercoil. These domains are present in both, eukaryotic and prokaryotic organisms, but are found mainly in eukaryotes. Moreover long coiled-coil domains (more than 250 amino acids) are absent from bacterial genomes but present in archaea and eukaryotes [26]. Therefore, coiled-coil domains longer than 250 amino acids can be considered as typical eukaryotic motifs. Several of the currently known Dot/Icm T4SS substrates possess long coiled-coil regions [13,

27, 28]. As proteins with coiled-coil domains are involved in molecular recognition systems and protein refolding processes or can form ion channels [29], these proteins might be secreted by the Dot/Icm T4SS and help *L. pneumophila* to subvert host functions. This hypothesis has been confirmed recently for three of these coiled-coil proteins. Lpp1666/Lpg1701, YflA/Lpg2298/Lpp2246 and YflB/Lpg1884/Lpp1848 have been shown to be Dot/Icm T4SS effectors that contribute to the intracellular trafficking of *L. pneumophila* [30].

Eukaryotic-Like Proteins of *L. pneumophila* Implicated in Virulence and Host Cell Modulation

After adhesion to a phagocytic cell, it is thought that *L. pneumophila* is uptaken by a host-driven phagocytosis [7]. Once *L. pneumophila* has entered the eukaryotic host, it is able to modulate trafficking so that the *Legionella*-containing phagosome or *Legionella* containing vacuole (LCV) is completely isolated from the host endocytic pathway and the lysosome [31]. Shortly after bacterial internalization, LCVs are found associated with endoplasmatic reticulum-derived vesicles [32, 33]. After replication and depletion of nutrients the LCVs undergo maturation following a pathway similar to the autophagy pathway [34–36]. The egress of bacteria following completion of replication is probably due to the formation, in addition to the Dot/Icm transporter pore, of a second pore required for host lysis [37, 38]. To date it is only partly understood how *L. pneumophila* is able to subvert host functions to replicate inside eukaryotic cells like aquatic protozoa but also human alveolar macrophages thus provoking pneumonia.

According to predictions from genome analysis, the ELPs and EPDs identified in the *L. pneumophila* genomes are good candidates for acting at all the different steps of the intracellular cycle [8, 12]. Indeed, the role for roughly 15 of them has meanwhile been investigated confirming their implication in virulence and host cell modulation. Most of these proteins are also candidates for being secreted by the Dot/Icm T4SS, as they must be translocated to the host cytoplasm to be able to affect the eukaryotic cell.

Entry and Blocking of Phagosomal-Lysosomal Fusion

A eukaryotic-like protein of *L. pneumophila,* a predicted ecto-nucleoside triphosphate diphosphohydrolases (ecto-NTPDases) (Lpp1880/Lpg1905) that shares similarities with human CD39 and other eukaryotic ecto-NTPDases, has been shown to play a role during uptake of *L. pneumophila* into the host cell. In humans, CD39 is located on the surface of endothelial cells and it controls extracellular levels of ATP by converting it in its diphosphate and monophosphate forms. In this way it plays a major role in maintaining vascular fluidity by regulating platelet aggregation [39]. CD39/NTPDases are found in a wide range of pathogens such as in protozoan parasites, but their role in infection is poorly understood. One of the two predicted ecto-NTPDases in *L. pneumophila* is secreted into the host cell and its activity is required for successful infection. This defect was not correlated with the ability to recruit the ER or avoiding phago-lysosomal fusion but mainly to a less efficient entry [40]. Recently, it was shown that the enzyme catalyzed the hydrolysis of ATP and ADP, and also of GTP and GDP but had only limited activity against CTP, CDP, UTP, and UDP. Furthermore, mutational analysis revealed, that all five apyrase domains are necessary for infection following intratracheal inoculation of A/J mice [41].

The Dot/Icm-translocated proteins VipA, VipD, VipF are thought to participate in blocking lysosomal fusion. They have been identified in a yeast screen as *L. pneumophila* proteins able to cause vacuolar missorting and to inhibit yeast lysosomal protein trafficking [42]. Two of them (VipA and VipD) contain eukaryotic-like domains. VipA contains a large coiled-coil region. These regions usually form highly versatile structures involved in protein-protein interactions commonly found in trafficking components such as soluble N-ethylmaleimide-sensitive fusion attachment receptor proteins (SNARE) and early endosomal antigen 1 (EEA1). VipD is characterized by a patatin domain with strong homology to eukaryotic phospholipase A2 proteins. As suggested by its trafficking defect in yeast, VipD is thought to be involved in the intracellular infection process of *L. pneumophila* [42, 43].

Additional eukaryotic domain proteins shown to be implicated in modulating trafficking in the host cell are proteins that contain the eukaryotic Sel-1 domains. Sel-1 repeats represent a subfamily of tetratrico peptide repeats (TPRs) which are degenerated repeated motifs that form a scaffold to mediate protein-protein interactions [44]. Three of the five Sel-1 domain containing *L. pneumophila* proteins, LpnE, EnhC and LidL interact with the host cell to modulate early trafficking events that determine the fate of *Legionella* right after internalization and in growth within the host cell [45–48].

Establishment of an ER-Derived Replicative Vacuole

To promote the fusion to ER membranes, *L. pneumophila* recruits host factors to the surface of the LCVs like Arf-1 and Rab-1, important cell signaling proteins involved in the regulation of the ER-Golgi traffic [31, 49, 50]. The *L. pneumophila* gene *ralF* encodes a protein with a Sec-7 domain. These domains are found in eukaryotes as components of Arf-specific guanine nucleotide exchange factors (GEFs). GEFs catalyze the nucleotide exchange of Arfs thereby converting them from an inactive state (GDP-bound) to the active one (GTP-bound). Following secretion by T4SS, RalF recruits Arf-1 and then functions like an Arf-1 specific GEF [51].

Another Dot/Icm translocated effector DrrA or SidM is able to interact with Rab1 [52, 53]. GDP-bound Rabs are kept inactive by a GDP association inhibitor (GDI) that prevents their spontaneous activation. Rabs are released from GDI by a guanine nucleotide dissociation inhibitor displacement factor (GDF) before their recruitment to the membrane and activation by GEFs. DrrA/SidM is characterized by two distinct regions: the N-terminal part recruits Rab1 to LCV membranes and functions as a GDF while the C-terminal part, characterized by highly specific Rab1-GEF activity, activates Rab1 [54].

Another interesting example of eukaryotic domain containing-proteins of *L. pneumophila* are the twenty ankyrin proteins. The ankyrin domain is a 33-residue L-shaped motif containing two antiparallel alpha-helices connected by a short loop [55]. The modular architecture and variable modular surfaces generated by the assembly of multiple compatible repeats render ankyrin proteins highly versatile in protein binding. This versatility and the multiple associated roles make the prediction of their function difficult. Ankyrin proteins are involved in cell signaling, cytoskeleton integrity and regulation, transcription and cell cycle regulation, inflammatory response and oncogenesis [56]. *L. pneumophila* single mutants for eleven of the thirteen ankyrin proteins of *L. pneumophila* Philadelphia, have been generated and analysed. Two of them, called AnkH (Lpp2248) and AnkJ (Lpp0503), play a role in intracellular replication during protozoan host infection [57]. Furthermore, the AnkX (Lpp0750) protein was shown to prevent microtubule-dependent vesicular transport to interfere with fusion of the LCV with late endosomes after infection of macrophages [58]. It is not known yet whether the redundant effect of the ankyrin proteins or of other bacterial effectors mask a possible role in virulence of the remaining ankyrin proteins or if those are not involved in protozoan host tropism.

Replication in the LCV and Egress from the Host

During bacterial replication unidentified ubiquitinated proteins are recruited to the LCV in a Dot/Icm-dependent manner [59]. Although the presence of these ubiquitinated proteins seems to be very important for bacterial replication the mechanism of their recruitment is unknown. Interestingly, the *L. pneumophila* genome encodes proteins containing domains with high similarity to F-box and U-box domains of eukaryotic proteins [8]. F-box and U-box domains are found in eukaryotic E3-ubiquitin ligases where they act by recognizing the targets of the ubiquitination process to lead them to proteasomal degradation. It has been shown that the *L. pneumophila* U-box containing effector, called LubX (Lpp2887), possesses in vitro ubiquitin ligase activity specific for the Cdc2-like kinase Clk1. While pharmacological inhibition of Clk1 inhibits bacterial replication, indicating its implication during intracellular replication of *L. pneumophila*, a *lubX* mutant was neither impaired in replication, nor in any step of the intracellular cycle [60].

After completion of intracellular replication, bacteria must exit the exhausted host cell in order to infect a new one. The egress process is not well understood but the formation of an egress pore has been hypothesized [61]. Two Dot/Icm effectors have been shown to be implicated in an active but non-lytic egress of *L. pneumophila* from protozoa, but not mammalian cells. These two effectors are LepA and LepB: both have weak homology to eukaryotic SNAREs. SNAREs are protein receptors that mediate vesicle-membrane fusions [62]. LepB has also Rab-GAP activity involved in the formation of LCVs, but it may contain also other functional domains involved in *L. pneumophila* host escape.

Evolutionary Origin of Eukaryotic-Like Proteins and Proteins with Eukaryotic Domains

ELPs and EPDs are clearly implicated in modulating cellular activities of the host, revealing that molecular mimicry is an important strategy of *L. pneumophila* to exploit host cell functions to its advantage. How did *L. pneumophila* acquire these proteins? Two hypotheses may explain their origin: (i) horizontal gene transfer (HGT) or (ii) convergent evolution. The close co-evolution of *Legionella* with the eukaryotic host has probably led to a constant cross talk between bacterial and protozoan proteins. The selective advantage of *Legionella* that acquired these proteins allowing them to manipulate the host cells may explain a successful incorporation in the genome through HGT. This hypothesis is supported by the fact that most of these genes show a G+C bias as compared to other *L. pneumophila* genes [13]. At least for one protein, RalF, it has been suggested that it was acquired through interdomain HGT [51]. Structural studies have shown that the three-dimensional structure of this protein resembles the well-known eukaryotic Sec7 domain fold [63]. However, the current number of completed eukaryotic genomes available is small, so it is difficult to predict the flow of horizontal gene transfer.

On the other hand the possible origin of these proteins through convergent evolution cannot be ruled out. This process implies changes in the amino acid sequence of the protein during evolution in order to become similar to the eukaryotic effector. However convergent evolution is perhaps the more intriguing of the two ways, as it involves sculpting genes already present in the bacteria to perform a new function. In some cases the bacterial proteins possess a structural architecture that differs markedly from that of their functional homologs of the host. However, the molecular surfaces that interact with their targets, the true level at which natural selection ultimately acts, are seen as excellent mimicry of proteins that operate normally in the cell. Therefore, in this second case the detection of the similarity to eukaryotic counterparts becomes difficult since normally it is restricted to a specific region of the protein and not over the whole length.

The two possibilities, horizontal transfer and convergent evolution are not exclusive; both of them can have taken place depending on the protein. Only future studies combining phylogenetic and structural information for each of these proteins together with the access to more completed protozoan genome sequences, will help to reveal the origin of each eukaryotic like gene.

Conclusions

L. pneumophila is able to modulate, manipulate and subvert many eukaryotic host cell functions to its advantage, in order to enter, replicate and evade protozoa or human alveolar macrophages during disease. Many studies have shown, that eukaryotic like proteins and proteins encoding eukaryotic like domains play an important role. Thus, molecular mimicry seems to be one of the main characteristics of *L. pneumophila* host cell infection. Future studies will elucidate the contribution of additional eukaryotic-like factors for their ability helping *L. pneumophila* to invade, replicate and finally exit human and protozoan hosts thereby providing new insights into *L. pneumophila* pathogenesis.

Acknowledgements

We would like to thank many of our colleagues who have contributed in different ways to this research. Work in the authors laboratory received financial support from the Institut Pasteur, the Centre National de la Recherche (CNRS) the Institut Carnot and the Network of Excellence 'Europathogenomics' LSHB-CT-2005–512061. M. Lomma is holder of a Marie Curie fellowship (Early stage training in infectious diseases) financed by the European Commission in the framework of the INTRAPTAH project MEST-CT-2005–020715 coordinated by Institut Pasteur and L. Gomez-Valero is holder of a Roux postdoctoral research Fellowship financed by the Institut Pasteur.

References

1 Fleischmann RD, Adams MD, White O, Clayton RA, Kirkness EF, et al: Whole-genome random sequencing and assembly of *Haemophilus influenzae* Rd. Science 1995;269:496–512.
2 Fields BS, Benson RF, Besser RE: *Legionella* and Legionnaires' disease: 25 years of investigation. Clin Microbiol Rev 2002;15:506–526.
3 Steinert M, Hentschel U, Hacker J: *Legionella pneumophila*: an aquatic microbe goes astray. FEMS Microbiol Rev 2002;26:149–162.
4 Diederen BM: *Legionella* spp. and Legionnaires' disease. J Infect 2008;56:1–12.
5 Berger KH, Isberg RR: Two distinct defects in intracellular growth complemented by a single genetic locus in *Legionella pneumophila*. Mol Microbiol 1993;7:7–19.
6 Marra A, Blander SJ, Horwitz MA, Shuman HA: Identification of a *Legionella pneumophila* locus required for intracellular multiplication in human macrophages. Proc Natl Acad Sci USA 1992;89:9607–9611.
7 Shin S, Roy CR: Host cell processes that influence the intracellular survival of *Legionella pneumophila*. Cell Microbiol 2008;10:1209–1220.
8 Cazalet C, Rusniok C, Bruggemann H, Zidane N, Magnier A, et al: Evidence in the *Legionella pneumophila* genome for exploitation of host cell functions and high genome plasticity. Nat Genet 2004;36:1165–1173.
9 Chien M, Morozova I, Shi S, Sheng H, Chen J, et al: The genomic sequence of the accidental pathogen *Legionella pneumophila*. Science 2004;305:1966–1968.
10 Steinert M, Heuner K, Buchrieser C, Albert-Weissenberger C, Glöckner G: *Legionella* pathogenicity: genome structure, regulatory networks and the host cell response. Int J Med Microbiol 2007;297:577–587.
11 Cazalet C, Jarraud S, Ghavi-Helm Y, Kunst F, Glaser P, et al: Multigenome analysis identifies a worldwide distributed epidemic *Legionella pneumophila* clone that emerged within a highly diverse species. Genome Res 2008;18:431–441.

12 Brüggemann H, Cazalet C, Buchrieser C: Adaptation of *Legionella pneumophila* to the host environment: role of protein secretion, effectors and eukaryotic-like proteins. Curr Opin Microbiol 2006;9:86–94.
13 de Felipe KS, Pampou S, Jovanovic OS, Pericone CD, Ye SF, et al: Evidence for acquisition of *Legionella* type IV secretion substrates via interdomain horizontal gene transfer. J Bacteriol 2005;187: 7716–7726.
14 Albert-Weissenberger C, Cazalet C, Buchrieser C: *Legionella pneumophila* – a human pathogen that co-evolved with fresh water protozoa. Cell Mol Life Sci 2007;64:432–448.
15 Nichols FC: Novel ceramides recovered from *Porphyromonas gingivalis*: relationship to adult periodontitis. J Lipid Res 1998;39:2360–2372.
16 Bandhuvula P, Saba JD: Sphingosine-1-phosphate lyase in immunity and cancer: silencing the siren. Trends Mol Med 2007;13:210–217.
17 Zhang K, Pompey JM, Hsu FF, Key P, Bandhuvula P, et al: Redirection of sphingolipid metabolism toward de novo synthesis of ethanolamine in *Leishmania*. EMBO J 2007;26:1094–1104.
18 Li G, Foote C, Alexander S, Alexander H: Sphingosine-1-phosphate lyase has a central role in the development of *Dictyostelium discoideum*. Development 2001;128:3473–3483.
19 Oggioni MR, Memmi G, Maggi T, Chiavolini D, Iannelli F, Pozzi G: Pneumococcal zinc metalloproteinase ZmpC cleaves human matrix metalloproteinase 9 and is a virulence factor in experimental pneumonia. Mol Microbiol 2003;49:795–805.
20 Seshadri R, Paulsen IT, Eisen JA, Read TD, Nelson KE, et al: Complete genome sequence of the Q-fever pathogen *Coxiella burnetii*. Proc Natl Acad Sci USA 2003;100:5455–5460.
21 Wu M, Sun LV, Vamathevan J, Riegler M, Deboy R, et al: Phylogenomics of the reproductive parasite *Wolbachia pipientis* wMel: A streamlined genome overrun by mobile genetic elements. PLoS Biol 2004;2:E69.
22 Ogata H, La Scola B, Audic S, Renesto P, Blanc G, et al: Genome sequence of *Rickettsia bellii* illuminates the role of amoebae in gene exchanges between intracellular pathogens. PLoS Genet 2006;2:e:76.
23 Walburger A, Koul A, Ferrari G, Nguyen L, Prescianotto-Baschong C, et al: Protein kinase G from pathogenic mycobacteria promotes survival within macrophages. Science 2004;304:1800–1804.
24 Fernandez P, Saint-Joanis B, Barilone N, Jackson M, Gicquel B, et al: The Ser/Thr protein kinase PknB is essential for sustaining mycobacterial growth. J Bacteriol 2006;188:7778–7784.
25 Greenstein AE, MacGurn JA, Baer CE, Falick AM, Cox JS, Alber T: *M. tuberculosis* Ser/Thr protein kinase D phosphorylates an anti-anti-sigma factor homolog. PLoS Pathog 2007;3:e49.
26 Rose A, Schraegle SJ, Stahlberg EA, Meier I: Coiled-coil protein composition of 22 proteomes – differences and common themes in subcellular infrastructure and traffic control. BMC Evol Biol 2005;16:66.
27 Chen J, Reyes M, Clarke M, Shuman HA: Host cell-dependent secretion and translocation of the LepA and LepB effectors of *Legionella pneumophila*. Cell Microbiol 2007;9:1660–1671.
28 Luo ZQ, Isberg RR: Multiple substrates of the *Legionella pneumophila* Dot/Icm system identified by interbacterial protein transfer. Proc Natl Acad Sci USA 2004;101:841–846.
29 Burkhard P, Stetefeld J, Strelkov SV: Coiled coils: a highly versatile protein folding motif. Trends Cell Biol 2001;11:82–88.
30 de Felipe KS, Glover RT, Charpentier X, Anderson OR, Reyes M, et al: *Legionella* eukaryotic-like type IV substrates interfere with organelle trafficking. PLoS Pathog 2008;4:e1000117.
31 Kagan JC, Roy CR: *Legionella* phagosomes intercept vesicular traffic from endoplasmic reticulum exit sites. Nat Cell Biol 2002;4:945–954.
32 Tilney LG, Harb OS, Connelly PS, Robinson CG, Roy CR: How the parasitic bacterium *Legionella pneumophila* modifies its phagosome and transforms it into rough ER: implications for conversion of plasma membrane to the ER membrane. J Cell Sci 2001;114:4637–4650.
33 Horwitz MA: The Legionnaires' disease bacterium (*Legionella pneumophila*) inhibits phagosome-lysosome fusion in human monocytes. J Exp Med 1983; 158:2108–2126.
34 Dubuisson JF, Swanson MS: Mouse infection by *Legionella*, a model to analyze autophagy. Autophagy 2006;2:179–182.
35 Amer AO, Swanson MS: Autophagy is an immediate macrophage response to *Legionella pneumophila*. Cell Microbiol 2005;7:765–778.
36 Sturgill-Koszycki S, Swanson MS: *Legionella pneumophila* replication vacuoles mature into acidic, endocytic organelles. J Exp Med 2000;192:1261–1272.
37 Molmeret M, Bitar DM, Han L, Kwaik YA: Disruption of the phagosomal membrane and egress of *Legionella pneumophila* into the cytoplasm during the last stages of intracellular infection of macrophages and *Acanthamoeba polyphaga*. Infect Immun 2004;72:4040–4051.

38 Alli OA, Gao LY, Pedersen LL, Zink S, Radulic M, et al: Temporal pore formation-mediated egress from macrophages and alveolar epithelial cells by *Legionella pneumophila*. Infect Immun 2000;68: 6431–6440.
39 Marcus AJ, Broekman MJ, Drosopoulos JH, Olson KE, Islam N, et al: Role of CD39 (NTPDase-1) in thromboregulation, cerebroprotection, and cardioprotection. Semin Thromb Hemost 2005;31:234–246.
40 Sansom FM, Newton HJ, Crikis S, Cianciotto NP, Cowan PJ, et al: A bacterial ecto-triphosphate diphosphohydrolase similar to human CD39 is essential for intracellular multiplication of *Legionella pneumophila*. Cell Microbiol 2007;9:1922–1935.
41 Sansom FM, Riedmaier P, Newton HJ, Dunstone MA, Müller CE, et al: Enzymatic properties of an ecto-nucleoside triphosphate diphosphohydrolase from *Legionella pneumophila*: substrate specificity and requirement for virulence. J Biol Chem 2008;283:12909–12918.
42 Shohdy N, Efe JA, Emr SD, Shuman HA: Pathogen effector protein screening in yeast identifies *Legionella* factors that interfere with membrane trafficking. Proc Natl Acad Sci USA 2005;102:4866–4871.
43 Banerji S, Aurass P, Flieger A: The manifold phospholipases A of *Legionella pneumophila* – identification, export, regulation, and their link to bacterial virulence. Int J Med Microbiol 2008;298:169–181.
44 Goebl M, Yanagida M: The TPR snap helix: a novel protein repeat motif from mitosis to transcription. Trends Biochem Sci 1991;16:173–177.
45 Liu M, Conover GM, Isberg RR: *Legionella pneumophila* EnhC is required for efficient replication in tumor necrosis factor alpha-stimulated macrophages. Cell Microbiol 2008;10:1906–1923.
46 Cirillo SL, Lum J, Cirillo JD: Identification of novel loci involved in entry by *Legionella pneumophila*. Microbiology 2000;146:1345–1359.
47 Newton HJ, Sansom FM, Bennett-Wood V, Hartland EL: Identification of *Legionella pneumophila*-specific genes by genomic subtractive hybridization with *Legionella micdadei* and identification of *lpnE*, a gene required for efficient host cell entry. Infect Immun 2006;74:1683–1691.
48 Newton HJ, Sansom FM, Dao J, McAlister AD, Sloan J, et al: Sel1 repeat protein LpnE is a *Legionella pneumophila* virulence determinant that influences vacuolar trafficking. Infect Immun 2007;75:5575–5585.
49 Kagan JC, Stein MP, Pypaert M, Roy CR: *Legionella* subvert the functions of Rab1 and Sec22b to create a replicative organelle. J Exp Med 2004;199:1201–1211.
50 Derré I, Isberg RR: *Legionella pneumophila* replication vacuole formation involves rapid recruitment of proteins of the early secretory system. Infect Immun 2004;72:3048–3053.
51 Nagai H, Kagan JC, Zhu X, Kahn RA, Roy CR: A bacterial guanine nucleotide exchange factor activates ARF on *Legionella* phagosomes. Science 2002; 295:679–682.
52 Machner MP, Isberg RR: Targeting of host Rab GTPase function by the intravacuolar pathogen *Legionella pneumophila*. Dev Cell 2006;11:47–56.
53 Murata T, Delprato A, Ingmundson A, Toomre DK, Lambright DG, Roy CR: The *Legionella pneumophila* effector protein DrrA is a Rab1 guanine nucleotide-exchange factor. Nat Cell Biol 2006;8: 971–977.
54 Ingmundson A, Delprato A, Lambright DG, Roy CR: *Legionella pneumophila* proteins that regulate Rab1 membrane cycling. Nature 2007;450:365–369.
55 Sedgwick SG, Smerdon SJ: The ankyrin repeat: a diversity of interactions on a common structural framework. Trends Biochem Sci 1999;24:311–316.
56 Mosavi LK, Minor DL Jr, Peng ZY: Consensus-derived structural determinants of the ankyrin repeat motif. Proc Natl Acad Sci USA 2002;99:16029–16034.
57 Habyarimana F, Al-Khodor S, Kalia A, Graham JE, Price CT, et al: Role for the Ankyrin eukaryotic-like genes of *Legionella pneumophila* in parasitism of protozoan hosts and human macrophages. Environ Microbiol 2008;10:1460–1474.
58 Pan X, Lührmann A, Satoh A, Laskowski-Arce MA, Roy CR: Ankyrin repeat proteins comprise a diverse family of bacterial type IV effectors. Science 2008; 320:1651–1654.
59 Dorer MS, Kirton D, Bader JS, Isberg RR: RNA interference analysis of *Legionella* in *Drosophila* cells: exploitation of early secretory apparatus dynamics. PLoS Pathog 2006;2:e34.
60 Kubori T, Hyakutake A, Nagai H: *Legionella* translocates an E3 ubiquitin ligase that has multiple U-boxes with distinct functions. Mol Microbiol 2008;67:1307–1319.
61 Molmeret M, Abu Kwaik Y: How does *Legionella pneumophila* exit the host cell? Trends Microbiol 2002;10:258–260.

62 Sutton RB, Fasshauer D, Jahn R, Brunger AT: Crystal structure of a SNARE complex involved in synaptic exocytosis at 2.4 A resolution. Nature 1998;395:347–353.

63 Amor JC, Swails J, Zhu X, Roy CR, Nagai H, et al: The structure of RalF, an ADP-ribosylation factor guanine nucleotide exchange factor from *Legionella pneumophila*, reveals the presence of a cap over the active site. J Biol Chem 2005;280:1392–1400.

Carmen Buchrieser
Biologie des Bactéries Intracellulaires, Institut Pasteur
25, rue du Dr. Roux
FR–75724 Paris Cedex 15 (France)
Tel. +33 1 45 68 83 72, Fax +33 1 45 68 87 86, E-Mail cbuch@pasteur.fr

de Reuse H, Bereswill S (eds): Microbial Pathogenomics.
Genome Dyn. Basel, Karger, 2009, vol 6, pp 187–197

A Proteomics View of Virulence Factors of *Staphylococcus aureus*

S. Engelmann · M. Hecker

Institut für Mikrobiologie, Ernst-Moritz-Arndt-Universität, Greifswald, Germany

Abstract

The pathogenicity of *Staphylococcus aureus* is determined by its ability to express multiple virulence factors. Thus far the virulence potential of *S. aureus* isolates has been described by the virulence gene repertoire, which, in part, varies considerably among the different isolates. Extracellular proteins constitute a reservoir of virulence factors and have been shown to play an important role in the pathogenicity of bacteria. Analyses of the expression of these virulence factors and elucidation of regulatory networks involved in *S. aureus* virulence by using gel based proteomics can yield information important for our understanding of the virulence potential of this pathogen and its interaction with the host. In addition, these approaches are critical for a comprehensive understanding of secretion and modification of virulence factors.

Staphylococcus aureus – A Commensal and Versatile Human Pathogen

S. aureus is a human commensal that asymptomatically colonizes the anterior nares of at least one third of the human population. On the other hand, it is also one of the main causes of nosocomial infections, which are often difficult to treat because of the increased prevalence of strains resistant to multiple antibiotics. The types of infection induced by *S. aureus* range from mild skin infections to more severe systemic infections, including pneumonia, endocarditis, osteomyelitis, and sepsis. One cause of the pathogenic diversity of *S. aureus* is its ability to produce a large variety of virulence factors. An impressive number of extracellular and surface associated proteins, e. g. α-toxin, coagulase, lipases, hemolysins, enterotoxins, protein A, and fibronectin binding proteins are already known to contribute to the virulence of *S. aureus*. These proteins have overlapping functions and can act either in concert or alone. Staphylococcal virulence factors can be divided according to their function into at least four groups: (i) proteins, usually localized on the bacterial cell surface, which are involved in adhesion to and invasion of host cells; (ii) proteins mediating the degradation of host cells for both bacterial nutrition and spreading; (iii) proteins that enable the bacteria to evade the immune response and (iv) proteins required for degradation of nutrients selectively found within the host. In most cases, an *S. aureus* infection is initiated by a breach of the skin or mucosal barrier. The course of these infections largely depends on the complex and poorly understood interplay of bacterial virulence determinants with each other and with components of the host. Patients implanted with foreign bodies, such as catheters, have an increased likelihood of infection due to the capacity of *S. aureus* to form biofilms on these materials [for review see 1–4].

S. aureus Isolates Show Extraordinary Diversity in the Genome Sequence

Sequencing of several *S. aureus* strains has uncovered marked heterogeneity of the species. The number of open reading frames ranges from 2,600 to 2,700 in these strains. Homologous analysis revealed that about 75% of the genome sequences of all isolates seem to be conserved and this portion mainly codes for proteins with house keeping functions. Interestingly, there are also some virulence associated genes belonging to the core genome, such as *spa, aur, hla, lip, clfAB, map/eap, fnbA* and *coa* [5, 6]. However, most virulence factors are located on highly variable regions of the staphylococcal genome, such as pathogenicity islands and lysogenic bacteriophages, or even on plasmids [7, 8]. The extensive genetic diversity of those genomic regions, in which virulence genes accumulate, might explain the broad spectrum of clinical symptoms observed in *S. aureus* infections. Clonal isolates of the same epidemic strains can differ significantly in their carriage of highly variable regions [8]. It is well understood that hypervariation of virulence genes is due to selection imposed by interaction with the host immune system and/or to the fact that they are not critical for basic metabolism. It remains to be established whether different isolates of the same clone are indeed equally pathogenic. Studies performed by von Eiff et al. [9] showed that patients suffering from *S. aureus* infections were usually infected with the same strain found as a commensal in their nose. Evidence has recently been provided that the *S. aureus* virulence gene pattern necessary for invasive diseases may also be important for nasal colonization [10]. Some diseases are related to certain *agr* groups. For example, *agr*III is associated with menstrual shock syndrome and Panton-Valentine leukocidin (PVL) induced necrotizing pneumonia, *agr*IV with exfoliatin production and *agr*I and II with reduced vancomycin susceptibility [11–13]. It is highly probable that the genome of a particular *agr* group has specific gene combinations that give rise to a specific phenotype. However, hospital infections with *S. aureus* are not restricted to a few highly virulent strains. On the contrary one *S. aureus* isolate can behave either as a commensal or as a pathogen.

Studies aimed to find a correlation between virulence gene repertoire and virulence potential of different *S. aureus* strains have not been very promising to date. Apart from the genomic variability of the bacterium, differences in the activity of

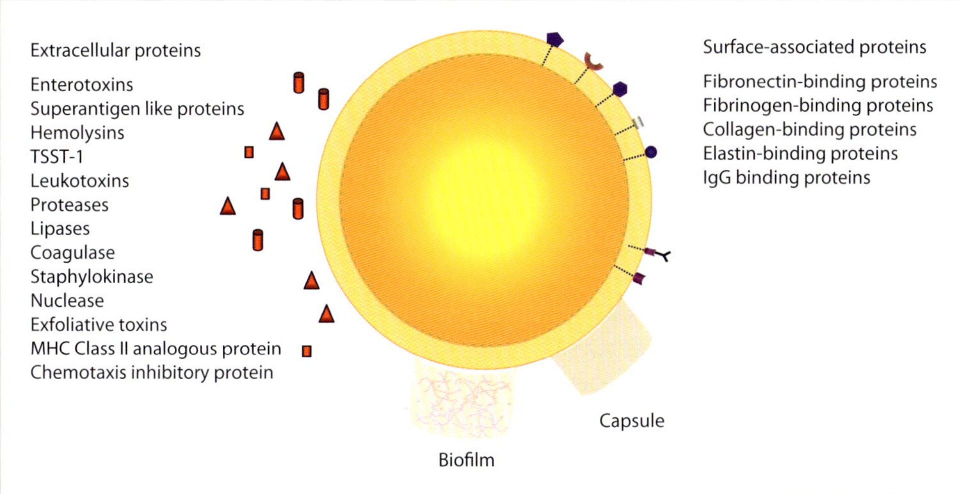

Fig. 1. Schematic presentation of known virulence factors in *S. aureus* and their localization.

virulence associated regulators have been reported to lead to variations in the amount of some virulence factors produced in different clinical isolates [14–16]. Consequently, genomic studies of clinical isolates, particularly of the distribution patterns of virulence genes on the genome (by PCR and DNA arrays) [8, 10] can only be the first step towards an understanding of these complex phenomena. Beyond the information about the presence or absence of virulence genes, the investigation of the proteome can provide information about the expression of individual virulence factors and their possible posttranslational modifications. Moreover, by using appropriate mutants the mechanism by which these proteins are secreted can be studied.

Extracellular and Surface Associated Proteins as Potential Virulence Factors

Expression of virulence factors might be analyzed either by transcriptomics or by proteomics. However, the amount of a factor at its appropriate location might be crucial for virulence activity and, since most of the virulence factors are cell surface molecules or are released into the extracellular milieu (fig. 1), secretion processes and post-translational modifications have to be taken into account when analyzing virulence activity. Components involved in different mechanisms of protein transport, including the Sec-, Tat-, Com-, and ESAT-pathways, as well as various ABC-transporters, are encoded in the *S. aureus* genome [17]. Proteins to be transported from the cytoplasm to the extra-cytoplasmic compartment of the cell or into the extracellular milieu need to contain specific signal peptides and these can be classified by the transport and modification pathway which they require. Comparative secretomics of six *S. aureus* genomes (COL, MRSA252, MSSA476, Mu50, N315, MW2) revealed an extreme heterogeneity among secreted proteins. While 58 proteins which possess a signal sequence for translocation via the Sec-system are encoded in all strains, and therefore belong to the core exoproteome, 61 proteins comprise the variant exoproteome. A similar situation was observed for the lipoproteome. By searching for typical lipoboxes in the genome sequences of the six *S. aureus* strains, 43 proteins were predicted to form the core lipoproteome and a further 43 to form the variant lipoproteome [17].

Since extracellular proteins constitute a reservoir of virulence factors and thereby play an important role in the pathogenicity of bacteria (fig. 1), the comprehensive analysis of the extracellular proteome of *S. aureus* offers the chance to identify new virulence factors and elucidate their regulation [18–21]. The proteomic approach is a useful tool for the analysis of the extracellular protein patterns of different clinical isolates and may in the future allow the correlation of the different staphylococcal disease types with the gene expression and protein secretion patterns of the causative infectious strains. The comparison of *S. aureus* extracellular protein patterns has revealed a marked diversity among different isolates [17, 22–24]. Interestingly, some strains produce only a few extracellular proteins. For example, 'small colony variants', which are involved in persistent *S. aureus* infections, are characterized by very low expression of toxins and proteases [25, unpublished data].

The extracellular proteome consists of all *S. aureus* proteins that are actively secreted via different secretion pathways. The theoretical extracellular proteome map of *S. aureus,* which considers the proteins that are actively secreted via the Sec-pathway, indicates that most of the proteins that belong to the core and the variable exoproteome [17] can be allocated to the pI region of 3.5–10.4 (fig. 2). Based on this calculation, 106 of these proteins should be detected on gels with a pI range of 3–10 and a molecular weight range of 10–140 kDa. Consequently, the 2-dimensional (2D) gel electrophoresis technique combined with mass spectrometry represents a very efficient tool to identify all of the proteins which are present in the extracellular milieu and to analyze the secreted protein pattern under different growth conditions and in different strains.

The extracellular proteome of *S. aureus* COL has been extensively characterized. These data show that only the combination of proteomics and genomics gives a complete picture of the virulence gene expression of a strain [18, 22]. The genome of *S. aureus* COL encodes 2615 proteins [26]. Among these are 83 proteins which possess a typical Sec-signal sequence and thus belong to the predicted exoproteome of this strain [17]. Nine of these proteins contain an LPXTG motif and should be covalently linked to the cell wall by a sortase dependent mechanism after secretion. At high cell densities in complex medium, 42 different proteins were identified by mass spectrometry [22]. 29 of these proteins were predicted to be secreted via the Sec-system and 21 belong to the defined core exoproteome of *S. aureus.* Interestingly, eight proteins identified among the extracellular proteins of *S. aureus* COL contain a typical

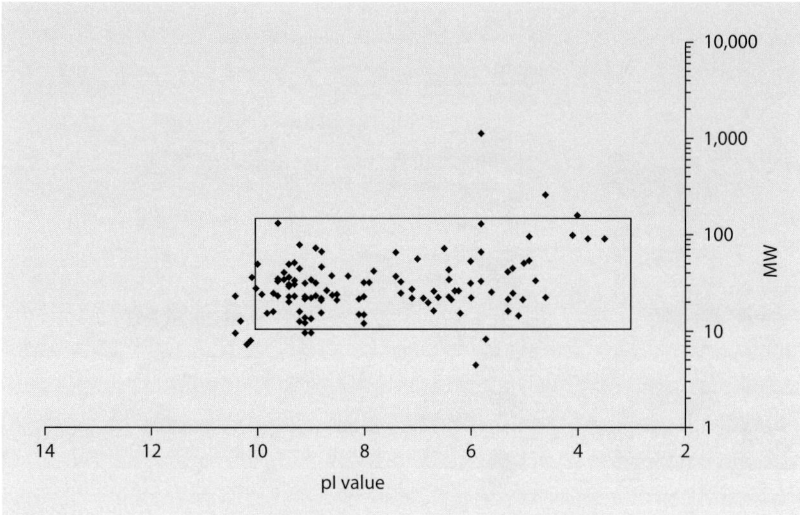

Fig. 2. The theoretical reference gel of the exoproteome of *S. aureus* predicted by Sibbald et al. [17]. The theoretical pI and molecular weight (MW) of the native proteins (without signal sequences) derived from the genome sequences of *S. aureus* COL, N315 or MW2 was obtained from NCBI database (www.ncbi.nlm.nih.gov). The region which is represented on 2D gels of fig. 3 is framed.

Sec-signal sequence [22], but are absent from the predicted exoproteome [17]. As expected, many of the extracellular proteins were already known to play a role in the virulence of *S. aureus*. However, additional proteins of unknown function were identified and these merit detailed characterization of their potential roles in virulence (e.g. Aly, IsaA, SceD, SsaA, YfnI).

A detailed comparison of the extracellular proteome of strain COL with that of *S. aureus* Newman showed an extremely heterogeneous extracellular protein pattern that cannot only be explained by differences in the variable regions of the genome sequences. Of the 29 possibly Sec-translocated proteins identified in strain COL, 21 were also found in the supernatant of *S. aureus* Newman (fig. 3). Although these 21 proteins were detected in both strains, some of them differed significantly in amount. Eight proteins were unique to *S. aureus* COL and fourteen were only detected in supernatants of *S. aureus* Newman (fig. 3). Why are these 22 proteins strain specific? There are at least three potential explanations: (i) the respective genes are unique to strain Newman or strain COL, (ii) the respective genes are pseudogenes in one of the strains or (iii) the proteins are synthesized in very low amounts in COL or Newman, and thus remain below the level of detection on protein gels. Surprisingly, only two of the 14 genes were missing in *S. aureus* COL and two of the eight genes were missing in strain Newman. Studies on the activity of virulence associated regulators implicate a higher level of activity of SaeRS, σ^B, and *agr* in strain Newman [22]. This observation strongly suggests that in addition to genomic diversity, the variability of gene

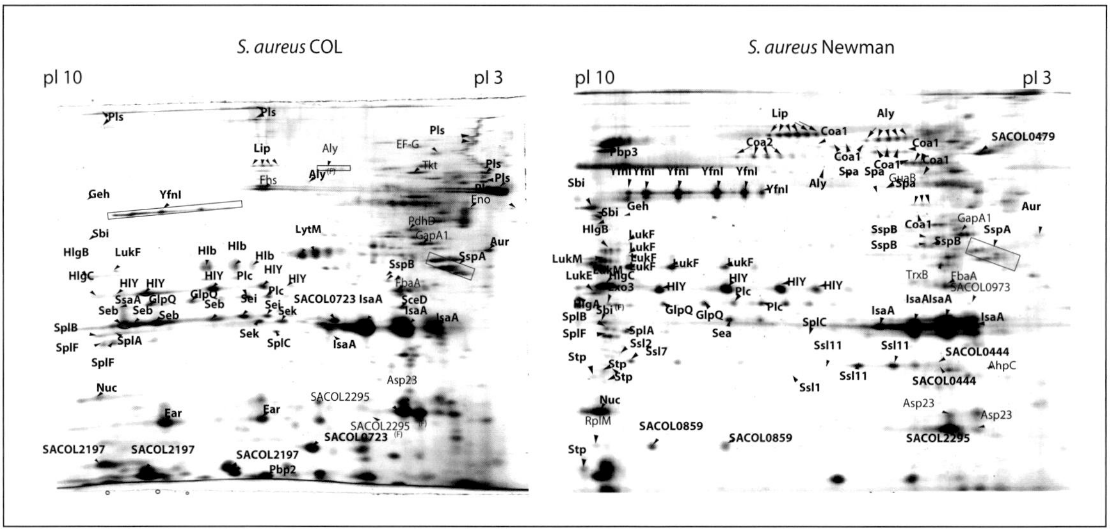

Fig. 3. Extracellular proteins of *S. aureus* COL and *S. aureus* Newman. Proteins (100 µg) isolated from the supernatant of *S. aureus* COL and *S. aureus* Newman grown in TSB medium to an OD_{540} of 10 were separated on 2D gels. The identified proteins are assigned to the open reading frame number as defined in the *S. aureus* COL, N315, and Mu50 genome sequencing projects [22].

regulation significantly contributes to the marked differences between the patterns of virulence factors in individual *S. aureus* strains.

A very similar phenomenon was also observed by Burlak and co-workers [24] who performed a comprehensive study on the exoproteomes of two community-associated MRSA (caMRSA) strains, MW2 and LAC. Altogether, the authors identified 250 distinct proteins in the supernatant of these strains. 11 of these proteins are known virulence factors and display marked differences in amount in both strains.

Regulation of Virulence Factors

The expression of virulence genes is regulated in a coordinated fashion during the growth cycle by a very complex network of regulators. As a result, the production of extracellular proteins takes place mainly at high population density during the late exponential and post-exponential phase of growth [23], and at the same time the synthesis of surface associated proteins is down-regulated. The so far best characterized regulators of virulence gene expression are Agr (**a**ccessory **g**ene **r**egulator) and SarA (Staphylococcal accessory regulator) [for review see 27]. The *sarA* locus encodes a DNA-binding protein that influences the amount of fibronectin- and fibrinogen-binding protein as well as immunodominant antigen IsaA, protein A, β-hemolysin, autolysin Aly, aureolysin, staphopain, V8 protease, and lipases Lip and Geh [18]. SarA may mediate its effects by (i) binding to the target gene promoters, (ii) indirect downstream effects on other global regulators, or (iii) degradation of proteins by *sarA*-dependent proteases. The *sar*-locus is believed to be necessary for the activation of the *agr* locus [28, 29]. The *agr* operon in turn acts as a quorum sensing system and enhances the synthesis of extracellular proteins, while simultaneously the synthesis of cell wall adhesins is repressed. RNAIII appears to be the major effector molecule of the *agr* system. It is thought to regulate most target genes at the level of transcription, but has also been shown to affect the translation of some genes [30–32]. Recent studies indicate that the alternative sigma factor σ^B may also contribute to virulence gene expression in Gram-positive bacteria by interfering the SarA and the RNAIII activity [23, 33, 34]. This pathogenicity network, however, is not confined to the interactions between SarA, RNAIII, SaeR, ArlR or σ^B. Many additional global regulators appear to be encoded in the genome sequence [27]. The network, therefore, consists of many overlapping regulons, which are expressed in a time-dependent manner to ensure an optimal mix of virulence factors at optimal concentrations during interactions with the host [18, 22, 23].

Interestingly, under in vivo conditions (e.g. in an animal model) the level of RNAIII did not influence virulence gene expression significantly [35, 36]. This was very surprising and shows once again that our knowledge of the signals that influence staphylococcal virulence gene expression within the host is still very preliminary. In particular, two component systems involved in signal perception might play important roles in modulating virulence gene expression at different sites within the host. The genome sequence of *S. aureus* codes for at least 15 two component systems and for most of these the signal detected is unknown and the structure of the respective regulons has been characterized in only a few cases (e. g. ArlSR, SaeRS, VicR) [22, 37, 38]. For a more comprehensive understanding of the regulatory network of virulence gene expression in *S. aureus,* a detailed characterization of each of these two component systems will be an important goal for future studies.

The Specific Immune Response as a Mirror of *S. aureus* Interaction with its Host

Community acquired invasive diseases caused by *S. aureus* are strongly dependent on host factors, especially on whether the host is immune compromised or not. Antibodies with specificity for *S. aureus* antigens are known to be prevalent in the human population and are thought to confer some degree of protection against *S. aureus* infections. Studies analyzing the antibody response of adult blood donors against superantigens have shown that carriers develop an immune response highly specific for the antigens of the colonizing strain [39]. Nevertheless, 80% of *S. aureus* infections in the hospital settings are caused by the colonizing strain [9]. This strongly indicates that the specific immune response directed to the colonizing strain does not fully protect against an infection. However, in cases of *S. aureus* bacteraemia, carriers have a much better prognosis to overcome this infection than non carriers [40]. Thus the *S. aureus* specific immune reaction cannot fully prevent an *S. aureus* infection, but has a decisive influence on the development of an infection, its outcome and probably also on the carrier status [41]. The characterization of the antibody response will provide us with new insights into the proteins expressed by *S. aureus* during interactions with the host and should therefore complement the analyses of virulence potential by genome, transcriptome, and proteome analysis of the bacterial strain. Moreover, these studies are a prerequisite for the development of new vaccine strategies aimed at mitigating or preventing *S. aureus* infections. Until now, studies addressing the humoral response have mostly been performed by using selected *Staphylococcus* antigens expressed in vitro [41]. However, these studies ignore the large diversity of antigens possibly produced within the host. Analyses of the immune response of carriers and patients against their own strain using gel based and gel free techniques will provide a comprehensive picture of the diversity of immunogenic *S. aureus* antigens. Moreover, proteins may be identified that rarely induce a specific immune response and this might pinpoint gaps in the humoral anti-staphylococcal immune defense. By using extracellular proteins of strain COL, a large variation in the specificities of antibodies in sera from different patients has been shown [42]. This might reflect the different composition of antigens expressed within the host by the respective carrier strain.

Concluding Remarks

Analyzing secreted proteins by gel based proteomics provides a valuable tool for identifying potential virulence genes in *S. aureus*. The theoretical extracellular proteome map of *S. aureus* indicates that most of the secreted proteins can be allocated to the pI region 3 to 10 and to the MW range of 10 to 140 kDa. If secreted in detectable amounts, about 90% of the predicted extracellular proteins should thus be present on 2D gels in a pI range of 3–10. Protein expression profiling of extracellular proteins by 2D gel analyses not only reveals the overall pattern of protein expression under given environmental conditions, but also provides additional information on post translational modifications and on the fate of the proteins. Identification of extracellular proteins showed that about 60% of the proteins secreted via the Sec pathway appear as multiple spots on 2D gels. Such multiple spots may be due to charge alteration (e. g. SEB, SEK, SEQ, Hla, Ear, IsaA, Lip, YfnI, Aly) or to fragmentation (e. g. Aly, Coa, LukF, LukM, Pls, Ssl11, SspA, SspB, Stp). Proteins with such deviations in pI and molecular mass are candidates for posttranslational modifications. To fully understand the pathogenicity of *S. aureus*, studies on the protein expression profiling of virulence factors have to be combined with detailed studies on protein modifications (such as disulfide formation, lipid modifications), as well as determination of protein stability and processing.

However, there are limitations to 2D protein gels that make certain groups of virulence associated proteins non accessible. Proteins with extreme pIs and molecular weights, very low abundant proteins and hydrophobic proteins escape the gel based proteomic approach. For this reason alternative techniques are certainly required. MS based approaches, which rely on separation of complex protein or peptide mixtures by liquid chromatography or 1D SDS gel electrophoresis, allow the identification of proteins in complex protein mixtures and circumvent the limitations of 2D gels. However, modified and processed proteins cannot currently be adequately distinguished by these approaches. Because of this a combination of 2D gel based and gel free (or semi gel free) approaches may be required to adequately target the extracellular and cell wall associated proteome. For analyses of membrane proteins, however, the use of gel free or semi gel free approaches will be essential [43].

The combination of proteomics with comparative genomics and transcriptomics together with an analysis of the host's immune response will provide new insights into the pathogenicity and virulence of *S. aureus* and will open the way towards new strategies to prevent and to treat infections caused by this pathogen. Moreover, genomic and proteomic data of clinical isolates may provide diagnostic information of value in selecting and tailoring clinical treatment regimes.

Acknowledgements

We are very grateful to Robert S. Jack for critical review of the manuscript and to Kathrin Rogasch and Christian Kohler for preparing the figures. This work was supported by grants of the BMBF (031U107A/-207A; 031U213B), the DFG (GK212/3–00, SFB/TRR34, FOR 585), the EU (Staphdynamics), the Land MV and the Fonds der Chemischen Industrie.

References

1 Dinges MM, Orwin PM, Schlievert PM: Exotoxins of *Staphylococcus aureus*. Clin Microbiol Rev 2000; 13:16–34.

2 Foster TJ, Hook M: Surface protein adhesins of *Staphylococcus aureus*. Trends Microbiol 1998;6: 484–488.

3 Foster TJ: Immune evasion by staphylococci. Nat Rev Microbiol 2005;3:948–958.

4 Lowy FD: *Staphylococcus aureus* infections. N Engl J Med 1998;339:520–532.

5 Peacock SJ, Moore CE, Justice A, Kantzanou M, Story L, et al: Virulent combinations of adhesin and toxin genes in natural populations of *Staphylococcus aureus*. Infect Immun 2002;70:4987–4996.

6 Lindsay JA, Holden MT: *Staphylococcus aureus:* superbug, super genome? Trends Microbiol 2004;12: 378–385.

7 Novick RP: Mobile genetic elements and bacterial toxinoses: the superantigen-encoding pathogenicity islands of *Staphylococcus aureus*. Plasmid 2003;49: 93–105.

8 Witney AA, Marsden GL, Holden MT, Stabler RA, Husain SE, et al: Design, validation, and application of a seven-strain *Staphylococcus aureus* PCR product microarray for comparative genomics. Appl Environ Microbiol 2005;71:7504–7514.

9 von Eiff C, Becker K, Machka K, Stammer H, Peters G: Nasal carriage as a source of *Staphylococcus aureus* bacteremia. Study Group. N Engl J Med 2001; 344:11–16.

10 Lindsay JA, Moore CE, Day NP, Peacock SJ, Witney AA, et al: Microarrays reveal that each of the ten dominant lineages of *Staphylococcus aureus* has a unique combination of surface-associated and regulatory genes. J Bacteriol 2006;188:669–676.

11 Gillet Y, Issartel B, Vanhems P, Fournet JC, Lina G, et al: Association between *Staphylococcus aureus* strains carrying gene for Panton-Valentine leukocidin and highly lethal necrotising pneumonia in young immunocompetent patients. Lancet 2002; 359:753–759.

12 Jarraud S, Lyon GJ, Figueiredo AM, Gerard L, Vandenesch F, et al: Exfoliatin-producing strains define a fourth *agr* specificity group in *Staphylococcus aureus*. J Bacteriol 2000;182:6517–6522.

13 Sakoulas G, Eliopoulos GM, Moellering RC Jr, Wennersten C, Venkataraman L, et al: Accessory gene regulator *(agr)* locus in geographically diverse *Staphylococcus aureus* isolates with reduced susceptibility to vancomycin. Antimicrob Agents Chemother 2002;46:1492–1502.

14 Blevins JS, Beenken KE, Elasri MO, Hurlburt BK, Smeltzer MS: Strain-dependent differences in the regulatory roles of *sarA* and *agr* in *Staphylococcus aureus*. Infect Immun 2002;70:470–480.

15 Li S, Arvidson S, Mollby R: Variation in the *agr*-dependent expression of alpha-toxin and protein A among clinical isolates of *Staphylococcus aureus* from patients with septicaemia. FEMS Microbiol Lett 1997;152:155–161.

16 Karlsson A, Arvidson S: Variation in extracellular protease production among clinical isolates of *Staphylococcus aureus* due to different levels of expression of the protease repressor *sarA*. Infect Immun 2002;70:4239–4246.

17 Sibbald MJ, Ziebandt AK, Engelmann S, Hecker M, de Jong A, et al: Mapping the pathways to staphylococcal pathogenesis by comparative secretomics. Microbiol Mol Biol Rev 2006;70:755–788.

18 Ziebandt AK, Weber H, Rudolph J, Schmid R, Höper D, et al: Extracellular proteins of *Staphylococcus aureus* and the role of SarA and sigma B. Proteomics 2001;1:480–493.

19 Bernardo K, Fleer S, Pakulat N, Krut O, Hunger F, Krönke M: Identification of *Staphylococcus aureus* exotoxins by combined sodium dodecyl sulfate gel electrophoresis and matrix-assisted laser desorption/ ionization-time of flight mass spectrometry. Proteomics 2002;2:740–746.

20 Kawano Y, Ito Y, Yamakawa Y, Yamashino T, Horii T, et al: Rapid isolation and identification of staphylococcal exoproteins by reverse phase capillary high performance liquid chromatography-electrospray ionization mass spectrometry. FEMS Microbiol Lett 2000;189:103–108.

21 Kawano Y, Kawagishi M, Nakano M, Mase K, Yamashino T, et al: Proteolytic cleavage of staphylococcal exoproteins analyzed by two-dimensional gel electrophoresis. Microbiol Immunol 2001;45:285–290.

22 Rogasch K, Rühmling V, Pané-Farré J, Höper D, Weinberg C, et al: Influence of the two-component system SaeRS on global gene expression in two different *Staphylococcus aureus* strains. J Bacteriol 2006;188:7742–7758.

23 Ziebandt AK, Becher D, Ohlsen K, Hacker J, Hecker M, Engelmann S: The influence of *agr* and sigmaB in growth phase dependent regulation of virulence factors in *Staphylococcus aureus*. Proteomics 2004;4: 3034–3047.

24 Burlak C, Hammer CH, Robinson MA, Whitney AR, McGavin MJ, et al: Global analysis of community-associated methicillin-resistant *Staphylococcus aureus* exoproteins reveals molecules produced in vitro and during infection. Cell Microbiol 2007;9: 1172–1190.

25 Moisan H, Brouillette E, Jacob CL, Langlois-Begin P, Michaud S, Malouin F: Transcription of virulence factors in *Staphylococcus aureus* small-colony variants isolated from cystic fibrosis patients is influenced by SigB. J Bacteriol 2006;188:64–76.

26 Gill SR, Fouts DE, Archer GL, Mongodin EF, Deboy RT, et al: Insights on evolution of virulence and resistance from the complete genome analysis of an early methicillin-resistant *Staphylococcus aureus* strain and a biofilm-producing methicillin-resistant *Staphylococcus epidermidis* strain. J Bacteriol 2005; 187:2426–2438.

27 Novick RP: Autoinduction and signal transduction in the regulation of staphylococcal virulence. Mol Microbiol 2003;48:1429–1449.

28 Morfeldt E, Tegmark K, Arvidson S: Transcriptional control of the *agr*-dependent virulence gene regulator, RNAIII, in *Staphylococcus aureus*. Mol Microbiol 1996;21:1227–1237.

29 Chien Y, Manna AC, Cheung AL: SarA level is a determinant of *agr* activation in *Staphylococcus aureus*. Mol Microbiol 1998;30:991–1001.

30 Janzon L, Arvidson S: The role of the delta-lysin gene *(hld)* in the regulation of virulence genes by the accessory gene regulator *(agr)* in *Staphylococcus aureus*. EMBO J 1990;9:1391–1399.

31 Morfeldt E, Taylor D, von Gabain A, Arvidson S: Activation of alpha-toxin translation in *Staphylococcus aureus* by the trans-encoded antisense RNA, RNAIII. EMBO J 1995;14:4569–4577.

32 Novick RP, Ross HF, Projan SJ, Kornblum J, Kreiswirth B, Moghazeh S: Synthesis of staphylococcal virulence factors is controlled by a regulatory RNA molecule. EMBO J 1993;12:3967–3975.

33 Bischoff M, Entenza JM, Giachino P: Influence of a functional *sigB* operon on the global regulators *sar* and *agr* in *Staphylococcus aureus*. J Bacteriol 2001; 183:5171–5179.

34 Horsburgh MJ, Aish JL, White IJ, Shaw L, Lithgow JK, Foster SJ: SigmaB modulates virulence determinant expression and stress resistance: characterization of a functional *rsbU* strain derived from *Staphylococcus aureus* 8325-4. J Bacteriol 2002;184: 5457–5467.
35 Goerke C, Campana S, Bayer MG, Döring G, Botzenhart K, Wolz C: Direct quantitative transcript analysis of the *agr* regulon of *Staphylococcus aureus* during human infection in comparison to the expression profile in vitro. Infect Immun 2000; 68:1304–1311.
36 Goerke C, Fluckiger U, Steinhuber A, Zimmerli W, Wolz C: Impact of the regulatory loci *agr*, *sarA* and *sae* of *Staphylococcus aureus* on the induction of alpha-toxin during device-related infection resolved by direct quantitative transcript analysis. Mol Microbiol 2001;40:1439–1447.
37 Fournier B, Klier A, Rapoport G: The two-component system ArlS-ArlR is a regulator of virulence gene expression in *Staphylococcus aureus*. Mol Microbiol 2001;41:247–261.
38 Dubrac S, Boneca IG, Poupel O, Msadek T: New insights into the WalK/WalR (YycG/YycF) essential signal transduction pathway reveal a major role in controlling cell wall metabolism and biofilm formation in *Staphylococcus aureus*. J Bacteriol 2007;189: 8257–8269.
39 Holtfreter S, Roschack K, Eichler P, Eske K, Holtfreter B, et al: *Staphylococcus aureus* carriers neutralize superantigens by antibodies specific for their colonizing strain: a potential explanation for their improved prognosis in severe sepsis. J Infect Dis 2006;193:1275–1278.
40 Wertheim HF, Vos MC, Ott A, van Belkum A, Voss A, et al: Risk and outcome of nosocomial *Staphylococcus aureus* bacteraemia in nasal carriers versus non-carriers. Lancet 2004;364:703–705.
41 Clarke SR, Brummell KJ, Horsburgh MJ, McDowell PW, Mohamad SA, et al: Identification of in vivo-expressed antigens of *Staphylococcus aureus* and their use in vaccinations for protection against nasal carriage. J Infect Dis 2006;193:1098–1108.
42 Vytvytska O, Nagy E, Bluggel M, Meyer HE, Kurzbauer R, et al: Identification of vaccine candidate antigens of *Staphylococcus aureus* by serological proteome analysis. Proteomics 2002;2:580–590.
43 Wolff S, Hahne H, Hecker M, Becher D: Complementary analysis of the vegetative membrane proteome of the human pathogen *Staphylococcus aureus*. Mol Cell Proteomics 2008;7: 1460–1468.

Susanne Engelmann
Institut für Mikrobiologie
Jahnstrasse 15
DE–17487 Greifswald (Germany)
Tel. +49 3834 864227, Fax +49 3834 864202, E-Mail Susanne.Engelmann@uni-greifswald.de

de Reuse H, Bereswill S (eds): Microbial Pathogenomics.
Genome Dyn. Basel, Karger, 2009, vol 6, pp 198–210

Pathogenomics of Mycobacteria

M.C. Gutierrez[a] · P. Supply[b] · R. Brosch[c]

[a]Institut Pasteur, Department Infection and Epidemiology, Paris, [b]INSERM U629 and Institut Pasteur de Lille, Lille, [c]Institut Pasteur, UP Pathogénomique Mycobactérienne Intégrée, Paris, France

Abstract

Among the 130 species that constitute the genus *Mycobacterium,* the great majority are harmless saprophytes. However, a few species have very efficiently adapted to a pathogenic lifestyle. Among them are two of the most important human pathogens, *Mycobacterium tuberculosis* and *Mycobacterium leprae*, and one emerging pathogen, *Mycobacterium ulcerans*. Their slow growth, virulence for humans and particular physiology make these organisms very difficult to work with, however the need to develop new strategies in the fight against these pathogens requires a clear understanding of their genetic and physiological repertoires and the mechanisms that have contributed to their evolutionary success. The rapid development of mycobacterial genomics following the completion of the *Mycobacterium tuberculosis* genome sequence provides now the basis for finding the important factors distinguishing pathogens and non-pathogens. In this chapter we will therefore present some of the major insights that have been gained from recent studies, with focus on the roles played by various evolutionary processes in shaping the structure of mycobacterial genomes and pathogen populations.

The genus *Mycobacterium* was an early focus of medical interest as it includes the agents of two devastating human diseases, leprosy and tuberculosis. *Mycobacterium* is the single genus in the family Mycobacteriaceae, which belongs to the order Actinomycetales and the phylum Actinobacteria [1]. Within this widespread class, mycobacteria present an unusual, waxy cell envelope containing specifically long-chained mycolic acids. This cell envelope helps pathogenic mycobacteria to resist dehydration, antimicrobial drugs and host defenses. Mycolic acids confer the characteristic ability to resist decolorization by acidic ethanol following staining with basic fuchsin to mycobacteria and some closely related actinomycetes, a property (still) widely used for the fast recognition of mycobacteria [2].

Mycobacteria are ubiquitous and enormously abundant in soil and untreated water, supposedly linked to early colonization of terrestrial environments by their ancestors billions of years ago [3]. Their evolution has resulted in a wide biological diversity, with highly complex lifestyles ranging from environmental saprophytes to intracellular parasites. As mammals and humans evolved in or close to terrestrial and water environments, their exposure to mycobacteria was inevitable since the beginning of their evolution [4]. This constant exposure and co-evolution is suggested by the presence of CD1-restricted T-cell subsets that appear to recognize only mycobacterial lipids and glycolipids [5]. Since the discovery of *Mycobacterium leprae* (Armauer Hansen, 1873) and *M. tuberculosis* (Robert Koch, 1883) more than one century ago, 130 mycobacterial species have been validly described [6] (see also: List of Prokaryotic Names with Standing in Nomenclature, URL: http://www.bacterio.net). The majority can be isolated from the environment and are collectively called nontuberculous mycobacteria (NTM). Although mycobacteria are in general not components of the normal human bacterial flora, many NTM species are occasionally isolated from skin and mucosa of asymptomatic individuals, and half of them may have clinical relevance under certain circumstances. The nature and level of environmental exposure depend upon human lifestyle and habitat localization. For instance, domestic water supplies in developing countries can contain as many as 10^9 mycobacteria per liter and therefore generally evoke immune responses among the residents that may have an influence on vaccine efficacy, whereas such responses are much less common in developed countries [4].

The major human mycobacterial pathogens have been recently subjected to analyses of their complete genome sequences. At the time of writing this chapter, whole genome sequences of 40 mycobacterial strains are determined or at various stages of completion (http://www.ncbi.nlm.nih.gov/genomes/lproks.cgi, http://www.tbdb.org), and selected genetic sequences of many other thousands have already been well characterized. Analysis of this huge amount of data provides unique opportunities to explain and reconstruct the biology and pathogenomic evolution of the genus *Mycobacterium*, for which some selected examples will be given in the following paragraphs.

Evolution of Pathogenicity within the Genus *Mycobacterium*

Taxonomic studies early recognized the natural division that exists between slowly and rapidly growing species of mycobacteria. Slow- and rapid-growers require more than seven days or less than seven days, respectively, to produce colonies on solid media. There is greater than 94.3% of 16S rRNA gene sequence similarity found within the genus sequences. Genetic relationships inferred from comparison of these sequences supported the traditional division of mycobacteria into two branches and suggested that the slow-growers constitute a sub-group that evolved from a fast-growing ancestor [7]. This partition is also supported by more robust phylogenetic reconstructions using concatenated sequences of house-keeping genes [8]. An explanation for the growth rate difference is still awaited, although differences in the number of

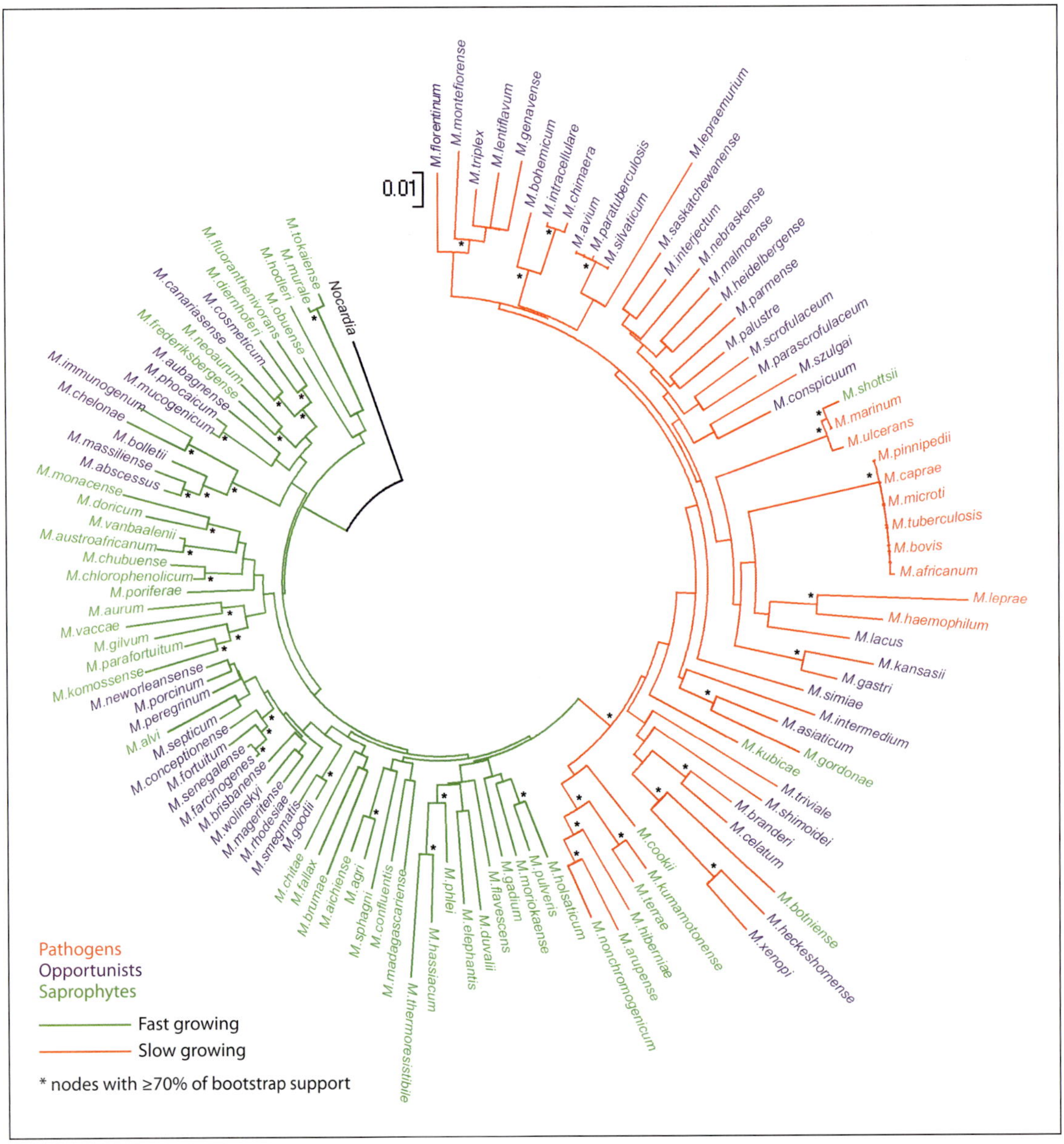

Fig. 1. Evolutionary relationships of 119 mycobacterial species based on 16S rRNA, *hsp65* and *rpoB* genes. The evolutionary history was inferred using the Neighbor-Joining method [62]. The bootstrap test was performed using 1000 replications. The optimal tree with the sum of branch length = 2.82884649 is shown. The tree is drawn to scale, with branch lengths in the same units as those of the evolutionary distances used to infer the phylogenetic tree. The evolutionary distances were computed using the Maximum Composite Likelihood method [63] and are in the units of the number of base substitutions per site. All positions containing gaps and missing data were eliminated from the dataset (Complete deletion option). There were a total of 738 positions in the final dataset. Phylogenetic analyses were conducted using MEGA4 [64].

rRNA *(rrn)* operons have been suggested to play a role, as fast-growers have generally two operons instead of one for slow-growers [9]. Importantly, most of fast-growing mycobacteria are harmless saprophytic organisms, whereas major human pathogens like *M. tuberculosis*, *M. leprae* or *M. ulcerans* are slow-growers. Apparent evolution towards slow growth rate and reduced number of *rrn* operons seems therefore of great importance, because it is associated with increased pathogenic potential of mycobacteria. This association is clearly reflected in figure 1, which shows the phylogenetic relationship of 119 mycobacterial species as determined by combined use of 16S rRNA, *rpoB* and *hsp65* gene sequences. Interestingly, the strict pathogens *M. tuberculosis* complex, *M. leprae, M. haemophilum, M. ulcerans*, and *M. marinum* (shown in red in figure 1) form a tight cluster among the slow-growers, indicating a common ancestry.

However, it should be noted that the notion of pathogenicity of mycobacteria also strongly depends on the proper functionality of the host's immune system. As many as 90% of persons infected with the strict pathogen *M. tuberculosis* remain lifelong asymptomatic carriers. The opposite situation is for instance seen with the normally harmless fast-growing *M. smegmatis*, which can nevertheless cause fatal disseminated disease in the case of inherited interferon gamma receptor deficiency [10]. Likewise, another fast-grower, *M. abscessus*, is an emerging pathogen in patients with underlying medical disorders like cystic fibrosis [11].

Genomics of *M. tuberculosis*

The first mycobacterial genome to be sequenced was that of the agent of human tuberculosis, *M. tuberculosis* H37Rv. This paradigm strain of tuberculosis research is used in various laboratories all over the world and has kept its virulence in spite of numerous passages since its original isolation in 1905 [12]. The complete genome sequence obtained from this strain in a pioneering collaborative project between the Institut Pasteur in Paris, France, and the Sanger Institute in Hinxton, UK, became available about 10 years ago and consists of 4,411,532 bp [13]. There are about 4,000 proteins and 50 RNAs encoded in the genome of this strain that can be consulted via a regularly updated genome browser (http://genolist.pasteur.fr/TubercuList/) that is part of the GenoList browsers developed at the Institut Pasteur.

In depth analyses of the *M. tuberculosis* H37Rv genome sequence revealed that 3.4% of the genome is composed of insertion sequences (IS) and prophages (phiRv1, phiRv2). Among the 56 loci harboring IS elements belonging to various families (e.g. IS*3*, IS*5*, IS*21*, IS*30*, IS*110*, IS*256*, and IS*L3*), IS*6110*, a member of the IS*3* family, is the most abundant insertion element. IS*6110* is a useful epidemiological tool as it transposes frequently, thereby generating restriction fragment length polymorphisms that can be exploited for molecular epidemiological studies [14].

The information provided by the genome sequence led to valuable insight into the biology of the tubercle bacilli. It was noted that about 8% of the genome of *M. tuberculosis* H37Rv is encoding proteins involved in lipid metabolism, which highlights the importance of this class of molecules for the particular lifestyle of *M. tuberculosis*. These findings are in good agreement with the presence of a wide range of lipids, glycolipids, lipoglycans and polyketides in the cell wall of *M. tuberculosis*, but also suggest that numerous proteins show lipolytic functions that enable *M. tuberculosis* to use host-cell lipids and sterols as energy sources. *M. tuberculosis* presents the prototype β-oxidation cycle required for lipid catabolism, and also encodes more than 100 enzymes potentially involved in alternative lipid oxidation pathways in which degradation products of host cell membranes could be metabolized [13]. The resulting acetyl-CoA can then either be used for the synthesis of mycobacterial cell wall components or fed into central metabolic pathways.

Another major finding of the genome project was the identification of novel gene and protein families, which were either previously unknown or poorly understood. The most notable of these were the PE and PPE families, which were named according to their characteristic N-terminal motifs ProGlu (PE), or ProProGlu (PPE), and consist of at least 100 and 67 members, respectively. The PE family proteins show a conserved N-terminal domain of ~110 amino acid residues, which is often followed by a glycin-rich domain encoded by a polymorphic G+C-rich sequence (PGRS). PPE family members have a 180-amino acid conserved N-terminal part and often contain major polymorphic tandem repeats (MPTR). There has been recent progress in the characterization of these proteins in the cell envelope where they are surface-exposed [15–17], and on the evolutionary history of these proteins [18]. However, although they are postulated to be involved in antigenic variation and pathogenesis, the actual biological functions of these two protein families remain still mostly unknown.

Another unusual gene family that was identified by genome analysis are the Mce proteins. This family, whose first representative was described as a protein promoting mammalian cell entry [19], consists of four genomic loci of eight genes each, which are organized in a highly similar manner and comprise two genes that resemble YrbE from *Escherichia coli* and six *mce* genes that encode Mce proteins with hydrophobic stretches at the N-terminus, probably representing signal or anchor sequences [13]. This molecular organization suggests that the Mce proteins are exposed on the cell surface and likely play an important role in the infection process with *M. tuberculosis*.

Similarly, the ESX-1 protein family is also playing a key role in the host-pathogen interaction. The prototype protein of this family, the 6-kDa early secreted antigenic target (ESAT-6), was first identified in the supernatant of *M. tuberculosis* cultures [20] and is encoded together with its protein partner CFP-10, the 10-kDa culture filtrate protein of *M. tuberculosis*, in proximity to the origin of replication. This genomic region has been termed region of difference 1 (RD1). Most interestingly, overlapping portions of RD1 are missing from the attenuated Bacille de Calmette et Guérin (BCG) vaccine [21] and the attenuated *M. microti* vole bacillus that was as well used as an attenuated live vaccine [22]. Complementation of RD1 in BCG and/or *M. microti* via RD1 knock-in constructs partially restored the virulence of the recombinant strains. Recombinant BCG::RD1 strains which have the RD1 encoded ESAT-6 secretion system (ESX-1) restored persist to a greater degree in organs of immuno-competent mice, and reproducibly induce better protection against disseminated tuberculosis in the mouse and the guinea pig model [23]. The ESX-1 system has most recently been proposed to be named type VII secretion system [24]. In total agreement with these RD1 knock-in studies, knock-out of RD1 from *M. tuberculosis* results in attenuation [25, 26]. Lack of ESAT-6 secretion linked to a point mutation in a two component regulator PhoP was also identified as one of the factors contributing to the attenuation of the widely used laboratory strain H37Ra ('a' for avirulent) [27–29]. In accordance with these observations, several genes involved in ESAT-6 secretion located inside and outside the RD1 region are part of the 194 candidate genes that were identified by a global, genome wide transposon site hybridization (TraSH) study as being specifically required for growth of *M. tuberculosis* under in vivo conditions (in the mouse) [30]. This very interesting work confirmed and extended previous signature tagged mutagenesis studies and showed that about 5% of the genes in *M. tuberculosis* were directly involved in the survival of the bacterium in the host (table 1) [30, and references therein]. The genes contained in this list represent candidates for further functional work in order to understand the global network of factors involved in mycobacterial pathogenicity.

Mycobacterial Pathogenomic Specialization

Studies of closely related mycobacterial species, often grouped into species complexes, provide an illustration of mycobacterial pathogenomic evolution. Members of species or complex can greatly differ in phenotypic, pathogenic, habitat and/or host range properties but still share more than 98% gene sequence identity and show identical or almost identical 16S rRNA sequences. This situation reflects the existence of differentially specialized clones, originating from a wider initial mycobacterial pool that have passed through recent evolutionary bottlenecks and have adapted to new ecological niches. This specialization is characterized by genomic signatures such as acquisition of novel genes via horizontal gene transfer (HGT), genome downsizing and rearrangements, accumulation of pseudogenes, and/or proliferation of insertion sequences.

For example, *M. ulcerans* and *M. marinum* exhibit high genomic similarity but differ greatly in their pathogenic potential [31] due to the ability of *M. ulcerans* to produce mycolactone, an unusual polyketide with strong cytotoxic potential that leads to cell necrosis [32] and immune suppression [33]. In contrast to *M. ulcerans* and some very closely related mycolactone producing mycobacteria (MPM), *M. marinum*

Table 1. Predicted functional classification of genes identified by transposon site hybridization (TraSH) analysis (after Sassetti and Rubin [30]) as being essential for growth of *M. tuberculosis* H37Rv under in vivo conditions

Functional classification	No. of genes	Percent of category[a]
Lipid metabolism	15	7.5
Carbohydrate transport and metabolism	9	8.4
Inorganic ion transport and metabolism	8	8.0
Cell envelope biogenesis, outer membrane	8	7.3
Amino acid transport and metabolism	8	4.3
Transcription	7	5.4
Coenzyme metabolism	7	6.0
DNA replication, recombination and repair	5	4.6
Translation, ribosomal structure	5	3.9
Signal transduction mechanisms	4	5.2
Secretion	3	13.6
Energy production and conversion	3	1.6
Cell division and chromosome partitioning	2	8.7
Posttranslational modification, chaperones	2	2.5
Nucleotide transport and metabolism	1	1.5
Unknown	107	4.7
Total	194	5.0

[a]Refers to the fraction of genes of the particular functional class.

does not produce mycolactone, but causes granulomatous lesions in fish and other ectotherms, and, only sporadically, limited granulomatous skin lesions in humans. *M. marinum* has a 10-fold faster growth and a more diverse metabolism than *M. ulcerans*. Despite these marked phenotypic differences, it has been shown by multi-locus sequence analysis and comparative genomics that MPM, including *M. ulcerans*, have recently evolved from a common *M. marinum* ancestral clone. The key event in this evolution has probably been the acquisition of the pMUM plasmid by HGT, harboring polyketide synthase genes required for mycolactone biosynthesis [34, 35]. MPM have subsequently diverged into at least two distinct lineages, one including *M. ulcerans* and the other one comprising the ectotherm-infecting MPM. Massive amplification of IS elements in the *M. ulcerans* lineage had a major impact on the genome of this organism, by generating pseudogenes via intragenic insertions, and marking chromosomal inversions and deletions. Accumulated deletions account for >1-Mb downsizing relative to the *M. marinum* genome. Gene lesions and deletions principally affect PE and PPE gene families, and paralogs involved in essential cell wall biosynthesis, nitrogen metabolism and solute transport. The resulting loss of genetic redundancy might contribute to slow growth via reduced gene dosage, which in turn may reflect relaxed selection due to adaptation from a free-living to a more stable, possibly arthropod host-adapted niche [31]. Interestingly, PE and PPE genes are also poorly represented in the *M. avium* subsp. *paratuberculosis* genome [36]. Other gene loss or inactivation more directly suggest specialization to a protected niche, such as a gene normally involved in pigment synthesis in *M. marinum* and protecting it from sunlight. Of direct relevance for pathogenesis, deletions of the *esx-1* locus might contribute to the predominantly extracellular infection cycle of *M. ulcerans*, in conjunction with the antiphagocytic properties of mycolactone. However, this deletion and deletions in other genomic regions are not universally found among geographically diverse *M. ulcerans* strains [37]. These results indicate that the *M. ulcerans* genome is at an intermediate stage of reductive evolution between that of a more generalist mycobacterium such as *M. marinum*, and the extreme genome contraction of the highly host-specialized *M. leprae* [38, 39].

The *M. tuberculosis* complex (MTBC) represents an example of refined host-adapted evolution, probably on a very recent evolutionary scale. Despite their high genetic relatedness, some complex members exclusively infect humans (e.g. *M. tuberculosis*, *M. africanum*) or rodents (*M. microti*), whereas others differentially infect a variety of mammals (e.g. *M. bovis*, *M. pinnipedii*). *M. tuberculosis* itself is composed of different phylogeographic lineages, which seem to differ in their pathogenic potential and are associated with specific, sympatric human populations [40, 41]. These observations suggest that MTBC lineages may even have adapted to populations of a particular host species. This intra-complex differentiation probably reflects recent divergence from a single ancestor clone, resulting from an evolutionary bottleneck estimated to have occurred 35,000 to 40,000 years ago [42–45, Wirth et al., unpublished data]. Importantly, the highly clonal structure of the MTBC [45–48] implies that HGT had little, if any, impact on divergence at this evolutionary scale. Instead, relatively limited genomic insertion-deletions and pseudogene accumulation, as well as single nucleotide polymorphisms and variation in genes encoding PE and PPE protein families appear as potential driving forces of the observed clonal specialization [49, 50]. Functional interpretation of most of these polymorphisms is not straightforward. Nevertheless, cell wall components and secreted proteins show the greatest variation between *M. bovis* and *M. tuberculosis*, suggesting roles in differentiated host-pathogen interaction and immune evasion [49]. At a more focused level, *M. tuberculosis* lineage(s)-specific polymorphisms, such as a deletion affecting the *Rv1519* gene of unknown function [51] or a 7-bp polymorphism in the *pks*15/1 gene required for synthesis of phenolic glycolipid [52–54], have also been associated with immune subversion and epidemic potential of clinical strains.

The contribution of HGT and reductive evolution to the long-term tuberculosis bacillus evolution becomes logically apparent at a higher evolutionary scale. Recent comparisons of *M. tuberculosis, M. marinum, M. ulcerans, M. avium* subsp. *paratuberculosis* and *M. smegmatis* genomes confirmed the close genetic relationship between *M. tuberculosis* and the *M. marinum*/*M. ulcerans* group, supported by 16S rRNA sequence analysis [55, and refs therein]. As *M. marinum* has a 50% bigger genome than *M. tuberculosis*, this analysis also indicated how the two species diverged from a common environmental mycobacterium with *M. tuberculosis* undergoing a dominantly reductive evolution compatible with its host-adapted lifestyle. Nevertheless, 630 coding DNA sequences (CDS) are specifically possessed by *M. tuberculosis*, of which 360 distributed into 80 genomic regions appear to have been acquired by HGT [55, and refs. therein]. The latter CDSs are involved in proven or potentially important functions, such as the direct repeat locus potentially conferring immunity to phage infection [56], an ABC transporter putatively involved in virulence [57], and the *virS* virulence locus [58]. Overall, as in the case of the *M. tuberculosis* complex, the major genome differences among relatively distant mycobacterial species are interestingly again found in genes encoding cell wall components and the PE and PPE protein families [59]. This is consistent with a key role of these components localized at the interface between the pathogen and its host, their variation probably contributing in primary pathogenesis differences between these pathogens.

Most interestingly, genomic analysis of '*M. prototuberculosis*' tuberculosis bacilli may provide missing links to further define the impact of reductive evolution and HGT on the tuberculosis bacillus evolution. Genetic analysis of these bacilli, isolated from immuno-competent tuberculous patients, indicated that they represent extant derivatives from a larger and non-clonal bacterial species, including *M. canettii*, from which the MTBC recently emerged [60, 61]. In these tubercle bacilli, detection of mosaic gene sequences, whose individual elements are retrieved in classical *M. tuberculosis* complex strain genomes, suggests that the present highly clonal framework of the MTBC is actually a composite assembly of genetic sequences resulting from multiple remote HGTs (fig. 2). The genomes of four most divergent '*M. prototuberculosis*' strains are presently being sequenced. Together with biological characterization of these strains, the resulting data will certainly provide new exciting insights into the pathogenomic adaptation of the tuberculosis bacillus, and the actual contribution of HGT and reductive evolution to this process.

Applications and Perspectives

Mycobacterial evolution to pathogenicity is obviously the result of a long evolutionary process, starting from generalist environmental bacteria to produce the breadth of highly host-adapted and sometimes highly successful pathogens that we see today. Expected increase of available genome sequences of mycobacterial strains from pathogenic and non-pathogenic species will probably permit a quantum leap in our understanding of the evolutionary forces and the genetic determinants that are driving this course. The new genomic data will help identify HGT- or genome decay-associated gene clusters at different pathoadaptive steps among the mycobacteria. Large-scale comparative genomics of both environmental and host-adapted mycobacteria will

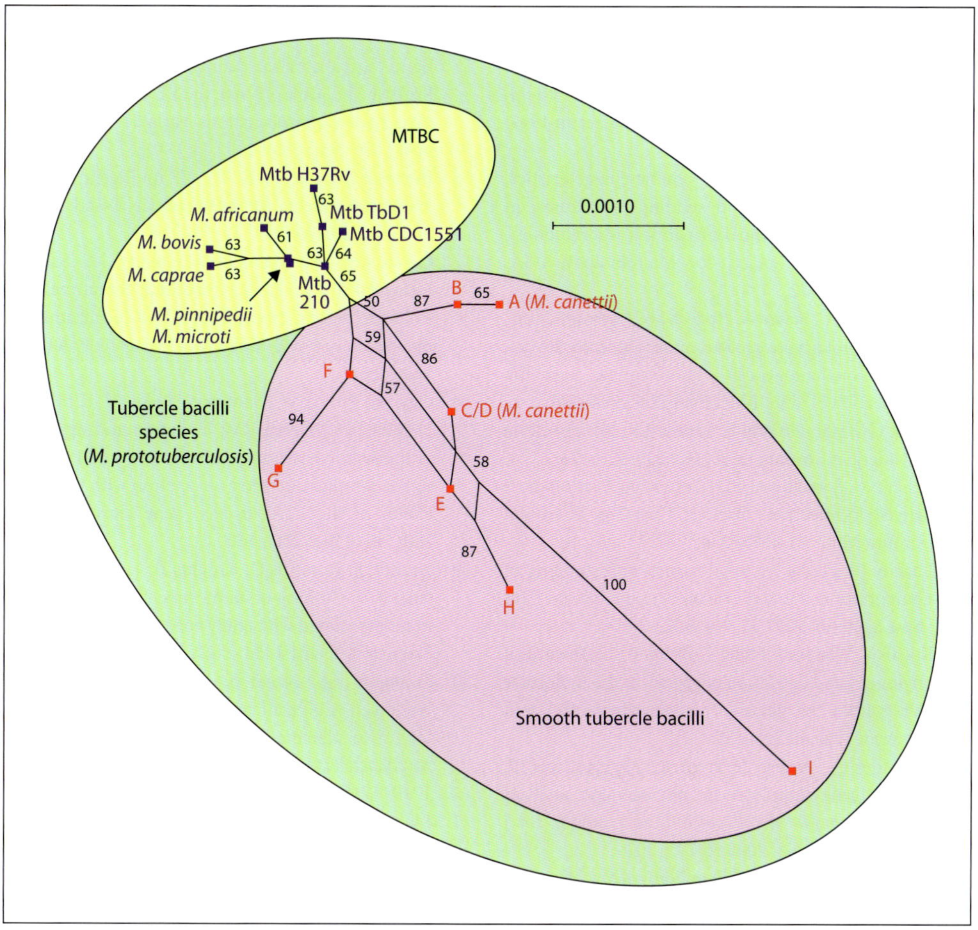

Fig. 2. Phylogenetic analysis of the tuberculosis bacilli species using a split decomposition graph (reprinted from Gutierrez and colleagues [60]. The MTBC forms a single compact bifurcating branch, rooted within the much larger array constituted by the smooth '*M. prototuberculosis*' tuberculosis bacilli.

shed light on metabolic evolution under the selection pressures of different environments. Together these data will provide new therapeutic, diagnostic and vaccine targets for combating all mycobacterial diseases.

Acknowledgements

We thank Faranoush Doustdar for help with data management for figure 1. This work was supported by the European Union (contracts LHSP-CT-2005–018923, HEALTH-F3–2007–201762), and the Institut Pasteur (PTR202). P.S. is a Researcher of the Centre National de la Recherche Scientifique (CNRS).

References

1 Stackebrandt E, Rainey FA, Ward-Rainey NL: Proposal for a new hierarchic classification system, *Actinobacteria* classis nov. Int J Syst Bacteriol 1997; 47:479–491.

2 Pfyffer GE: *Mycobacterium*: general characteristics, laboratory detection, and staining procedures; in Murray PR (ed): Manual of Clinical Microbiology, ed 9. American Society for Microbiology, USA, 2007, pp 543–572.

3 Battistuzzi FU, Feijao A, Hedges SB: A genomic timescale of prokaryote evolution: insights into the origin of methanogenesis, phototrophy, and the colonization of land. BMC Evol Biol 2004;4:44.

4 Rook GA, Hamelmann E, Brunet LR: Mycobacteria and allergies. Immunobiol 2007;212:461–473.

5 Behar SM, Porcelli SA: CD1-restricted T cells in host defense to infectious diseases. Curr Top Microbiol Immunol 2007;314:215–250.

6 Euzeby JP: List of bacterial names with standing in nomenclature: a folder available on the Internet. Int J Syst Bacteriol 1997;47:590–592.

7 Rogall T, Wolters J, Flohr T, Böttger EC: Towards a phylogeny and definition of species at the molecular level within the genus *Mycobacterium*. Int J Syst Bacteriol 1990;40:323–330.

8 Devulder G, Perouse de Montclos M, Flandrois JP: A multigene approach to phylogenetic analysis using the genus *Mycobacterium* as a model. Int J Syst Evol Microbiol 2005;55:293–302.

9 Goodfellow M, Magee JG: Taxonomy of Mycobacteria; in Gangadharam PRJ, Jenkins PA (eds): Mycobacteria: Basic Aspects. Chapman and Hall Medical Microbiology Series, USA, 1998, pp 1–71.

10 Pierre-Audigier C, Jouanguy E, Lamhamedi S, Altare F, Rauzier J, et al: Fatal disseminated *Mycobacterium smegmatis* infection in a child with inherited interferon gamma receptor deficiency. Clin Infect Dis 1997;24:982–984.

11 Sermet-Gaudelus I, Le Bourgeois M, Pierre-Audigier C, Offredo C, Guillemot D, et al: *Mycobacterium abscessus* and children with cystic fibrosis. Emerg Infect Dis 2003;9:1587–1591.

12 Manca C, Tsenova L, Barry CE 3rd, Bergtold A, Freeman S, et al: *Mycobacterium tuberculosis* CDC1551 induces a more vigorous host response in vivo and in vitro, but is not more virulent than other clinical isolates. J Immunol 1999;162:6740–6746.

13 Cole ST, Brosch R, Parkhill J, Garnier T, Churcher C, et al: Deciphering the biology of *Mycobacterium tuberculosis* from the complete genome sequence. Nature 1998;393:537–544.

14 Mathema B, Kurepina NE, Bifani PJ, Kreiswirth BN: Molecular epidemiology of tuberculosis: current insights. Clin Microbiol Rev 2006;19:658–685.

15 Banu S, Honoré N, Saint-Joanis B, Philpott D, Prévost MC, Cole ST: Are the PE-PGRS proteins of *Mycobacterium tuberculosis* variable surface antigens? Mol Microbiol 2002;44:9–19.

16 Brennan MJ, Delogu G: The PE multigene family: a 'molecular mantra' for mycobacteria. Trends Microbiol 2002;10:246–249.

17 Cascioferro A, Delogu G, Colone M, Sali M, Stringaro A, et al: PE is a functional domain responsible for protein translocation and localization on mycobacterial cell wall. Mol Microbiol 2007;66: 1536–1547.

18 Gey van Pittius NC, Sampson SL, Lee H, Kim Y, van Helden PD, Warren RM: Evolution and expansion of the *Mycobacterium tuberculosis* PE and PPE multigene families and their association with the duplication of the ESAT-6 (esx) gene cluster regions. BMC Evol Biol 2006;6:95.

19 Arruda S, Bomfim G, Knights R, Huima-Byron T, Riley LW: Cloning of an *M. tuberculosis* DNA fragment associated with entry and survival inside cells. Science 1993;261:1454–1457.

20 Sørensen AL, Nagai S, Houen G, Andersen P, Andersen AB: Purification and characterization of a low-molecular-mass T-cell antigen secreted by *Mycobacterium tuberculosis*. Infect Immun 1995;63: 1710–1717.

21 Mahairas GG, Sabo PJ, Hickey MJ, Singh DC, Stover CK: Molecular analysis of genetic differences between *Mycobacterium bovis* BCG and virulent *M. bovis*. J Bacteriol 1996;178:1274–1282.

22 Brodin P, Eiglmeier K, Marmiesse M, Billault A, Garnier T, et al: Bacterial artificial chromosome-based comparative genomic analysis identifies *Mycobacterium microti* as a natural ESAT-6 deletion mutant. Infect Immun 2002;70:5568–5578.

23 Pym AS, Brodin P, Majlessi L, Brosch R, Demangel C, et al: Recombinant BCG exporting ESAT-6 confers enhanced protection against tuberculosis. Nat Med 2003;9:533–539.

24 Abdallah AM, Gey van Pittius NC, Champion PA, Cox J, Luirink J, et al: Type VII secretion–mycobacteria show the way. Nat Rev Microbiol 2007;5:883–891.

25 Lewis KN, Liao R, Guinn KM, Hickey MJ, Smith S, et al: Deletion of RD1 from *Mycobacterium tuberculosis* mimics bacille Calmette-Guérin attenuation. J Infect Dis 2003;187:117–123.

26 Hsu T, Hingley-Wilson SM, Chen B, Chen M, Dai AZ, et al: The primary mechanism of attenuation of bacillus Calmette-Guerin is a loss of secreted lytic function required for invasion of lung interstitial tissue. Proc Natl Acad Sci USA 2003;100:12420–12425.

27 Frigui W, Bottai D, Majlessi L, Monot M, Josselin E, et al: Control of *M. tuberculosis* ESAT-6 secretion and specific T cell recognition by PhoP. PLoS Pathog 2008;4:e33.
28 Lee JS, Krause R, Schreiber J, Mollenkopf HJ, Kowall J, et al: Mutation in the transcriptional regulator PhoP contributes to avirulence of *Mycobacterium tuberculosis* H37Ra strain. Cell Host Microbe 2008; 14:97–103.
29 Zheng H, Lu L, Wang B, Pu S, Zhang X, et al: Genetic basis of virulence attenuation revealed by comparative genomic analysis of *Mycobacterium tuberculosis* strain H37Ra versus H37Rv. PLoS ONE 2008;11: e2375.
30 Sassetti CM, Rubin EJ: Genetic requirements for mycobacterial survival during infection. Proc Natl Acad Sci USA 2003;100:12989–12894.
31 Stinear TP, Seemann T, Pidot S, Frigui W, Reysset G, et al: Reductive evolution and niche adaptation inferred from the genome of *Mycobacterium ulcerans,* the causative agent of Buruli ulcer. Genome Res 2007;17:192–200.
32 George KM, Chatterjee D, Gunawardana G, Welty D, Hayman J, et al: Mycolactone: a polyketide toxin from *Mycobacterium ulcerans* required for virulence. Science 1999;283:854–857.
33 Coutanceau E, Decalf J, Martino A, Babon A, Winter N, et al: Selective suppression of dendritic cell functions by *Mycobacterium ulcerans* toxin mycolactone. J Exp Med 2007;204:1395–1403.
34 Yip MJ, Porter JL, Fyfe JA, Lavender CJ, Portaels F, et al: Evolution of *Mycobacterium ulcerans* and other mycolactone-producing mycobacteria from a common *Mycobacterium marinum* progenitor. J Bacteriol 2007;189:2021–2029.
35 Stinear TP, Mve-Obiang A, Small PL, Frigui W, Pryor MJ, et al: Giant plasmid-encoded polyketide synthases produce the macrolide toxin of *Mycobacterium ulcerans.* Proc Natl Acad Sci USA 2004; 101:1345–1349.
36 Li L, Bannantine JP, Zhang Q, Amonsin A, May BJ, et al: The complete genome sequence of *Mycobacterium avium* subspecies *paratuberculosis.* Proc Natl Acad Sci USA 2005;102:12344–12349.
37 Käser M, Rondini S, Naegeli M, Stinear T, Portaels F, et al: Evolution of two distinct phylogenetic lineages of the emerging human pathogen *Mycobacterium ulcerans.* BMC Evol Biol 2007;7:177.
38 Cole ST, Eiglmeier K, Parkhill J, James KD, Thomson NR, et al: Massive gene decay in the leprosy bacillus. Nature 2001;409:1007–1011.
39 Gómez-Valero L, Rocha EP, Latorre A, Silva FJ: Reconstructing the ancestor of *Mycobacterium leprae*: the dynamics of gene loss and genome reduction. Genome Res 2007;17:1178–1185.
40 Gagneux S, DeRiemer K, Van T, Kato-Maeda M, de Jong BC, et al: Variable host-pathogen compatibility in *Mycobacterium tuberculosis.* Proc Natl Acad Sci USA 2006;103:2869–2873.
41 Caws M, Thwaites G, Dunstan S, Hawn TR, Lan NT, et al: The influence of host and bacterial genotype on the development of disseminated disease with *Mycobacterium tuberculosis.* PLoS Pathog 2008; 4:e1000034.
42 Sreevatsan S, Pan X, Stockbauer KE, Connell ND, Kreiswirth BN, et al: Restricted structural gene polymorphism in the *Mycobacterium tuberculosis* complex indicates evolutionarily recent global dissemination. Proc Natl Acad Sci USA 1997;94:9869–9874.
43 Gutacker MM, Smoot JC, Migliaccio CA, Ricklefs SM, Hua S, et al: Genome-wide analysis of synonymous single nucleotide polymorphisms in *Mycobacterium tuberculosis* complex organisms: resolution of genetic relationships among closely related microbial strains. Genetics 2002;162:1533–1543.
44 Hughes AL, Friedman R, Murray M: Genomewide pattern of synonymous nucleotide substitution in two complete genomes of *Mycobacterium tuberculosis.* Emerg Infect Dis 2002;8:1342–1346.
45 Brosch R, Gordon SV, Marmiesse M, Brodin P, Buchrieser C, et al: A new evolutionary scenario for the *Mycobacterium tuberculosis* complex. Proc Natl Acad Sci USA 2002;99:3684–3689.
46 Supply P, Warren RM, Bañuls AL, Lesjean S, Van Der Spuy GD, et al: Linkage disequilibrium between minisatellite loci supports clonal evolution of *Mycobacterium tuberculosis* in a high tuberculosis incidence area. Mol Microbiol 2003;47:529–538.
47 Smith NH, Dale J, Inwald J, Palmer S, Gordon SV, et al: The population structure of *Mycobacterium bovis* in Great Britain: clonal expansion. Proc Natl Acad Sci USA 2003;100:15271–15275.
48 Hirsh AE, Tsolaki AG, DeRiemer K, Feldman MW, Small PM: Stable association between strains of *Mycobacterium tuberculosis* and their human host populations. Proc Natl Acad Sci USA 2004;101:4871–4876.
49 Garnier T, Eiglmeier K, Camus JC, Medina N, Mansoor H, et al: The complete genome sequence of *Mycobacterium bovis.* Proc Natl Acad Sci USA 2003;100:7877–7882.
50 Fleischmann RD, Alland D, Eisen JA, Carpenter L, White O, et al: Whole-genome comparison of *Mycobacterium tuberculosis* clinical and laboratory strains. J Bacteriol 2002;184:5479–5490.

51 Newton SM, Smith RJ, Wilkinson KA, Nicol MP, Garton NJ, et al: A deletion defining a common Asian lineage of *Mycobacterium tuberculosis* associates with immune subversion. Proc Natl Acad Sci USA 2006;103:15594–15598.
52 Constant P, Perez E, Malaga W, Lanéelle MA, Saurel O, et al: Role of the *pks15/1* gene in the biosynthesis of phenolglycolipids in the *Mycobacterium tuberculosis* complex. Evidence that all strains synthesize glycosylated p-hydroxybenzoic methyl esters and that strains devoid of phenolglycolipids harbor a frameshift mutation in the *pks15/1* gene. J Biol Chem 2002;277:38148–38158.
53 Reed MB, Domenech P, Manca C, Su H, Barczak AK, et al: A glycolipid of hypervirulent tuberculosis strains that inhibits the innate immune response. Nature 2004;431:84–87.
54 Sinsimer D, Huet G, Manca C, Tsenova L, Koo MS, et al: The phenolic glycolipid of *Mycobacterium tuberculosis* differentially modulates the early host cytokine response but does not in itself confer hypervirulence. Infect Immun 2008;76:3027–3036.
55 Stinear TP, Seemann T, Harrison PF, Jenkin GA, Davies JK, et al: Insights from the complete genome sequence of *Mycobacterium marinum* on the evolution of *Mycobacterium tuberculosis*. Genome Res 2008;18:729–741.
56 Barrangou R, Fremaux C, Deveau H, Richards M, Boyaval P, et al: CRISPR provides acquired resistance against viruses in prokaryotes. Science 2007; 315:1709–1712.
57 Rosas-Magallanes V, Deschavanne P, Quintana-Murci L, Brosch R, Gicquel B, Neyrolles O: Horizontal transfer of a virulence operon to the ancestor of *Mycobacterium tuberculosis*. Mol Biol Evol 2006;23:1129–1135.
58 Singh R, Singh A, Tyagi AK: Deciphering the genes involved in pathogenesis of *Mycobacterium tuberculosis*. Tuberculosis 2005;85:325–335.
59 Marri PR, Bannantine JP, Golding GB: Comparative genomics of metabolic pathways in *Mycobacterium* species: gene duplication, gene decay and lateral gene transfer. FEMS Microbiol Rev 2006;30:906–925.
60 Gutierrez MC, Brisse S, Brosch R, Fabre M, Omaïs B, et al: Ancient origin and gene mosaicism of the progenitor of *Mycobacterium tuberculosis*. PLoS Pathog 2005;1:e5.
61 Fabre M, Koeck JL, Le Flèche P, Simon F, Hervé V, et al: High genetic diversity revealed by variable-number tandem repeat genotyping and analysis of *hsp65* gene polymorphism in a large collection of '*Mycobacterium canettii*' strains indicates that the *M. tuberculosis* complex is a recently emerged clone of '*M. canettii*'. J Clin Microbiol 2004;42:3248–3255.
62 Saitou N, Nei M: The neighbor-joining method: A new method for reconstructing phylogenetic trees. Mol Biol Evol 1987;4:406–425.
63 Tamura K, Nei M, Kumar S: Prospects for inferring very large phylogenies by using the neighbor-joining method. Proc Natl Acad Sci USA 2004;101: 11030–11035.
64 Tamura K, Dudley J, Nei M, Kumar S: MEGA4: Molecular Evolutionary Genetics Analysis (MEGA) software version 4.0. Mol Biol Evol 2007;24:1596–1599.

M. Cristina Gutierrez
Institut Pasteur, Department Infection and Epidemiology
28, rue du Dr Roux
FR–75015 Paris (France)
Tel. +33 145688360, Fax +33 145688837, E-Mail crisgupe@pasteur.fr

Author Index

Subject Index